T0327617

Physics of Solar Energy and Energy Storage

Physics of Solar Energy and Energy Storage

Second Edition

C. Julian Chen

Columbia University

New York, USA

Published by John Wiley & Sons, Inc., Hoboken, New Jersey.
Published simultaneously in Canada.

For general information on our other products and services or for technical
support, please contact our Customer Care Department within the United
States at (800) 762-2974, outside the United States at (317) 572-3993 or fax
(317) 572-4002.

Wiley also publishes its books in a variety of electronic formats. Some content
that appears in print may not be available in electronic formats. For more
information about Wiley products, visit our web site at www.wiley.com.

Library of Congress Cataloging-in-Publication Data Applied for:
Hardback ISBN: 9781394203611

Cover Design: Wiley
Cover Image: © bilanol/Adobe Stock Photos

Contents

List of Figures

List of Tables

Preface to the Second Edition

Twelve years have elapsed since the publication of the first edition of Physics of Solar Energy. During the one dozen years, dramatic changes have emerged in solar energy and related fields. A new edition is needed to reflect those advances.

First, from 2010 to 2022, the price of solar panels has dropped from $2 per watt to $0.2 per watt, a full order of magnitude. The cost reduction made solar electricity the least expensive of all energy sources in most places on the Earth. In the future, solely based on economics, solar energy will become a main source of energy.

Second, around 2010, lithium-ion rechargeable battery was almost exclusively applied to portable electronics, such as laptop computers, cellular phones, digital cameras, hand-held tools, etc. In 2020s, lithium-ion battery has been massively applied to automobiles and utility-scale energy storage. Because solar radiation is intermittent, to make solar energy the main source, energy storage is a necessity. Lithium-ion battery is the most versatile energy storage device. Logically, the 2019 Nobel Prize for chemistry was awarded to John B. Goodenough, M. Stanley Whittingham, and Akira Yoshino for their contributions to the development of the lithium-ion battery. The Nobel Prize press release concluded with "It can also store significant amounts of energy from solar and wind power, making possible a fossil fuel-free society."

Third, around 2010, light-emitting diodes (LED) accounted for less than 1% of the illumination market. In 2022, it grew to more than 50%. It is expected to reach almost 100% in 2025. As a reverse process of solar cells, LED is also based on a semiconductor pn-junction. The efficiency of LED lights is more than 10 folds higher than that incandescent light, and its lifetime is more than 10 times longer. Logically, the 2014 Nobel Prize in Physics was awarded to Isamu Akasaki, Hiroshi Amano and Shuji Nakamura "for the invention of efficient blue light-emitting diodes which has enabled bright, long-lasting, and energy-saving white light sources".

Last but not the least, in 2015, the United Nations Framework Convention on Climate Change (UNFCCC) held the 21st Conference of the Parties in Paris, where a legally binding international treaty on climate change was reached by 196 parties, known as the "Paris Agreement". Accordingly, all countries or regions signed up to that treaty should propose and implement their nationally determined contributions to reduce their greenhouse gas emissions in order to limit the increase of global average temperature following the goals of the Paris Agreement. The implementation of the above technological advances becomes a globally coordinated endeavor.

To reflect those advances, in the second edition of the book, several new sections are added, including Section 1.1 entitled *Shaping a More Livable World*, Section 1.4 entitled *A Rechargeable-Battery Primer*, and Section 8.4 entitled *Light-Emitting Diodes for Illumination*. Because of the importance and richness of contents, a new Chapter *Rechargeable Batteries* is added. Accordingly, the title of the second edition of the book is changed to *Physics of Solar Energy and Energy Storage*.

Besides adding new contents, in view of the recent advances in fundamental physics, the presentation is also modernized. The common theoretical basis for solar cells, LEDs, and rechargeable batteries is *quantum mechanics*. It is the centerpiece of modern physics. In the first edition, essential quantum mechanics was presented in Chapter 7 and Appendix C. It follows the traditional formulation based on Hilbert space, where the dynamic variables of point particles are represented by Hermitian operators. The crown jewelry of the traditional formulation, Wolfgang Pauli's 1925 algebraic solution of the hydrogen atom problem, was presented in detail because of its mathematical beauty. Nevertheless, my teaching experience showed that the traditional formulation of quantum mechanics was difficult to understand, and not useful to explain solid-state physics and chemistry, as related to solar energy and energy storage.

The 2023 Nobel Prize in Physics elables a natural view of wavefuncions, the central concept in quantum mechanics. Using attosecond light pulses, atomic and molecular wavefunctions, also called as orbitals, were imaged experimentally in real space [74, 43]. It showed that Schrödinger's wavefunction is a physical field, the same as electromagnetic fields. Teaching quantum mechanics becomes much easier. The confusing and paradoxical concepts such as Hilbert space, von Neumann axioms, complex operators, uncertainty relations, Born statistical rule, and wave-particle duality, are eliminated. All wavefunctions are real and they never collapse. Complex number is an optional mathematical tool for time-dependent phenomena, same as in electromagnetics. In Chapter 7, *A Quantum Mechanics Primer*, a conceptually and logically consistent elementary quantum mechanics is presented at a sophomore level for all natural scientists and engineers. It is the foundation for the understanding of atomic physics, molecular physics, solid-state physics, chemistry, molecular biology, including the interaction of radiation with atomic systems. More mathematical details of Section 7.3, The Chemical Bond, can be found in Chapter 4 of the third edition of Introduction to Scanning Tunneling Microscopy, Oxford University Press 2021, entitled Atomic Forces. In Appendix G, entitled Quantum Measurement in Light of Experiments, some basic concepts concerning the understanding of quantum mechanics are presented.

In Spring 2009, I started to teach a graduate-level course Physics of Solar Energy at Columbia University. I sincerely thank Professors Irving Herman, Cevdet Noyan, and Richard Osgood for assistance to establish the new course. The first edition of this book was based on the lecture notes of that course. I especially thank Marina Zamalin, the Associate Dean of Online Education of the School of Engineering and Applied Science, to establish a Columbia Video Network (CVN) course. I sincerely thank Professor Marc Spiegelman for assistance to establish an updated course, Physics of Solar Energy and Energy Storage, for the Fall semester 2023 and on. For 14 consecutive years, the Physics of Solar Energy course has attended by several hundreds of students, mostly graduates. I heartily thank their valuable feedback to improve the course.

C. Julian Chen

Columbia University
in the City of New York

August 2023

Preface to the First Edition

One of the greatest challenges facing mankind in the twenty-first century is energy. Starting with the industrial revolution in the 18th century, fossil fuels such as coal, petroleum, and natural gas have been the main energy resources for everything vital for human society: from steam engines to Otto and diesel engines, from electricity to heating and cooling of buildings, from cooking and hot-water making, from lighting to various electric and electronic gadgets, as well as for most of the transportation means. However, fossil fuel resources as stored solar energy accumulated during hundreds of millions of years are being rapidly depleted by excessive exploration. In addition, the burning of fossil fuels has caused and is causing damage to the environment of Earth.

It is understandable that alternative or renewable energy resources, other than fossil fuels, have been studied and utilized. Hydropower, a derivative of solar energy, currently supplies about 2% of the world's energy consumption. The technology has matured, and the available resources are already heavily explored. Wind energy, also a derivative of solar energy, is being utilized rapidly. The resource of such highly intermittent energy is also limited. Nuclear energy is not renewable. The mineral resource of uranium is limited. The problems of accident prevention and nuclear waste management are still unresolved.

The most abundant energy resource available to human society is solar energy. At 4×10^6 EJ/year, it is ten thousand times the energy consumption of the world in 2007. For example, if 50% of the sunlight shining on the state of New Mexico is converted into useful energy, it can satisfy all the energy needs of the United States.

The utilization of solar energy is as old as human history. However, to date, among various types of renewable energy resources, solar energy is the least utilized. Currently, it only supplies about 0.1% of the world's energy consumption, or 0.00001% of the available solar radiation. Nevertheless, as a result of intensive research and development, the utilization of solar energy, especially solar photovoltaics, is enjoying an amazingly rapid progress. Therefore, it is reasonable to expect that in the latter half of the 21st century solar energy will become the main source of energy, surpassing all fossil fuel energy resources.

Similar to other fields of technology, the first step to achieve success in solar energy utilization is to have a good understanding of its basic science. Three years ago, Columbia University launched a master's degree program in solar energy science and engineering. I was asked to give a graduate-level course on the physics of solar energy. In the spring semester of 2009, when the first course was launched, 46 students registered. Columbia's CVN (Columbia Video Network) decided to record the lectures and distribute them to outside students. Because of the high demand, the lectures series for regular students repeated for two more semesters, and the CVN course on the physics of solar energy was repeated for seven consecutive semesters. This book is a compilation of lecture notes.

The basic design of the book is as follows. The first chapter summarizes the energy problem and compares various types of renewable energy resources, including hydropower and wind energy, with solar energy. Chapter 2, "Nature of Solar Radiation," presents the electromagnetic wave theory of Maxwell as well as the photon theory of Einstein. Understanding of blackbody radiation is crucial to the understanding of solar radiation, which is described in detail. Chapter 3, "Origin of Solar Energy," summarizes the astrophysics of solar energy, including the basic parameters and structure of the Sun. The gravitational contraction theory of Lord Kelvin and the nuclear fusion theory of Hans Bethe for the origin of stellar energy are presented. Chapter 4, "Tracking Sunlight," is a self-contained but elementary treatment of the positional astronomy of the Sun for nonastronomy majors. It includes an elementary derivation of the coordinate transformation formulas. It also includes a transparent derivation of the equation of time, the difference between solar time and civil time, as the basis for tracking sunlight based on time as we know it. This chapter is supplemented with a brief summary of spherical trigonometry in Appendix B. The accumulated daily direct solar radiation on various types of surfaces over a year is analyzed with graphics. Chapter 5, "Interaction of Sunlight with Earth," presents both the effect of the atmosphere and the storage of solar energy in the ground, the basis for the so-called shallow geothermal energy. A simplified model for scattered or diffuse sunlight is presented. Chapter 6, "Thermodynamics of Solar Energy," starts with a summary of the basics of thermodynamics followed by several problems of the application of solar energy, including basics of heat pump and refrigeration. Chapters 7–10 deal with basic physics of solar photovoltaics and solar photochemistry. Chapter 7, "Quantum Transition," presents basic concepts of quantum mechanics in Dirac's format, with examples of organic molecules and semiconductors, with a full derivation of the golden rule and the principle of detailed balance. Chapter 8 is dedicated to the essential concept in solar cells, the pn-junction. Chapter 9 deals with semiconductor solar cells, including a full derivation of the Shockley–Queisser limit, with descriptions of the detailed structures of crystalline, thin-film, and tandem solar cells. Chapter 10, "Solar Photochemistry," presents an analysis of photosynthesis in plants as well as research in artificial photosynthesis. Various organic solar cells are described, including dye-sensitized solar cells and bilayer organic solar cells. Chapter 11 deals with solar thermal applications, including solar water heaters and solar thermal electricity generators. The vacuum tube collector and the thermosiphon solar heat collectors are emphasized. Concentration solar energy is also presented, with four types of optical concentrators: trough, parabolic dish, heliostat, and especially the compact linear Fresnel concentrator. Chapter 12 deals with energy storage, including sensible and phase-change thermal energy storage systems and rechargeable batteries, especially lithium–ion batteries. The last chapter, "Building with Sunshine," introduces architectural principles of solar energy utilization together with civil engineering elements.

Experience in teaching the course has shown me that the student backgrounds are highly diversified, including physics, chemistry, electrical engineering, mechanical engineering, chemical engineering, architecture, civil engineering, environmental science, materials science, aerospace engineering, economy, and finance. Although it is a senior undergraduate and beginning graduate-level course, it must accommodate a broad spectrum of student backgrounds. Therefore, necessary scientific background knowledge is part of the course. The book is designed with this in mind. For example, background knowledge in positional astronomy, thermodynamics, and quantum mechanics is included. For students who have already taken these courses, the background material serves as a quick review and as a reference for the terminology and symbols used in this book. The presentation of the background science is for the purpose of solar energy utilization only, along a "fast track." For example, quantum mechanics is presented using an "empirical" approach, starting from direct perception of quantum states by a scanning tunneling microscope; thus, the quantum states are not merely a mathematical tool but a perceptible reality. The scanning tunneling microscope is also an important tool in the research for novel devices in solar energy conversion.

At an insert of the book, a gallery of color graphics and photographs is constructed and compiled. It serves as a visual introduction to the mostly mathematical presentation of the materials, which is useful for intuitive understanding of the concepts.

During the course of giving lectures and writing the lecture notes, I have encountered many unexpected difficulties. Solar energy is a multidisciplinary topic. The subject fields comprise astronomy, thermodynamics, quantum mechanics, solid-state physics, organic chemistry, solid-state electronics, environmental science, mechanical engineering, architecture, and civil engineering. As a unified textbook and reference book, a complete and consistent set of terminology and symbols must be designed which should be as consistent as possible with the established terminology and symbols of the individual fields, but yet be concise and self-consistent. A list of symbols is included toward the end of the book.

I sincerely thank Professors Irving Herman, Richard Osgood, and Vijay Modi for helping me setting up the solar energy course. I am especially grateful to many business executives and researchers in the field of solar energy who provided valuable information: Steve O'Rourke, then Managing Director and Research Analyst of Deutsch Bank, currently Chief Strategy Officer of MEMC Electronics, for detailed analysis of solar photovoltaic industry. John Breckenridge, Managing Director of investment bank Good Energies, for information on renewable energy investment in the world. Robert David de Azevedo, Executive Director of Brazilian American Chamber of Commerce, for information and contacts of renewable energy in Brazil. Loury A. Eldada, Chief Technology Officer of HelioVolt, for manufacture technology of CIGS thin-film solar cells. Ioannis Kymissis, a colleague professor at Columbia University, for two guest lectures in the Solar Energy Course about organic solar cells. Section 10.5 is basically based on literature suggested by him. Vasili Fthenakis, also a colleague professor at Columbia University, for valuable information about economy and environment issues of solar cells. John Perlin, a well-known solar energy historian, for kindly sending me electronic versions of his two books. George Kitzmiller, owner of Miami Pluming and Solar Heating Company, for showing me a number of 80-years-old solar hot water heaters still working in Miami. Margaret O'Donoghue Castillo, President of American Institute of Architects, for introducing me to the geothermal heating and cooling system in AIA, New York City. Mitchell Thomashaw, President of Union College, Maine, for letting me eyewitness the history of solar energy in the United States through brokering the donation of a Carter-era White House solar panel to the Solar Energy Museum in

Dézhōu, China. Academician Hé Zuòxiū, a prominent advocate of renewable energy, for helping me establish contacts in renewable-energy research and industry in China. Lǐ Shēnshēng, Professor Emeritus of Beijing Normal Institute, for kindly gifted me an autographed copy of his out-of-print book Tàiyángnéng Wùlǐxué. Published in 1996, it is probably the first book about the physics of solar energy in any language. Mr. Huáng Míng, founder and CEO of Himin Solar Energy Group and Vice President of International Solar Energy Association, for many inspiring discussions and a visit to Himin Corp, including an impressive production line for vacuum tube solar collectors. Professor Huáng Xuéjié, a long-time researcher of lithium rechargeable batteries and the founder of Phylion Battery Co., for many discussions about electric cars and a tour to the production lines of Phylion. Mire Ma, Vice President of Yingli Green Energy Group, for valuable information and a tour to the entire manufacturing process of solar-grade silicon, solar cells and solar modules. Last but not least, the book could not be written without the patience and support of my wife Liching.

C. Julian Chen

Columbia University
in the City of New York

April 2011

Chapter 1

Introduction

1.1 Shaping a More Livable World

The progress of human civilization depends critically on the utilization of energy. The modern industrial revolution is intimately associated with the generation and conversion of energy, as evidenced by the sequence of inventions that greatly improved the living conditions of human society, see Table 1.1.

In the 18th century, the invention and improvement of steam engine not only drove up the modern industry, but also revolutionized transportation, especially trains and steamboats. The invention of light bulbs by Thomas Edison in 1878 started the era of centralized generation and transmission of electrical power. The steam turbine was invented. After the invention of the internal combustion engines (ICE), including the Otto engine (gasoline engine) and the Diesel engine, the efficiency is increased to more than 20%. And The ICE is significantly lighter than the steam engine. Transportation was revolutionized. Petroleum-based liquid fuels took central stage.

In the 20th century, the worldwide industrialization made an explosive expansion of energy generation and conversion. The invention of airplanes and the widespread application of electricity accelerates that trend. Figure. 1.1 shows the annual consumption of energy in different sectors, the data of 2015 in the United States. The total energy consumption is 102.9 EJ. The largest sector of energy consumption is electricity,

Table 1.1: Inventions in energy conversion

Year	Invention	Inventor
1712	Original steam engine	Thomas Newcomen
1764	Improved steam engine	James Watt
1807	Steam boat	Robert Fulton
1814	Steam locomotive	George Stephenson
1861	Gaseline engine	Nicolaus Otto
1884	Steam turbine	Charles Parsons
1898	Diesel engine	Rudolph Diesel

Figure 1.1 Energy consumption by sectors. The energy consumption in the United States in 2015 by sectors. The total energy consumption is 102.9 EJ. Therefore, the number in EJ is also approximately the percentage for different sectors. The largest consumer is electricity, accounts about 40%. The second largest consumer is transportation, accounts for about 29%. Residential and commercial consumption accounts for about 10%.

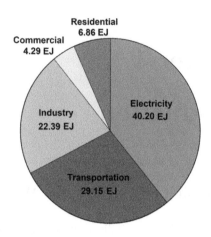

accounts almost 40%. The second largest is transportation, roughly 28%. Residential and commercial consumption account for about 10%.

1.1.1 Fossil Fuels and Beyond

In the early years of industrial revolution, up to late 19th century, coal was the main source of energy. For example, in 1750, Britian was producing 5.2 million tons of coal per year. By 1850, it was producing 62.5 million tons per year, more than 10 times greater than 1750. The efficiency of steam engine is much less than 10%. Currently, coal is still the main source of electrical energy in the world. Especially in many third-world countries, coal is the dominating source of electricity. In advanced industrialized countries, natural gas started to replace coal to become the main source of electrical power generation. After the invention of the internal combustion engines in late 19th century, an explosive growth of petroleum derivatives is seen.

Currently, fossil fuels are still the main source of energy, including coal, petroleum, and natural gas. Figure 1.2 shows the global consumption of fossil fuels from 1800 to 2021. The numbers are shown in Table 1.2. The surge of coal burning in early 21st century is because third-world countries like China and India are in an early stage of industrialization, and coal is indispensable. Nevertheless, the reliance in fossil fuels creates many problems detrimental to human civilization. Here is a short list.

1. The distribution of fossil fuel resource is extremely uneven over the world. According to a recent report by United Nations, about 80 percent of the global population lives in countries that are net importers of fossil fuels – that is about six billion people who are dependent on fossil fuels from other countries, which makes then vulnerable to geopolitical shocks and crises.

2. The burning of fossil fuels generates pollutions detrimental to environment and public health. One example is nitrogen oxides (NO_x). Because air contains about

Figure 1.2 History of fossil-fuels consumption In the 18th and 19th centuries, coal was the main source of energy. After the invention of internal combustion engines in late 19th century, petroleum takes the central stage. *Source:* U.S. Energy Information Administration (EIA/public domain).

80% of nitrogen, burning fossil fuel with air inevitably generates nitrogen oxides, which creates acid rain, hazy air, and causes respiratory problems.

3. The waste of fossil fuels, especially from coal, seriously damages the environment. For example, Duke Energy of North Carolina dumps more than 100 million tons of coal ash each year from generating electricity by burning coal. It seriously polluted the source of drinking water. In 2014, the state of North Carolina fined Duke Energy $25 million for the pollution. Oil tanker spills cause large-scale pollutions, for example the Exxon Valdez incident in 1989.

4. Coal mining and underwater oil drilling are among the most dangerous industrial activities. In early 20th century, in the United States, thousands of coal miners died of incidents every year. In third-world countries, coal mining incidents and deaths are still reported frequently. In 2010, the Deepwater Horizon oil drill rig exploded, causing disastrous prolusion in Gulf of Mexico, see Fig. 1.3.

5. The carbon dioxide emitted from burning fossil fuels causes global warming and

Table 1.2: Consumption of fossil fuels, in EJ

Year	1800	1850	1900	1950	2000	2020
Coal	0.29	2.05	20.62	45.37	97.84	159.7
Petroleum	0	0	0.65	19.6	159.3	191.7
Natural gas	0	0	0.23	7.53	86.4	138.7

Source: U.S. Energy Information Administration (EIA) / public domain

Figure 1.3 Deepwater Horizon oil rig explosion. On April 20, 2010, the Deepwater Horizon oil drill rig exploded, causing disastrous pollution in Gulf of Mexico, killed 11 workers and injured 17 others. The photo shows several fire extinguishing boats spraying water. *Source:* Reuters.

affecting Earth's climate system. According to the United Nations, as greenhouse gas emissions blanket the Earth, they trap the Sun's heat. This leads to global warming and climate change. The world is now warming faster than any point in recorded history. It should be amended immediately.

1.1.2 The Paris Agreement

The adverse impact of global warming and climate change due to the burning of fossil fuels has caught attentions of scientists and politicians over the world for many decades. In 1988, the Intergovernmental Panel on Climate Change (IPCC), an organization under the United Nations, started to publish annual reports that are the consensus of the world's scientists and experts AND the consensus of the world's governments too. In 2007, a Nobel Peace Prize was awarded to IPCC, "for their efforts to build up and disseminate greater knowledge about man-made climate change, and to lay the foundations for the measures that are needed to counteract such change;" because "global warming not only has negative consequences for human security, but can also fuel violence and conflict within and between states."

In 1992, the United Nations Framework Convention on Climate Change (UNFCCC) held an Earth Summit convention in Rio de Janeiro. A supreme decision-making body, The Conference of the Parties (COP), was established. The Convention was ratified by 194 countries in the world. Since 1995, the COP met each year to review the implementation of UNFCCC. Notable meetings include and COP21 in 2015, where a legally binding international treaty on climate change was reached by 196 parties, known as the "Paris Agreement".

The overarching goal of the Paris Agreement is "to hold the increase in the global average temperature to well below 2°C above pre-industrial levels" and pursue efforts "to limit global warming to 1.5°C above pre-industrial levels." According to the Paris agreement, all participating countries or regions should communicate actions they take to reduce their greenhouse gas emissions in order to reach the goals of the Paris agreement. This is called the nationally determined contributions (NDCs).

As reported by UNFCCC in 2022, in view of the Paris agreement, each country or region submits a plan to reduce the emission of greenhouse gases (GHG), especially a date to reach carbon neutrality, when the net emission of CO_2 will be zero. The commitments are reviewed in the subsequent COP conferences annually.

1.1.3 Phasing Out Coal-Generated Power

Science has made it clear that emissions from burning coal to generate electrical power is the single largest factor in the climate crisis. Currently, for many countries in the world, coal electricity generation still takes a substantial proportion, see Fig. 1.4.

Since the establishment of the Paris agreement, many countries and areas in the world initiated a Powering Past Coal Alliance (PPCA) to set up a date to end generating electric by burning coal. The United Kingdom and Canada were leading the way. In 2015, the United Kingdom was the first nation to commit to ending coal power by

Rank	Country	Coal generation (TWh)	Percentage of electricity production
1	China	4631	61
2	India	947	71
3	United States	774	19
4	Japan	274	29
5	South Korea	192	36
6	South Africa	191	86
7	Indonesia*	168	60
8	Russia	155	15
9	Vietnam	141	53
10	Australia	135	54

Figure 1.4 Volume and percentage of coal-generated electricity. Currently, in many countries in the world, coal electricity generation still takes a substantial proportion. According to the nationally determined contributions of the Paris Agreement, this situation is going to change.

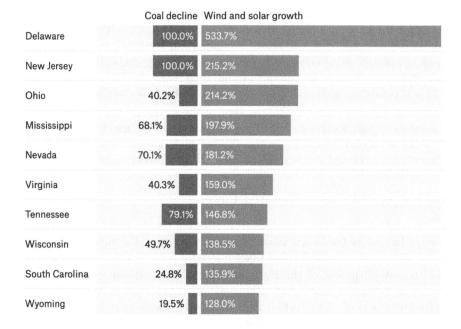

Figure 1.5 Phasing out coal electricity in the United States. In the United States, the commitment of phasing out coal electricity varies from state to state. Here is the plan for several states in 2029 to phase out coal electricity and increase renewable electricity, published in 2022.

2030. Subsequently, Canada announced in 2016 to eliminate coal power by 2030. In 2020, more than 40 countries joined PPCA. As a country, United States did not join. Nevertheless, since each state in the United States is as large as a mid-sized country in the world, and each has power to determine such a policy, 24 states in the United States joined PPCA. Figure 1.5 shows the plans of 10 states to reduce coal capacity and to increase wind and solar capacity by 2029.

1.1.4 Phasing Out ICE Vehicles

In accordance with the Paris Agreement, many countries in the world proposed dates to phase out internal combustion engine (ICE) vehicles. Here is a brief list.

2025 Norway.

2030 Germany, Israel, Netherlands, Sweden.

2035 Canada, China, Italy, Japan, Portugal, Thailand, United Kingdom, United States.

2040 Austria, India, Ireland, Mexico, New Zealand, Poland, Spain.

2050 Indonesia.

Note that in 2022, 80% of vehicles sold in Norway were electric, in the way to become the first nation to completely ban the sales of ICE vehicles. And even in 2022, the benefits of electrical vehicles are apparent. The streets in Norwegian cities become a lot quieter and cleaner. As reported on New York Times on May 8, 2023, in Oslo, previously, construction vehicles have been required to stop working when the children in nearby kindergartens napped. Now they can operate full time.

1.1.5 Economics of Renewable Energy

The inevitability of the transition from fossil-fuel energy to renewable energy is not only because of the harmful effects of burning fossil fuels. An even more practical reason is the constant reduction of the cost of renewable energy, especially solar PV electricity. Currently, solar PV is already the least expensive source of electricity due to the continuous improvement of technology. The trend will continue.

A major factor for the reduction of the cost of solar PV system is the decline of the cost of solar PV modules, see Fig. 1.6. In 2020, the cost of a watt pf PV module was about $3. In 2021, it dropped to about $0.2. Other costs such as inverters, other hardware including structural and electrical components, and soft costs are also declined mainly due to government policy and accumulation of experiences.

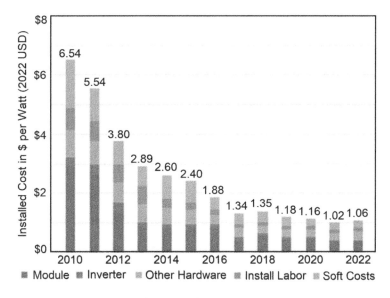

Figure 1.6 Cost decline of installed solar PV systems. According to The National Renewable Energy Laboratory (NREL), during the last decade, the total cost of installed solar PV system has declined by almost one order of magnitude. For example, the installed cost of utility-scale PV system with one-axis tracker per watt has declined from $6.54 per watt to $1.06 per watt.

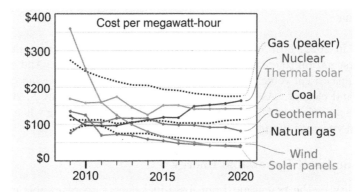

Figure 1.7 Cost of various energy resources. Levelized cost of electricity (LCOE) is a measure of the average net cost of electricity generation for a generator over its lifetime. Showing here is a chart of LCOE for several energy resources. See Wikipedia, levelized cost of electricity, 2022.

Figue 1.7 shows the variation of cost of various energy resources during the last decade. In 2009, solar PV was still a very expensive source of energy. During the single decade from 2010 to 2020, the cost of energy by solar PV dropped by a full order of magnitude, from about $350 to about $35 per megawatt-hour. For nuclear energy, the growing concern about safety and the related expenses cause a significant increase of its cost. On the other hand, due to mass production and technological advancement, the levelized cost of electricity generated from solar cells dropped to below coal-generated electricity in 2013, and dropped to below the wind-generated electricity in 2019, to become the least expensive source of electricity. The trend of further reduction of the cost of solar panels will continue in the foreseeable future. Therefore, from the point of view of pure economics, solar electricity will become the choice in the future.

In the following, we make a calculation of the cost of solar PV energy per kilowatt-hour (kWh), the unit of utility electricity. From Fig. 1.6, we can assume that in the future, the cost of installed solar PV system per kilowatt is $1000 or less. The lifetime of a solar PV system is no less than 25 years, which is 9125 days. By dividing the cost of installing solar PV system with the electricity generated over its lifetime in kWh, the cost of solar PV electricity per kWh can be calculated. Three typical cases are

Table 1.3: Production cost of solar PV electricity

Insolation	Lifetime kWh	Cost per kWh
5 h/day	45,625	$ 0.022
4 h/day	36,500	$ 0.027
3 h/day	27,375	$ 0.036

considered: places with average daily full sunshine of 3 h, 4 h, and 5 h, see Table 1.3. Apparently, at most places in the world, the production cost of a kWh of solar PV electricity is a lot less than its market price.

1.2 Solar Energy

According to well-established measurements, the average power density of solar radiation just outside the atmosphere of the Earth is 1366 W/m², known as the *solar constant*. After penetrating the atmosphere and reaches the surface of the Earth, the power density of sunlight is reduced, with average value of roughly 1 kW/m².

On the other hand, the original definition of the meter by the French Academy in 1791 is one ten-millionth of Earth's meridian, the length from the North Pole to the equator, see Fig. 1.8. This definition is still pretty accurate according to modern measurements. Therefore, the radius of Earth is $(2/\pi) \times 10^7$ m. Using the average value 1 kW/m², the total power of solar radiation reaching Earth is then

$$\text{Solar power} = 1 \times \frac{4}{\pi} \times 10^{17} \cong 1.26 \times 10^{17} \text{ W.} \qquad (1.1)$$

Each day has 86,400 s, and on average, each year has 365.24 days. The total energy of solar radiation reaching Earth per year is

$$\text{Annual solar energy} = 1.26 \times 10^{17} \times 86,400 \times 365.24 \cong 4 \times 10^{24} \text{ J.} \qquad (1.2)$$

For large amount of energy, the unit EJ, or 10^{18} J, is used. The annual solar energy reaching Earth is thus about 4 million EJ. As shown in Fig. 1.9, in 2020, the energy consumption of the entire world is about 635 EJ. A mere 0.016% of solar radiation energy reaching Earth can satisfy the energy need of the entire world.

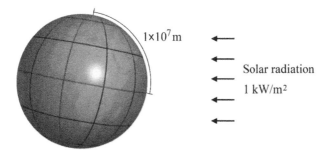

Figure 1.8 Annual solar energy arriving at surface of Earth. The average solar power reaching the Earth is 1 kW/m². The length of the meridian of Earth, according to the definition of the meter, is ten million meters. Using the value 1 kW/m², the total solar energy that arrives at the surface of Earth per year is about 4 million EJ.

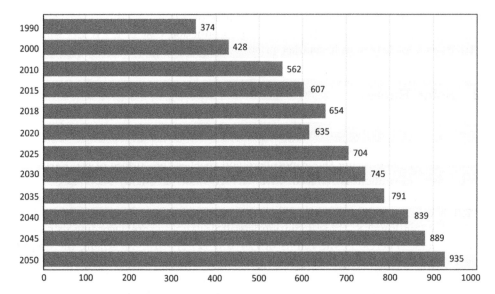

Figure 1.9 World marketed energy consumption. *Source:* Statistica. The unit is converted from quadrillion Btu to exajoule. The data for 2025 and beyond are projections.

It is interesting to compare the annual solar energy that reaches Earth with the proved total reserve of various fossil fuels, see Table 1.4. The numbers show that the total proved reserves of fossil fuel is approximately 1.5% of the solar energy that reaches the surface of Earth each year. Fossil fuels are solar energy stored as concentrated biomass over many millions of years. Actually, only a small percentage of solar energy was able to be preserved for mankind to explore. The current annual consumption of fossil fuel energy is approximately 450 EJ. If the current level of consumption of fossil fuel continues, the entire fossil energy reserve will be depleted in about 100 years.

In the 20th century, the cost of solar photovoltaic panels was very expensive. Prac-

Table 1.4: Proved resources of various fossil fuels

Item	Quantity	Unit Energy	Energy (EJ)
Crude oil	1.65×10^{11} tons	4.2×10^{10} J/ton	6,930
Natural gas	1.81×10^{14} m^3	3.6×10^7 J/m^3	6,500
High-quality coal	4.9×10^{11} tons	3.1×10^{10} J/ton	15,000
Low-quality coal	4.3×10^{11} tons	1.9×10^{10} J/ton	8,200
Total			36,600

Source: BP Statistical Review of World Energy, June 2007, British Petroleum.

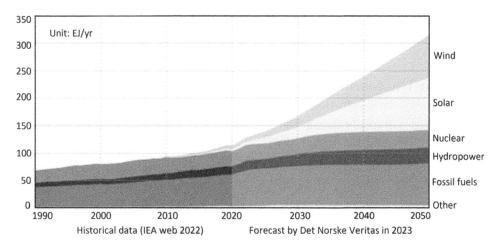

Figure 1.10 Sources of electricity: history and forecast. *Source:* Det Norske Veritas (DNV), a Norwegian institution. It includes the historical data published by International Energy Agency in 2022. Accordingly, in 2050, solar PV will become the largest source of electricity.

tically, the widespread use of solar electricity started in early 21st century. The new solar power installations reached 1 gigawatt (GW) in 2004, then reached 10 GW six years later in 2010, then reached 100 GW nine years later in 2019, then reached 150 GW in 2021. In 2022, the percentage of solar PV electricity already reached 3.5%. Because of the falling cost, globally, solar photovoltaic energy is the fastest–growing energy resource. A reasonable forecast is that around the middle of the 21st century, solar photovoltaics will become a dominant source of energy. Figure 1.10 is published by Det Norske Veritas, a Norwegian institution, that also includes the historical data published by International Energy Agency in 2022. According to that forecast, around 2050, solar PV will become the largest source of electricity.

The inevitability that fossil fuel will eventually be replaced by solar energy is simply a geological fact: The total recoverable reserve of crude oil is finite. For example, the United States used to be the largest oil producer in the world. By 1971, about one-half of the recoverable crude oil reserve in the continental United States (the lower 48 states) was depleted. Since then, crude oil production in this area started to decline. Therefore, crude oil production from more difficult geological and environmental conditions must be explored. Not only has the cost of oil drilling increased, but also the energy consumed to generate the crude oil has also increased. To evaluate the merit of an energy production process, the *energy return on energy invested* (EROI), also called *energy balance*, is often used: The definition is

$$\text{EROI} = \frac{\text{energy return}}{\text{energy invested}} = \frac{\text{energy in a volume of fuel}}{\text{energy required to produce it}}. \tag{1.3}$$

In the 1930's, the EROI value to produce crude oil was around 100. In 1970, it was 25. For deep-sea oil drilling, typical value is around 10. Shale oil, shale gas, and tar sands also have low EROI values. If the EROI of an energy production process is decreased to nearly 1, there is no value in pursuing in the process.

1.3   Solar Photovoltaics

It is recognized universally that in the first half of the twenty-first century, solar photovoltaics will becomes the largest source of electricity, replacing fossil fuels. In this section, we will present the history and an elementary overview of photovoltaics. Details will be presented in Chapters 7–10.

1.3.1   Birth of Modern Solar Cells

In 1953, Bell Labs set up a research project for providing electricity to remote parts of the world where no grid power was available. The leading scientist, Darryl Chapin, suggested using solar cells, and approved by his supervisors.

At that time, the photovoltaic effect in selenium, discovered in the 1870s, was commercialized for the measurement of light intensity for photography. Figure 1.11(a) is a schematic. A layer of Se is deposited on a copper substrate, then covered by a semitransparent film of gold. When the device is illuminated by visible light, a voltage is generated, which in turn generates a current. The electric current is proportional to the intensity of light. It has been a standard instrument in the first half of the 20th century for photographers to measure lighting conditions.

Chapin started his experiment with selenium photocells. He found that the efficiency, 0.5%, is too low to generate sufficient power for telephony applications. Then, there was a stroke of unbelievable luck. Two Bell Lab scientists involved in developing silicon transistors, Calvin Fuller and Gerald Pearson, joined Chapin in using the nascent silicon technology for solar cells; see Fig 1.12. In 1954, a solar cell with 5.7% efficiency was demonstrated [84]. A schematic is shown in Fig 1.11(b).

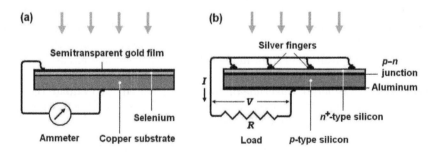

Figure 1.11 Selenium solar cell and silicon solar cell. (a) The selenium photovoltaic cell was discovered in the middle of the 19th century and was used for measuring light intensity in photography. (b) The silicon photovoltaic cell was invented at Bell Labs in 1954 using the technology for silicon transistors.

Figure 1.12 Inventors of silicon solar cells. Left to right: Gerald Pearson (1905–1987), Darryl Chapin (1906–1995), and Calvin Fuller (1902–1994). In 1953 Bell Labs set up a research project to provide energy sources for remote parts of the world where no grid power was available. Utilizing the nascent technology to make silicon transistors, in 1954, they designed and demonstrated the first silicon solar cells. The efficiency achieved, 5.7%, makes the solar cell a useful power source. Courtesy of AT&T Bell Labs.

The silicon solar cell was made from a single crystal of silicon. By judicially controlling the doping profile, a p–n junction is formed. The n-side of the junction is thin and highly doped to allow light to come to the p–n junction with very little attenuation, but the lateral electric conduction is high enough to collect the current to the front contact through an array of silver fingers. The back side of the silicon is covered with a metal film, typically aluminum. The basic structure of the silicon solar cell has remained almost unchanged until now.

The initial demonstration of the solar cell to the public in New York City was a fanfare. However, the cost of such solar cells was very high. From the mid 1950s to the early 1970s, photovoltaics research and development were directed primarily toward space applications. In 1976, the U.S. Department of Energy (DOE) was established. A Photovoltaics Program was created. The DOE, as well as many other international organizations, began funding research in photovoltaics aggressively. A terrestrial solar

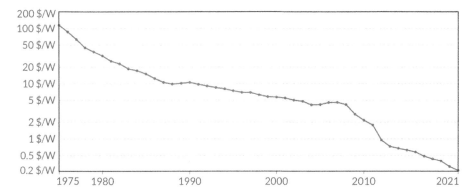

Figure 1.13 Average price of solar panels: 1975–2021. The average price of solar photovoltaic panels dropped from more than $100 per watt in 1975 to $0.2 per watt in 2021, a 500–fold reduction in 45 years. *Source*: Adapted from Our World in Data.

cell industry was established. Economies of scale and progress in technology reduced the price of solar panels dramatically. Figure 1.13 shows the evolution of price in a logarithmic scale from 1975 to 2021. As shown, the price of solar panels dropped 500 folds from more than \$100 per watt to \$0.2 per watt.

1.3.2 Basic Terms and Concepts on Solar Cells

Following is a list of key terms and concepts regarding solar cells.

Standard Illumination Conditions

The efficiency and power output of a solar module (or a solar cell) are tested under the following standard conditions: 1000 W/m^2 intensity, 25°C ambient temperature, and a spectrum that relates to sunlight that has passed through the atmosphere when the sun is at 42° elevation from the horizon, defined as air mass (AM) 1.5.

Fill Factor

The *open-circuit voltage* V_{op} is the voltage between the terminals of a solar cell under standard illumination conditions when the load has infinite resistance that is open. In this situation, the current is zero. The *short-circuit current* I_{sc} is the current of a solar cell under standard illumination conditions when the load has zero resistance. In this case, the voltage is zero. By using a resistive load R, the voltage V becomes smaller than V_{op}, and the current I becomes smaller than I_{sc}. The power $P = IV$. The maximum power output is determined by the condition

$$dP = d(IV) = IdV + VdI = 0. \tag{1.4}$$

Figure 1.14 shows the relation among these quantities. Denoting the point of maximum power by I_{mp} and V_{mp}, we have $P_{\text{max}} = I_{\text{mp}}V_{\text{mp}}$.

The fill factor of a solar cell FF is defined as

$$\text{FF} = \frac{P_{\text{max}}}{I_{\text{sc}}V_{\text{oc}}} = \frac{I_{\text{mp}}V_{\text{mp}}}{I_{\text{sc}}V_{\text{oc}}}. \tag{1.5}$$

Figure 1.14 Maximum power and fill factor. By connecting a load resistor to the two terminals of a solar cell, the solar cell supplies power to the load. The maximum power point occurs when $P = IV$ reaches maximum. At that point, $P_{\text{max}} = I_{\text{mp}}V_{\text{mp}}$. Obviously, there is always $I_{\text{mp}} < I_{\text{sc}}$ and $V_{\text{mp}} < V_{\text{oc}}$. The fill factor of a solar cell is defined as FF $= P_{\text{max}}/I_{\text{sc}}V_{\text{oc}} = I_{\text{mp}}V_{\text{mp}}/I_{\text{sc}}V_{\text{oc}}$.

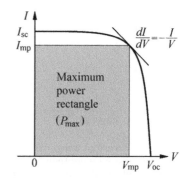

The typical value of the fill factor is between 0.8 and 0.9.

Efficiency

The efficiency of a solar cell is defined as the ratio of the output electric power over the input solar radiation power under standard illumination conditions at the maximum power point. For typical solar cells, the efficiency is between 20% and 25%.

Peak Watt

The "peak watt" (Wp) rating of a solar module is the power (in watts) produced by the solar module under standard illumination conditions at the maximum power point. The actual power output of a solar cell obviously depends on the actual illumination conditions. For a discussion of solar illumination, see Chapter 4.

1.3.3 Types of Solar Cells

The crystalline silicon solar cell was the first practical solar cell invented in 1954. Currently, the efficiency of such mass-produced solar cells is 20–25%, which is still the highest in single-junction solar cells. It also has a long life and a readiness for mass production. There are two versions of the crystalline silicon solar cell: monocrystalline and polycrystalline. To date, monocrystalline silicon solar cells account for more than 90% of the solar cell market. Thin film solar cells, especially CIGS (copper indium gallium selenide) and CdTe–CdS thin film solar cells, with a typical efficiency of around 15% and account for several percentages of the market. because the absorptivity of

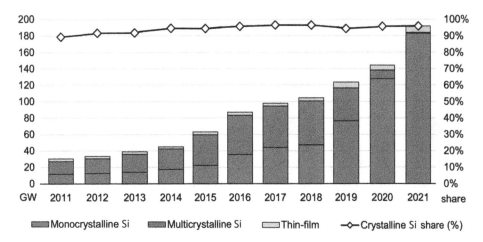

Figure 1.15 Volume and types of solar cells: 2011–2021. For the three major types of solar cells, the crystalline silicon solar cells take the greatest share and grew from 90% to 95% during a decade. Note that the monocrystalline silicon solar cell was the original type invented in 1954.

those materials are very high, little materials are needed to build a high-efficiency solar cells. Nevertheless, some necessary elements are rare and expensive. See Fig. 1.15 for a statistics of various types of solar cells over a decade.

1.4 A Rechargeable Battery Primer

Because solar radiation is intermittent, to make solar energy the main source, energy storage is essential. The most important energy storage device, lithium–ion rechargeable battery, is also revolutionizing transportation.

On October 9, 2019, the Royal Swedish Academy of Sciences announced that year's Nobel Prize in Chemistry was awarded to John B. Goodenough, M. Stanley Whittingham and Akira Yoshino for their contributions to the development of the lithium-ion battery. See Fig. 1.16. According to the Nobel Prize press release, "this lightweight, rechargeable and powerful battery is now used in everything from mobile phones to laptops and electric vehicles. It can also store significant amounts of energy from solar and wind power, making possible a fossil fuel-free society."

Lithium is the first metal element in the periodic table. It has just one electron in its outer electron shell, and that electron has a strong drive to leave lithium and move to elsewhere. When this happens, a positively charged – and more stable – lithium ion is formed. Lithium ion has a very small radius and a very small mass, enabling it to move freely from one electrode to another electrode to make intercalation.

Figure 1.16 Winners of the 2019 Nobel Prize in Chemistry. John B. Goodenough (left), M. Stanley Whittingham (middle), and Akira Yoshino (right) are the winners of the 2019 Nobel Prize in Chemistry for their contributions to the development of the lithium-ion battery. *Source*: Courtesy of The Royal Swedish Academy of Sciences.

1.4.1 Whittingham's Initial Invention

The first lithium–ion battery was demonstrated in 1978 by M. Stanley Whittingham, then a researcher at Exxon Mobile. After conducting a long-term systematic study of the intercalation of alkali metal atoms in layered compounds, a Eureka moment came up. Through many years of study, Whittingham realized that the intercalation process can be utilized to make high-energy-density rechargeable batteries. He started with a layered material titanium disulfide TaS_2. Lithium ions can be intercalated into the space between adjacent layers, without significantly alter its crystallographic structure, see Fig. 1.17. The intercalation of Li ions releases a significant amount of energy, which can be output as electrical energy. And most importantly, the intercalation action is *reversable*. By applying a voltage of an opposite polarity, the Li ions can be *deintercalated* from TaS_2 to become free Li ions, and move back to another electrode, made of metallic lithium. To complete the electrochemical system, an electrolyte is needed to allow the Li ion to move between the electrodes. To prevent short circuit, a separator made of porous plastic film is needed. Whittingham's team resolved all the problems and successfully built a prototype lithium-ion rechargeable battery.

After a successful demonstration, Whittingham convinced the executives of Exxon to develop a commercial battery based on his discovery. An experimental production line was established under the supervision of Whittingham. The first commercial rechargeable lithium-ion batteries appeared in the late 1970s.

Unexpectedly, during the test production, fire accidents occurred. In the recharging

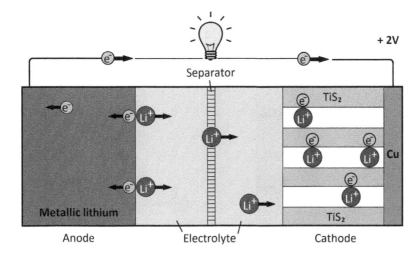

Figure 1.17 The first lithium-ion rechargeable battery of Stanley Whittingham. The cathode, or the positive electrode, is made of TiS_2, a layered material. The anode, or the negative electrode, is metallic lithium. Li ions can be intercalated into the space between adjacent layers of TiS_2 without significantly alter the crystalline structure, but releases a significant amount of energy, then output as electrical energy. By applying a voltage of opposite polarity, Li ions can deintercalated from TiS_2, and moved back to the metallic lithium anode.

process, Li ions move back from TiS_2 to the metallic Li electrode. However, instead of plating uniformly, dendrites grow across the separator to cause an internal short-circuit with incendiary and explosive consequences. Furthermore, the output voltage, 2V, is too small for valuable applications. As a result, the initial effort to commercialize a Li-ion rechargeable battery was suspended.

1.4.2 Goodenough's Improved Cathode

At that time, John B. Goodenough, an engineering professor at the University of Texas at Austin, learned about Whittingham's revolutionary battery. His specialized knowledge of matter's interior told him that the cathode could have a higher potential if it was built using metal oxide instead of a metal sulfide. In the 1980s, he experimented using lithium cobalt oxide Li_xCoO_2 as the cathode instead of TiS_2, which paid off: the battery doubled its energy potential to 4V instead of 2V. Note that the host material, CoO_2, is also a layered material. Each layer consists of three sheets, a sheet of cobalt atoms sandwiched between two sheets of oxygen atoms, see Fig. 1.18.

One key to this success was John Goodenough's realization that batteries did not have to be manufactured in a charged state. The rechargeable battery can be manufactured in a discharged state, and then being charged before using. In 1980, he published the discovery of this new, energy-dense cathode material which, resulted in powerful, high-capacity batteries. This was a decisive step towards the revolution in energy storage. Nevertheless, cobalt is a rare and expensive material, and the high energy density

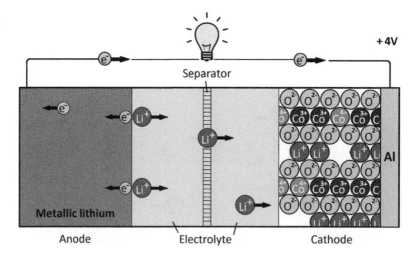

Figure 1.18 Improved Li-ion battery cathode materials of John Goodenough. Instead of using titanium disulphide as the cathode, John Goodenough discovered that by using an oxide of transition metal, for example Li_xCoO_2, the output voltage is increased from 2V to 4V. A key new idea is, that batteries could have been manufactured in a completely discharged state, then being charged using an external power supply before use. The cobalt-oxide based cathode enabled the commercialization of the Li-ion battery in 1991.

causes fire accidents. Other cathode materials are looked for.

After Li_xCoO_2 cathode material was commercialized, in 1997, John Goodenough discovered another cathode material, lithium iron phosphate Li_xFePO_4. Although the energy density of the Li_xFePO_4 cathode is lower, the battery is much safer. Furthermore, the raw materials are much less expensive. Currently, lithium iron phosphate is becoming the most commercialized cathode material.

1.4.3 Yoshino's Improved Anode

Akira Yoshino of Meijo University in Nagoya, Japan, made other breakthroughs. While the metallic lithium anode causes fire and explosion, in 1985, Yoshino discovered a carbon-based material as the anode to contain the Li ions by intercalation, see Fig 1.19. The carbonbased anode material prevented fire and explosion at the same time provided a high nominal voltage. It becomes the basis of later products. He also invented a novel multilayered battery structure and a continuous roll-to-roll processing method for mass production [122]. The novel structure and method of continuous manufacturing paved the way of large-scale manufacturing of the Li-ion battery.

The first commercialized Li-ion rechargeable battery was produced in a discharged state, with pure carbon anode and $LiCoO_2$ cathode. The charging process is:

$$C + xe^- + xLi^+ \Longrightarrow Li_xC \qquad \text{(anode)}. \qquad (1.6)$$

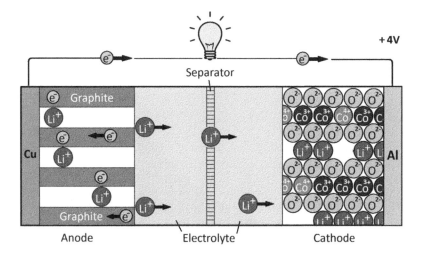

Figure 1.19 Improved Li-ion battery anode materials of Akira Yoshino. Rather than using metallic lithium as the anode, Akira Yoshino discovered that a carbon-based material like graphite can be used to store Li ions in the space between adjacent layers of graphene. He also invented a roll-to-roll continuous manufacturing method. The new anode and the new mass production method enabled the first commercial Li-ion rechargeable battery marketed by Sony in 1991.

$$LiCoO_2 - xe^- - xLi^+ \Longrightarrow Li_{1-x}CoO_2 \quad \text{(cathode).} \tag{1.7}$$

And the discharging process is:

$$Li_xC - xe^- - xLi^+ \Longrightarrow C \quad \text{(anode).} \tag{1.8}$$

$$Li_{1-x}CoO_2 + xe^- + xLi^+ \Longrightarrow LiCoO_2 \quad \text{(cathode).} \tag{1.9}$$

In 1991, a major Japanese electronics company started selling the nascent lithium-ion batteries, leading to a revolution in electronics. Mobile phones shrank, computers became portable, pocket MP3 players and tablets were developed.

1.4.4 Current Status

Since the commercialization of the Li-ion rechargeable batteries in 1991, the market has been expanding rapidly, see Fig 1.20. In late 20th century and the 2000s, the main market was electronic devices, such as smart phones, laptop computers, hand-held tools, and digital cameras. The second, and more important application of the lithium ion battery is in transportation and large-scale energy storage for implementation of wind and solar energy. In 2000, there was still no application of Li-ion battery on automobiles, partly because of the high cost. Since 2010, the use of Li-ion batteries in automobiles exploded. In 2016, more than one-half of the Li-ion batteries are installed in vehicles. The large-scale energy storage for wind and solar energy also started in 2010s, and it is expected to become significant in the future.

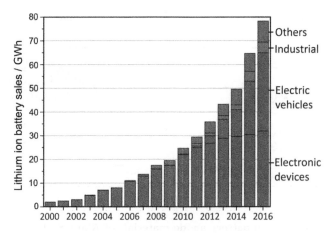

Figure 1.20 The expansion of Li-ion rechargeable battery market. Since the commercialization of the Li-ion batteries in 1991, the applications of those batteries expends rapidly. In late 20th century and the first decade of the 21st century, the main market was electronic devices. Since 2010, the use of Li-ion batteries in automobiles exploded. *Source*: [86] / Springer Nature.

Accompanying the market explosion, the design and technology are advanced rapidly. One serious problem is the mineral resources. Both lithium and cobalt are rare material, the source is limited and facing pressure from international geopolitical conflicts. To resolve such problems, various approaches are experimented. Instead of using pure cobalt, a mixture of transition metals are used instead of pure cobalt. The most successful example is the tri-metal cathode, with a mixture of nickel, manganese, and cobalt. A late discovery of John Goodenough, the iron phosphate-based cathodes, Li_xFePO_4, becomes popular. Furthermore, sodium ion rechargeable batteries are under rapid research and development. We will present those topics in Chapter 13.

1.5 Other Renewable Energy Resources

Although solar energy is by far the largest resource of renewable energy, other renewable energy resources, including hydropower, wind power, shallow geothermal energy, and geothermal energy, and tidal energy, have been extensively utilized. Except the last two items, they are derived from solar energy.

1.5.1 Hydroelectric Power

Hydroelectric power is a well-established technology. Since the late 19th century, it has been producing substantial amounts of energy at competitive prices. Currently, it produces about one-sixth of the world's electric output. As shown in Fig. 1.21, for many countries, hydropower accounts for a large percentage of total electricity. For example, Norway generates more than 98% of all its electricity from hydropower; in

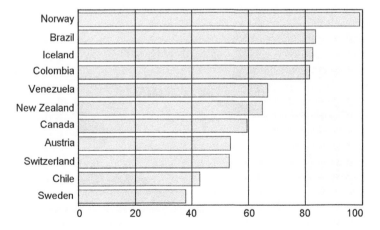

Figure 1.21 Percentage of electricity generation from hydropower in various countries.
Norway generates virtually all its electricity from hydropower; in Brazil, Iceland, and Colombia, more than 80% of electricity is generated by hydropower.

Table 1.5: Regional hydroelectric power potential and output

Region	Output (EJ/year)	Resource (EJ/year)	Percentage Explored
Europe	2.62	9.74	27%
North America	2.39	6.02	40%
Asia	2.06	18.35	11%
Africa	0.29	6.80	4.2%
South America	1.83	10.05	18%
Oceania	0.14	0.84	17%
World	9.33	51.76	18%

Source: Ref. [14], Chapter 5.

Brazil, Iceland, and Colombia, more than 80% of electricity is generated by hydropower. Table 1.5 lists the utilization of hydropower in various regions of the world.

The physics of hydropower is straightforward. A hydropower system is characterized by the *effective head*, the height H of the water fall, in meters; and the *flow rate*, the rate of water flowing through the turbine, Q, in cubic meters per second. The power carried by the water mass is given as

$$P(\text{kW}) = g \times Q \times H, \tag{1.10}$$

where g, 9.81 m/s^2, is the gravitational acceleration. Because a 2% error is insignificant,

Figure 1.22 Itaipu hydropower station at border of Brazil and Paraguay. With a capacity of 14.0 GW, the Itaipu hydropower station is one of the world's largest and generates about 20% of Brazil's electricity.

in the engineering community, it always takes $g \approx 10\,\mathrm{m/s}^2$. Thus, in terms of kilowatts,

$$P(\mathrm{kW}) = 10 \times Q \times H. \tag{1.11}$$

The standard equipment is the Francis turbine, invented by American engineer James B. Francis in 1848. With this machine, the efficiency η of converting water power to mechanical power is very high. Under optimum conditions, the overall efficiency of converting water power into electricity is greater than 90%, which makes it one of the most efficient machines. The electric power generated by the hydroelectric system is

$$P(\mathrm{kW}) = 10\,\eta\,QH. \tag{1.12}$$

A significant advantage over other renewable energy resources is that hydropower provides an energy storage mechanism of very high round-trip efficiency. The energy loss in the storage process is negligible. Therefore, the hydropower station together with the reservoir makes a highly efficient and economic energy storage system. Figure 1.22 is a photo of one of the world's largest hydropower station, the Itaipu hydropower station, which supplies about 20% of Brazil's electricity.

1.5.2 Wind Power

Generating electricity by wind turbines has been practiced since late 19th century, with potential just next to solar energy. The theory of wind power utilization is straightforward. The kinetic energy in a volume of air with mass m and velocity v is

$$\text{Kinetic energy} = \frac{1}{2}mv^2. \tag{1.13}$$

If the density of air is ρ, the mass of air passing through a surface of area A perpendicular to the velocity of wind per unit time is

$$\frac{dm}{dt} = \rho v A. \tag{1.14}$$

The wind power P_0, or the kinetic energy of air moving through an area A per unit time, is then

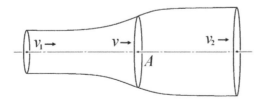

Figure 1.23 Derivation of Betz theorem of wind turbine. Wind velocity before the turbine rotor is v_1, and wind velocity after the turbine rotor is v_2. The velocity at the rotor is the average velocity, and the power generated by the rotor is related to the difference in kinetic energy.

$$P_0 = \frac{1}{2}\frac{dm}{dt}v^2 = \frac{1}{2}\rho v^3 A. \tag{1.15}$$

Under standard conditions (1 atm pressure and 18°C), the density of air is 1.225 kg/m³. If the wind speed is 10 m/s, the wind power density is

$$P_0 \approx 610 \text{ W/m}^2. \tag{1.16}$$

It is of the same order of magnitude as the solar power density.

However, the efficiency of a wind turbine is not as high as that of hydropower. Because the air velocity before the rotor, v_1, and the air velocity after the rotor, v_2, are different, see Figure 1.23, the air mass flowing through area A per unit time is determined by the *average wind speed* at the rotor,

$$\frac{dm}{dt} = \rho A \frac{v_1 + v_2}{2}. \tag{1.17}$$

Thus, the kinetic energy picked up by the rotor per unit time is

$$\text{Rate of kinetic energy loss} = \frac{1}{2}\frac{dm}{dt}v_1^2 - \frac{1}{2}\frac{dm}{dt}v_2^2. \tag{1.18}$$

Combining Eqs. 1.17 and 1.18, we obtain an expression of the wind power P picked up by the rotor,

$$P = \frac{1}{4}\rho A(v_1 + v_2)\left[v_1^2 - v_2^2\right]. \tag{1.19}$$

Rearranging Eq. 1.19, we can define the fraction C of wind power picked up by the rotor, or the *rotor efficiency*, as

$$P = \frac{1}{2}\rho v_1^3 A \left[\frac{1}{2}\left(1 + \frac{v_2}{v_1}\right)\left(1 - \frac{v_2^2}{v_1^2}\right)\right] = P_0\, C. \tag{1.20}$$

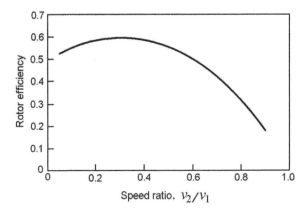

Figure 1.24 Efficiency of wind turbine. See Eq. 1.23. As shown, the maximum efficiency is 16/27, which occurs at a speed ratio of $v_2/v_1 = 1/3$.

Hence,

$$C = \frac{1}{2}\left(1 + \frac{v_2}{v_1}\right)\left(1 - \frac{v_2^2}{v_1^2}\right). \tag{1.21}$$

Let $x = v_2/v_1$ be the ratio of the wind speed after the rotor and the wind speed before the rotor. Then we have

$$C = \frac{1}{2}\left(1 + x\right)\left(1 - x^2\right). \tag{1.22}$$

The dependence of rotor efficiency C with speed ratio x is shown in Fig. 1.24. It is straightforward to show that the maximum occurs at $x = 1/3$ where $c = 16/27 = 59.3\%$. This result was first derived by Albert Betz in 1919 and is widely known as Betz's theorem or the Betz limit.

The estimate of worldwide available wind power varies. A conservative estimate shows that the total available wind power, 75 TW, is more than five times the world's total energy consumption. In contrast to hydropower, currently, only a small fraction of wind power has been utilized. However, it is growing very fast. From 2000 to 2009, total capacity grew nine fold to 158.5 GW. The Global Wind Energy Council expects that by 2014, total wind power capacity will reach 409 GW.

Because of a shortage of conventional energy resources, in the late 19th century, Denmark began to developed wind power and accelerated production after the 1970 energy crisis. Denmark is still the largest manufacturer of wind turbines, led by Vestas Cooperation, and it has about 20% of wind power in its electricity blend. Figure 1.25 is a photo the author took in Copenhagen. The little mermaid is staring at a dense array of wind turbines instead of the Prince.

Figure 1.25 Wind turbines in Copenhagen. A photo taken by the author in Copenhagen, Denmark, 2006. The statue of the little mermaid, a national symbol of Denmark, is staring at a dense array of wind turbines rather than the Prince.

Denmark's success in wind energy could not be achieved without its neighbors: Norway, Sweden, and Germany [50]. Because wind power is intermittent and irregular, a stable supply of electricity must be accomplished with a fast-responding power generation system as energy storage. Fortunately, almost 100% of the electricity in Norway is generated by hydropower, and the grids of the two countries share a 1000-MW interconnection. In periods of heavy wind, the excess power generated in Denmark is fed into the grid in Norway. By using the reversible turbine, the surplus electrical energy is stored as potential energy of water in the reservoirs. In 2005, the author visited the Tonstad Hydropower Station in Norway on a Sunday afternoon. I asked a Norwegian engineer why the largest turbine was sitting idle. He explained that one of the missions of that power station is to supply power to Denmark. On Monday morning, when the Danes brew their coffee and start to work, that turbine would run full speed.

1.5.3 Biomass and Bioenergy

Over many millennia in human history, until the industrial revolution when fossil fuels began to be used, the direct use of biomass was the main source of energy. Wood, straw, and animal waste were used for space heating and cooking. Candle (made of whale fat) and oil lamps were used for light. The mechanical power of horses was energized by feeding biomass. In many developed countries, this situation remains. Even in highly developed countries, direct use of biomass is still very common: for example, firewood for fireplaces and wood-burning stoves.

Biomass is created by photosynthesis from sunlight. For details, see Section 10.1. Although the efficiency of photosynthesis is only about 5% and land coverage by leaves is only a few percent, the total energy currently stored in terrestrial biomass is estimated to be 25,000 EJ, roughly equal to the energy content of the known fossil fuel reserve of the world, see Table 1.4. The energy content of the annual production of land biomass is about six times the total energy consumption of the world; see Table 1.6.

Currently, there is a well-established industry to generate liquid fuel using biomass for transportation. Two approaches are widely used: produce ethanol from sugar and produce biodiesel from vegetable oil or animal oil.

Table 1.6: Basic data of bioenergy

Item	In EJ/year	In TW
Rate of energy storage by land biomass	3000 EJ/year	95 TW
Total worldwide energy consumption	500 EJ/year	15 TW
Worldwide biomass consumption	56 EJ/year	1.6 TW
Worldwide food mass consumption	16 EJ/year	0.5 TW

Source: Ref. [14], page 107.

Figure 1.26 Costa Pinto Production Plant of sugar ethanol. The foreground shows the receiving operation of the sugarcane harvest; on the right in the background is the distillation facility where ethanol is produced. Courtesy of Mariordo.

Ethanol from Sugar Fermentation

The art of producing wine and liquor from sugar by fermentation has been known for thousands of years. Under the action of the enzymes in certain yeasts, sugar is converted into ethanol and CO_2:

$$C_6H_{12}O_6 \longrightarrow 2(C_2H_5OH) + 2(CO_2). \tag{1.23}$$

At the end of the reaction, the concentration of ethanol can reach 10–15% in the mixture, using specially cultured yeast, it can be up to 21%. Ethanol is then extracted by distillation.

One of the most successful examples is the production of ethanol from sugarcane in Brazil. An important number in the energy industry is *energy balance*, or EROI, the ratio of energy returned over energy invested; see Eq. 1.3. According to various studies, the energy balance in Brazil for sugar ethanol is over 8, which means that, in order to produce 1 J equivalent of ethanol, about 0.125 J of input energy is required. Also, the cost to produce 1 gal of ethanol in Brazil is about $0.83, much less than the cost of 1 gal of gasoline. This is at least partially due to the climate and topography of São Paulo, the south-east state of Brazil, a flat subtropical region with plenty of rainfall and sunshine. Since the advance of the flex-fuel automobiles in 2003 which can efficiently use an mixture of gasoline and ethanol of any proportion, the consumption of ethanol dramatically increased because of its low price. In 2008, as a world first, ethanol overtook gasoline as Brazil's most used motor fuel [55]. Figure 1.26 is a panoramic view of the Costa Pinto Production Plant for producing ethanol located in Piracicaba, São Paulo state. The foreground shows the receiving operation of the sugarcane harvest. On

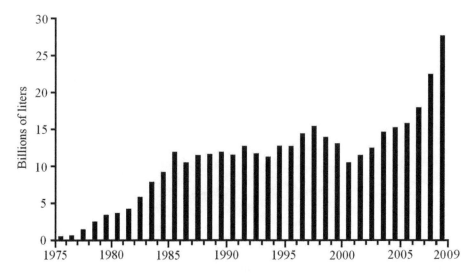

Figure 1.27 Annual production of ethanol in Brazil. *Source: Anuário Estatístico da Agroenergia* 2009, Ministério da Agricultura, Pecuária e Abastecimento, Brazil.

the right side, in the background, is the distillation facility where ethanol is produced. This plant produces all the electricity it needs from the bagasse of sugarcane left over by the milling process, and it sells the surplus electricity to public utilities.

Although the Brazil government makes no direct subsidy for the use of ethanol, there is continuous government-supported research to improve the efficiency of production and mechanization of the process. It is an important factor for the success of the Brazilian sugar-ethanol project. From 1975 to 2003, yield has grown from 2 to 6 m^3/ha. Recently, in the state of São Paulo, it has reached 9 m^3/ha. Figure 1.27 shows the annual production of fuel-grade ethanol in Brazil from 1975 to 2010. Although Brazil currently produces more than 50% of the fuel for domestic automobiles and about 30% of the world's traded ethanol, it only uses 1.5% of its arable land.

Biodiesel from Vegetable Oil or Animal Fat

Another example of using biomass for liquid fuel is the production of biodiesel from vegetable oil or animal fat. The chemical structures of vegetable oil and animal fat molecules are identical: triglyceride is formed from a single molecule of glycerol and three molecules of fatty acid; see Fig. 1.28. A fatty acid is a carboxylic acid (characterized by a —COOH group) with a long unbranched carbon hydride chain. With different types of fatty acids, different types of triglycerides are formed. Although vegetable oil can be used in diesel engines directly, the large molecule size and the resulting high viscosity as well as the tendency of incomplete combustion could damage the engine. Commercial biodiesel is made from reacting triglyceride with alcohol, typically methanol or ethanol. Using sodium hydroxide or potassium hydroxide as catalyst, the

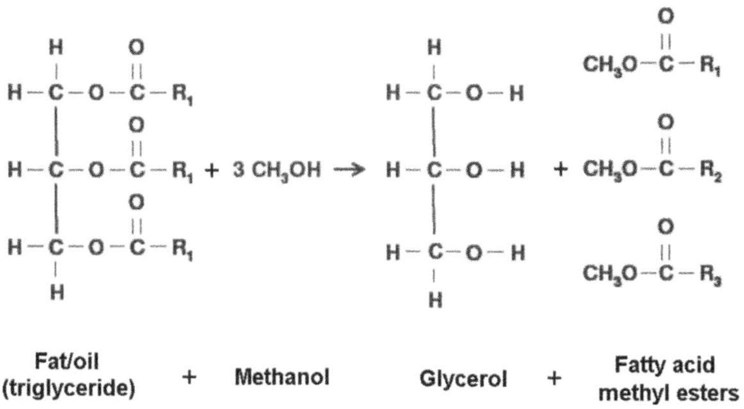

Figure 1.28 Production process of biodiesel. By mixing triglyceride with alcohol, using a catalyst, the triglyceride is transesterified to form three esters and a free glycerin. The ester, or the biodiesel, has a much smaller molecule size, which provides better lubrication to the engine parts.

triglyceride is transesterified to form three small esters and a free glycerin; see Fig. 1.28. The ester is immiscible with glycerin, and its specific gravity (typically 0.86–0.9 g/cm^3) is much lower than that of glycerin (1.15 g/cm^3). Therefore, the biodiesel can be easily separated from the mixture of glycerin and residuals.

The biodiesel thus produced has a much smaller molecule size than the triglycerides, which provides better lubrication to the engine parts. It was reported that the property of biodiesel is even better than petroleum-derived diesel oil in terms of lubricating properties and cetane ratings, although the calorific value is about 9% lower. Another advantage of biodiesel is the absence of sulfur, a severe environmental hazard of petroleum-derived diesel oil.

The cost and productivity of biodiesel depend critically on the yield and cost of the feedstock. Recycled grease, for example, used oil in making French fries and grease recovered from restaurant waste, is a primary source of raw materials. Byproducts of

Table 1.7: Yield of biofuel from different crop

For Ethanol	m^3/ha	For Biodiesel	m^3/ha
Sugar beet (France)	6.67	Palm oil	4.75
Sugarcane (Brazil)	6.19	Coconut	2.15
Sweet sorghum (India)	3.50	Rapeseed	0.95
Corn (United States)	3.31	Peanut	0.84

Source: Ref. [14], Chapter 4.

Figure 1.29 Oil palm fruit. The size and structure of oil palm fruit are similar to a peach or a plum. However, the soft tissue of the fruit contains about 50% palm oil. The yield of palm oil per unit plantation area is much higher than any other source of edible oil. The kernel is also rich in oil but of a different type. The oil palm kernel oil is critical ingredient of soap.

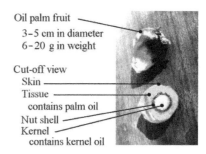

Oil palm fruit
3–5 cm in diameter
6–20 g in weight

Cut-off view
Skin
Tissue
  contains palm oil
Nut shell
Kernel
  contains kernel oil

the food industry, such as lard and chicken fat, often considered unhealthy for humans, are also frequently used. However, the availability of those handy resources is limited. Virgin oil is thus the bulk of the feedstock of biodiesel. The yield and cost of virgin oil vary considerably from crop to crop; see Table 1.7.

In Table 1.7, several crops for producing biofuels are listed, including those for producing ethanol. Two of them are sugar-rich roots (sugar beet and sweet sorghum). Harvesting these roots takes much more energy and is more labor intensive than harvesting sugarcane. Therefore, the energy balance (the ratio of energy produced versus the energy required to produce it) is often around 2, much lower than the case of sugarcane, which is higher than 8. The energy balance of corn is also lower (around 2), because the first step is to convert corn starch into sugar, which requires energy and

Figure 1.30 Wild oil palms in Africa. Oil palms are native trees in Africa which have supplied palm oil for centuries. Shown is a photo of wild oil palms taken by Marco Schmidt on the slopes of Mt. Cameroon, Cameroon, Africa.

labor. Palm oil, originally from Africa, has the highest yield per unit area of land. A photograph of the oil palm fruit is shown in Fig. 1.29. The fruit is typically 3–5 cm in diameter. The soft tissue of the fruit contains about 50% palm oil. The kernel contains another type of oil, the palm kernel oil, a critical ingredient for soap. Under favorable conditions, the yield of palm oil could easily reach 5 tons per hectare per year, far outstripping any other source of edible oil. Because it contains no cholesterol, it is also a healthy food oil. Currently, palm oil is the number one vegetable oil on the world market (48 million tons, or 30% of the world market share), with Malaysia and Indonesia, as the largest producers. Unlike other types of oil-producing plants (such as soybean and rapeseeds), which are annual, oil palms are huge trees; see Fig. 1.30. Once planted, an oil palm can produce oil for several decades.

1.5.4 Shallow Geothermal Energy

By definition, geothermal energy is the extraction of energy stored in Earth. However, there are two distinct types of geothermal energy depending on its origin: shallow geothermal energy and deep geothermal energy. Shallow geothermal energy is the solar energy stored in Earth, the origin of which will be described in Section 5.3. The temperature is typically some 10°C off that of the surface. The major application of shallow geothermal energy is to enhance the efficiency of the electrical heater and cooler (air conditioner) by using a vapor compression heat pump or refrigerator. Deep geothermal energy is the heat stored in the core and mantel of Earth. The temperature could be hundreds of degrees Celsius. It can be used for generating electricity and large-scale space heating. In this section, we will concentrate on shallow geothermal energy. Deep geothermal energy is presented in the following section.

The general behavior of the underground temperature distribution is shown in Fig. 1.31. At a great depth, for example, 20–30 m underground, the temperature is the annual average temperature of the surface, for example, $\overline{T} = 10°C$. At the

Figure 1.31 Shallow geothermal energy. Seasonal variation of underground temperature. On the surface, the summer temperature is much higher than in the winter temperature. Deeply underground, e.g., minus 20 m, the temperature is the annual average temperature of the surface. In the Summer, the temperature several meters underground is *lower* than the annual average; in the winter, the temperature several meters underground is *higher* than the annual average. The energy stored in Earth can be used for space heating and cooling, to make substantial energy savings.

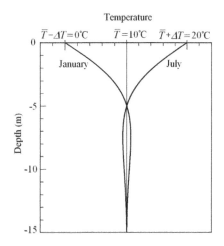

surface, the temperature varies with the seasons. In January, the temperature is the lowest, for example, $\overline{T} - \Delta T = 0°\text{C}$. In July, the temperature is the highest, for example, $\overline{T} + \Delta T = 20°\text{C}$. Because of the finite speed of heat conduction, at certain depth, typically -10 m below the surface, the temperature profile is *inverted*. In other words, in the summer, the temperature about 10 m underground is *lower* than the annual average; and in the winter, the temperature about 10 m underground is *higher* than the annual average.

The solar energy stored in Earth is universal and of very large quantity. In much of the temperate zone, it can be used directly for space cooling. By placing heat exchange structures underground and guiding the cool air through ducts to the living space, a virtually free air-conditioning system can be built. In areas with average temperature close to or slightly below 0°C, underground caves can be used as refrigerators, also virtually free of energy cost.

The major application of the shallow geothermal energy is the space heating and cooling systems using vapor-compression heat pump or refrigerator, taking the underground mass as a heat reservoir. Details will be presented in Chapter 6.

1.5.5   Deep Geothermal Energy

The various types of renewable energies presented in the previous sections are derivatives of solar energy. Deep geothermal energy, on the other hand, is the only major energy source not derived from solar energy. At the time Earth was formed from hot gas, the heat and gravitational energy made the core of Earth red hot. After Earth was formed, the radioactive elements continuously supplied energy to keep the core of Earth hot. Figure 1.32 is a schematic cross-section of Earth. The crest of Earth, a relatively cold layer of rocks with a relatively low density (2–3 g/cm^3), is divided into several *tectonic plates*. The thickness varies from place to place, from 0 to some 30 km. Underneath the crest is the *mantle*, a relatively hot layer of partially molten rocks

Figure 1.32   Deep geothermal energy. The origin of deep geothermal energy is the core of Earth. First, during the formation of Earth, gravitational contraction generated heat. Then, nuclear reactions in Earth continuously supplied energy. Because of the thickness of the tectonic plates, deep geothermal energy is economical only at the edges of the plate or near the volcanoes.

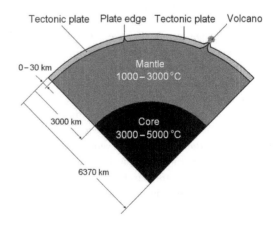

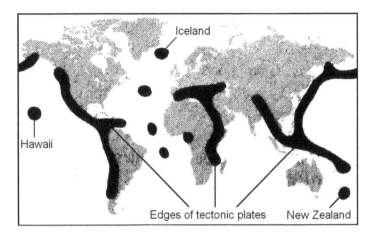

Figure 1.33 Regions for deep geothermal energy extraction. At the edges of tectonic plates and regions with active volcanoes, deep geothermal energy can be extracted economically.

with relatively high density (3–5.5 g/cm^3). It is the reservoir of magma for volcanic activities. From about 3000 km and down is the core of Earth, which is believed to be molten iron and nickel, with the highest density (10–13 g/cm^3).

The heat content of the mantle and the core is enormous. In principle, by drilling a deep well to the hot part of Earth, injecting water, superheated steam can be produced to drive turbines to generate electricity. In general, such operation is prohibitively expensive and difficult.

Figure 1.34 Nesjavellir geothermal power station, Iceland. Due the high concentration of volcanoes, Iceland has an unusual advantage of utilizing geothermal energy. Shown here is Nesjavellir Geothermal Power Station with a capacity of 120 MW.

Figure 1.35 The Rance Tidal Power Station, France. The barrage of the Rance Tidal Power Station, 750 m long, from Brebis point in the west to Briantais point in the east in France. In spite of the high development cost of the project, the costs have now been recovered.

Most current geothermal power stations are located either in the vicinity of edges of tectonic plates or in regions with active volcanoes, where the thickness of Earth's crest is less than a few kilometers and drilling to hot rocks is practical. Figure 1.33 shows the regions on Earth where deep geothermal energy can be extracted. Figure 1.34 shows a photo of the Nesjavellir geothermal power station, Iceland.

1.5.6   Tidal Energy

Tidal energy is a form of power produced by the natural rise and fall of tides caused by the gravitational interaction between Earth, the sun, and the moon. Tidal currents with sufficient energy for harvesting occur when water passes through a constriction, causing the water to move faster. Using specially engineered generators in suitable locations, tidal energy can be converted into useful forms of power, including electricity.

A successful example is the Rance Tidal Power Station located on the estuary of the Rance River in Brittany, France. Opened in 1966 as the world's first tidal power station, it is currently operated by Électricité de France. And for 45 years. its 24 turbines reach peak output at 240 megawatts (MW) and average 57 MW, a capacity factor of approximately 24%. See Fig. 1.35.

In spite of the high development cost of the project, the costs have now been recovered, and the current electricity production cost is 0.018 €/kWh, much less than the market price of electricity per kWh.

Problems

1.1. In the United States, the British thermal unit (Btu) is defined as the energy to raise the temperature of one pound water by one degree Fahrenheit. Show that to a good approximation, 1 Btu equals 1 kJ.

1.2. Approximately (to ±5%), how much energy is in one billion barrels of petroleum in gigajoules (GJ) and megawatt-hours (MWh)?

1.3. The area of New Mexico is 121,666 square miles. The average annual insolation (accumulated solar radiation energy density) is 2200 kWh/m². If one-half of the area of New Mexico is covered with solar panels of 10% efficiency, how much electricity can be generated per year? What percentage of United States energy needs can be satisfied? (The total energy consumption of the United States in 2007 was 100 EJ.)

1.4. The area of Tibet is 1,230,000 km². The average annual insolation (accumulated solar radiation energy density) is 3000 kWh/m². If one-half the area of Tibet is covered with solar panels of 10% efficiency, how much electricity per year can be generated? What percentage of the world's energy needs can be satisfied? (The total energy consumption of the world in 2007 was 500 EJ.)

1.5. A solar oven has a concentration mirror of 1 m² with a solar tracking mechanism. If the efficiency is 75%, on a sunny day, how long will it take to melt one kg of ice at 0°C at the same temperature? How long will it take to heat it to the boiling point? How long will it take to evaporate it at 100°C?

1.6. For wind speeds of 20, 30, 40, ..., up to 100 mph, calculate the wind power density (in watts per square meter).

1.7. The distance from the equator to the North Pole along the surface of Earth is 1.00×10^7 m. If the average solar radiation power density on Earth is one sun, how much energy is falling on Earth annually? If the annual energy consumption of the entire world in 2040 is 800 EJ, what percentage of solar energy is required to supply the world's energy need in 2040? (*Hint*: One day equals 24×60×60=86,400 seconds.)

1.8. Using a solar photovoltaic field of 1 square mile (2.59 km^2) with efficiency of 15%, how many kilowatt-hours will this field generate annually at locations of average daily insolation (on flat ground) of 3 h (Alaska), 4 h (New York), 5 h (Georgia), and 6 h (Arizona)? An average household consumes 1000 kWh per month. How many households can this field support in the four states, respectively?

Chapter 2

Nature of Solar Radiation

Solar energy comes to Earth in the form of radiation, or sunlight, with spectral components mostly in the visible, near infrared, and near ultraviolet. According to Maxwell's theory, radiation is an electromagnetic wave. Nevertheless, to understand the principles of solar cells and solar photochemistry, we must follow the theory of Planck and Einstein that the energy of radiation is *quantized*. Although the quantum of energy is called a *photon*, it does not mean that a photon is a geometrical point moving in space. For example, in the presentation of the blackbody radiation, the electromagnetic waves in a cavity are decomposed into standing waves in the cavity. Each standing wave spreads out through the entire space of the cavity. The energy of each such standing wave can only take discrete values. Nevertheless, each quantum of energy, or photon, is associated with the standing wave occupying the entire space of the cavity.

2.1 Light as Electromagnetic Waves

Up to the middle of the 19th century, electromagnetic phenomena and light have been considered as totally independent entities. In 1865, in a monumental paper *A Dy-*

Figure 2.1 James Clerk Maxwell. Scottish physicist (1831–1879), one of the most influential physicists along with Isaac Newton and Albert Einstein. He developed a set of equations describing electromagnetism, known as the *Maxwell's equations*. In 1865, based on those equations, he predicted the existence of electromagnetic waves and proposed that light is an electromagnetic wave [65]. He also pioneered the kinetic theory of gases, and created a science fiction character *Maxwell's demon*. Portrait courtesy of Smithsonian Museum.

namic Theory of the Electromagnetic Field, James Clerk Maxwell (Fig. 2.1) proposed that light is an electromagnetic wave [65]. In that paper, he developed a complete set of equations for the electromagnetic phenomena, now known as *Maxwell's equations*. Based on those equations, he predicted the existence of electromagnetic waves, propagating in free space with a speed that equals exactly the speed of light, which was then verified experimentally by Heinrich Hertz. Maxwell's bold postulation that light is an electromagnetic wave has since become one of the cornerstones of physics.

2.1.1 Maxwell's Equations

In vacuum, or free space, Maxwell's equations are

$$\nabla \cdot \mathbf{E} = \frac{\rho}{\varepsilon_0}, \tag{2.1}$$

$$\nabla \cdot \mathbf{B} = \mathbf{0}, \tag{2.2}$$

$$\nabla \times \mathbf{E} = -\frac{\partial \mathbf{B}}{\partial t}, \tag{2.3}$$

$$\nabla \times \mathbf{B} = \varepsilon_0 \mu_0 \frac{\partial \mathbf{E}}{\partial t} + \mu_0 \mathbf{J}. \tag{2.4}$$

Electric current cannot exist in free space. For linear, uniform, isotropic materials, the current density $\mathbf{J}$ is determined by the electric field intensity $\mathbf{E}$ through Ohm's law,

$$\mathbf{J} = \sigma \mathbf{E}. \tag{2.5}$$

The names, meanings, and units of the physical quantities in these equations are listed in Table 2.1. For example, the electric constant has an intuitive meaning as

Table 2.1: Quantities in Maxwell's equations

Symbol	Name	Unit	Meaning or Value
$\mathbf{E}$	Electric field intensity	V/m	
$\mathbf{B}$	Magnetic field intensity	T (tesla)	N/A·m
ρ	Electric charge density	C/m^3	
$\mathbf{J}$	Electric current density	A/m^2	
ε_0	Electric constant (permittivity of free space)	F/m	8.85×10^{-12} F/m
μ_0	Magnetic constant (permeability of free space)	H/m	$4\pi \times 10^{-7}$ H/m
σ	Conductivity	$(\Omega \cdot \text{m})^{-1}$	

follows. A capacitor made of two parallel conducting plates with area A and distance d has a capacitance $C = \varepsilon_0 A/d$ in farads. Similarly, the electric constant has an intuitive meaning as follows. An inductor made of a long solenoid of N loops with cross-sectional area A and length l has an inductance $L = \mu_0 N^2 A/l$ in henrys.

The constant $\varepsilon_0 \mu_0$ in Eq. 2.4 occurs very often. Defining a constant with the dimension of speed,

$$c = \frac{1}{\sqrt{\varepsilon_0 \mu_0}}, \tag{2.6}$$

Eq. 2.4 becomes

$$\nabla \times \mathbf{B} = \frac{1}{c^2} \frac{\partial \mathbf{E}}{\partial t} + \mu_0 \mathbf{J}. \tag{2.7}$$

From the values of ε_0 and μ_0 in Table 2.1,

$$c = 2.998 \times 10^8 \mathrm{m/s}. \tag{2.8}$$

We will show in Section 2.1.3 that c is the speed of electromagnetic waves in vacuum, that is the speed of light.

2.1.2 Vector Potential and Scalar Potential

To treat the electromagnetic field in space, a convenient method is to use the *vector potential*. From Eq. 2.2, it is possible to construct a vector field $\mathbf{A}$ which satisfies

$$\mathbf{B} = \nabla \times \mathbf{A}. \tag{2.9}$$

Then, Eq. 2.2 is automatically satisfied. Substituting Eq. 2.9 into Eq. 2.3, one obtains

$$\nabla \times \mathbf{E} = -\frac{\partial}{\partial t} \nabla \times \mathbf{A}. \tag{2.10}$$

For any function $\phi(\mathbf{r})$, $\nabla \times [\nabla \phi(\mathbf{r})] = \mathbf{0}$. Therefore, it is always possible to set up the vector potential $\mathbf{A}$ such that

$$\mathbf{E} = -\frac{\partial \mathbf{A}}{\partial t} - \nabla \phi, \tag{2.11}$$

where ϕ is the scalar potential of the electromagnetic field. The choice of the vector potential is not unique. By adding a gradient of an arbitrary function to it, values of the electric field and magnetic field do not change. This is called the *gauge invariance* of the vector potential and the scalar potential. One may require the potentials to satisfy gauge conditions. The Lorenz gauge is the most conveniant one,

$$\nabla \cdot \mathbf{A} + \frac{1}{c^2} \frac{\partial \phi}{\partial t} = 0. \tag{2.12}$$

Substitute Eq. 2.9 into Eq. 2.4, using Eq. C.13,

$$\nabla \times \mathbf{B} = \nabla \times (\nabla \times \mathbf{A})$$
$$= \nabla(\nabla \bullet \mathbf{A}) - \nabla^2 \mathbf{A}$$
$$= \mu_0 \mathbf{J} + \frac{1}{c^2} \frac{\partial \mathbf{E}}{\partial t} \qquad (2.13)$$
$$= \mu_0 \mathbf{J} + \frac{1}{c^2} \nabla \frac{\partial \phi}{\partial t} - \frac{1}{c^2} \frac{\partial^2 \mathbf{A}}{\partial t^2}.$$

Using Eq. 2.12, one reaches a wave equation for the vector potential,

$$\nabla^2 \mathbf{A} - \frac{1}{c^2} \frac{\partial^2 \mathbf{A}}{\partial t^2} = -\mu_0 \mathbf{J}. \qquad (2.14)$$

And using Eq. 2.11, we find another wave equation for the scalar potential,

$$\nabla^2 \phi - \frac{1}{c^2} \frac{\partial^2 \phi}{\partial t^2} = -\frac{\rho}{\varepsilon_0}. \qquad (2.15)$$

Those differential equations are equivalent to Maxwell's equations.

2.1.3   Electromagnetic Waves

In this section, we study the electromagnetic waves in free space, that is, where the electric charge ρ and current $\mathbf{J}$ are zero. Eqs. 2.14 and 2.15 become

$$\nabla^2 \mathbf{A} - \frac{1}{c^2} \frac{\partial^2 \mathbf{A}}{\partial t^2} = 0, \qquad (2.16)$$

$$\nabla^2 \phi - \frac{1}{c^2} \frac{\partial^2 \phi}{\partial t^2} = 0, \qquad (2.17)$$

which are wave equations with velocity c. Because of Eqs. 2.9 and 2.11, the electric field intensity and the magnetic field intensity also satisfy the same wave equation,

$$\nabla^2 \mathbf{E} - \frac{1}{c^2} \frac{\partial^2 \mathbf{E}}{\partial t^2} = 0 \qquad (2.18)$$

and

$$\nabla^2 \mathbf{B} - \frac{1}{c^2} \frac{\partial^2 \mathbf{B}}{\partial t^2} = 0, \qquad (2.19)$$

According to values of ε_0 and μ_0 coming from electromagnetic measurements in 1860s, the velocity of electromagnetic waves should be 3.1×10^8 m/s. On the other hand, experimental values of the speed of light at that time were 2.98×10^8–3.15×10^8 m/s. The difference was within experimental error. Maxwell proposed thusly [65]:

> The agreement of the results seems to show that light and magnetism are affections of the same substance, and that light is an electromagnetic disturbance propagated through the field according to electromagnetic laws.

Maxwell's theory of electromagnetic waves was experimentally verified by Heinrich Hertz in 1865. From recent electrical measurements, one finds $1/\sqrt{\varepsilon_0 \mu_0} = 2.998 \times 10^8$ m/s, which is exactly the speed of light in a vacuum, c.

2.1.4 Plane Waves and Polarization

The sunlight can be described with high accuracy as a plane wave. The mathematical form is much simpler than the general three-dimensional case, and the treatment can provide a good conceptual understanding.

Let the direction of propagation be z. All physical quantities only depend on z. Because there is no electric charge in vacuum, Eq. 2.1 becomes

$$\frac{\partial E_z}{\partial z} = 0. \tag{2.20}$$

The z-component of the electric field intensity is a constant. Because we are only looking at the wave, the constant electric field intensity can be omitted. Therefore, only the x-component and the y-component of the electric field exist. Similarly, Eq. 2.2 implies that the z-component of the magnetic field intensity is zero. In other words, the electromagnetic wave is *transverse*, where the intensity vectors **E** and **B** are perpendicular to the direction of propagation.

The non-zero components of Eq. 2.3 are

$$\frac{\partial E_y}{\partial z} = \frac{\partial B_x}{\partial t} \tag{2.21}$$

and

$$\frac{\partial E_x}{\partial z} = -\frac{\partial B_y}{\partial t}. \tag{2.22}$$

The non-zero components of Eq. 2.4 are

$$-\frac{\partial B_y}{\partial z} = \frac{1}{c^2}\frac{\partial E_x}{\partial t} \tag{2.23}$$

and

$$\frac{\partial B_x}{\partial z} = \frac{1}{c^2}\frac{\partial E_y}{\partial t}, \tag{2.24}$$

where Eq. 2.6 is applied. Differential Eq. 2.22 with z and differential Eq. 2.23 with t, we obtain

$$\frac{\partial^2 E_x}{\partial z^2} = \frac{1}{c^2}\frac{\partial^2 E_x}{\partial t^2}. \tag{2.25}$$

Similarly, differential Eq. 2.21 with z and differential Eq. 2.24 with t, we obtain

$$\frac{\partial^2 E_y}{\partial z^2} = \frac{1}{c^2}\frac{\partial^2 E_y}{\partial t^2}. \tag{2.26}$$

Equations 2.25 and 2.26 shows that the x-component and the y component of the electric field are *independent of each other*, which are two *polarizations* of the electromagnetic wave. We may study one of the polarizations at a time.

2.1.5 Sinusoidal Waves

Electromagnetic waves with a fixed frequency ν are particularly important. It is convenient to use the circular frequency $\omega = 2\pi\nu$ in mathematical expressions. For example, an x-polarized wave with frequency ν can be written as

$$E_x(z,t) = E_x(z)\, e^{-i\omega t}. \tag{2.27}$$

Insert Eq. 2.27 into Eq. 2.25, defining a *wavevector* k_z as

$$k_z = \frac{\omega}{c}. \tag{2.28}$$

one obtains a *Helmholtz equation* for $E_z(x)$:

$$\frac{d^2 E_x(z)}{dz^2} + k_z^2\, E_x(z) = 0. \tag{2.29}$$

The general solution is

$$E_x(z) = E_{x0}\, e^{i(k_z z - \omega t)}. \tag{2.30}$$

where E_{x0} is a constant.

Because any field variable follows the same space-and-time variation as Eq. 2.30, one can substitute the partial differentiation operator with a constant, such as

$$\frac{\partial}{\partial x} \longrightarrow ik_z, \tag{2.31}$$

and

$$\frac{\partial}{\partial t} \longrightarrow -i\omega. \tag{2.32}$$

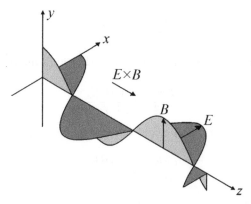

Figure 2.2 Electromagnetic wave. The electric field intensity $\mathbf{E}$ is perpendicular to the magnetic field intensity $\mathbf{B}$. The energy flux vector $\mathbf{S} = \mu_0^{-1}\mathbf{E} \times \mathbf{B}$ is formed from $\mathbf{E}$ and $\mathbf{B}$ by a right-hand rule.

Using the above expressions, we study the relation of the magnetic field and the electric field in a sinusoidal plane wave. For an x-polarized wave, Eq. 2.22 gives

$$ik_z E_x = i\omega B_y. \tag{2.33}$$

In other words, we have

$$B_y = \frac{k_z}{\omega} E_x = \frac{1}{c} E_x. \tag{2.34}$$

On the other hand, for a y-polarized wave, Eq. 2.21 gives

$$ik_z E_y = -i\omega B_x. \tag{2.35}$$

Similarly, we have

$$B_x = -\frac{k_z}{\omega} E_y = -\frac{1}{c} E_y. \tag{2.36}$$

In a summary, according to the electromagnetic theory of light, the electric field intensity vector is perpendicular to the direction of the propagation of light. The magnetic field intensity is perpendicular to both the direction of the electric field intensity vector and the direction of the propagation of light, and its magnitude is proportional to the electric field intensity. See Fig. 2.2.

2.2 Interface Phenomena

Maxwell's theory of light plays a critical role in the understanding of selective absorption films for solar thermal applications and antireflection films in photovoltaics. The general theory with an arbitrary incident angle is rather complicated. However, for applications related to solar energy, it suffices to study the case of normal incidence, which demonstrates most of the related physics. First, let us extend Maxwell's equations to dielectrics.

2.2.1 Relative Dielectric Constant and Refractive Index

Maxwell's equations, Eqs. 2.1–2.4 are used for the case of a vacuum. To describe electromagnetic phenomena in a nonmagnetic medium, the electric constant ε_0 is replaced by the electric constant of the medium, ε. Maxwell's equations are

$$\nabla \cdot \mathbf{E} = \frac{\rho}{\varepsilon}, \tag{2.37}$$

$$\nabla \cdot \mathbf{B} = \mathbf{0}, \tag{2.38}$$

$$\nabla \times \mathbf{E} = -\frac{\partial \mathbf{B}}{\partial t}, \tag{2.39}$$

$$\nabla \times \mathbf{B} = \varepsilon \mu_0 \frac{\partial \mathbf{E}}{\partial t} + \mu_0 \mathbf{J}. \tag{2.40}$$

Following the procedures in Section 2.1.1, in regions with no electric charge and no electric current, the wave equations for field intensities are:

$$\nabla^2 \mathbf{E} - \frac{1}{v^2}\frac{\partial^2 \mathbf{E}}{\partial t^2} = 0, \tag{2.41}$$

$$\nabla^2 \mathbf{B} - \frac{1}{v^2}\frac{\partial^2 \mathbf{B}}{\partial t^2} = 0, \tag{2.42}$$

where the velocity v is given as

$$v = \frac{1}{\sqrt{\varepsilon \mu_0}}. \tag{2.43}$$

Comparing with Eq. 2.6, the relation of v with c is

$$\frac{c}{v} = \sqrt{\frac{\varepsilon}{\varepsilon_0}}. \tag{2.44}$$

Defining the relative dielectric constant of the medium as

$$\varepsilon_r \equiv \frac{\varepsilon}{\varepsilon_0}, \tag{2.45}$$

the ratio of the speed of light in a vacuum and the speed of light in the medium, defined as the *refractive index n*, is

$$n \equiv \frac{c}{v} = \sqrt{\varepsilon_r}. \tag{2.46}$$

In general, the relative dielectric constant and the refractive index depend on the frequency or wavelength of the electromagnetic wave. For application in solar energy

Table 2.2: Dielectric constants and refractive indices

Material	Wavelength	ε_r	n
Silicon	1.39 μm	11.7	3.42
Germanium	2.1 μm	16.8	4.10
TiO_2	2.0 μm	5.76	2.4
SiO_2	Visible	2.40	1.55
Window glass	Visible	2.25	1.50
ZnS	Visible	5.43	2.33
CeO_2	Visible	3.81	1.953
CaF_2	Visible	2.06	1.435
MgF_2	Visible	1.91	1.383

Source: American Institute of Physics Handbook,
3rd Ed, McGraw-Hill, New York, 1982.

devices, the most relevant case is solar radiation in the visible or infrared. Table 2.2 shows the relative dielectric constant and refractive index of several materials often used in solar energy devices.

For electromagnetic waves propagating in the z-direction with wavevector k and electric field intensity in x, the nonzero components are

$$E_x = E_0 \, e^{i(kz - \omega t)} \tag{2.47}$$

$$B_y = \frac{k}{\omega} E_0 \, e^{i(kz - \omega t)}. \tag{2.48}$$

The wavevector k is given as

$$k = \frac{\omega}{v} = \frac{\omega n}{c}. \tag{2.49}$$

And, according to Eq. 2.46, the electric and magnetic fields are in phase and proportional,

$$B_y = \frac{1}{v} E_x = \frac{n}{c} E_x. \tag{2.50}$$

2.2.2 Energy Balance and Poynting Vector

Let us study the energy balance in an electromagnetic field by considering a unit volume with relatively uniform fields. If the current density is $\mathbf{J}$ and the electric field intensity is $\mathbf{E}$, the ohmic energy loss per unit time per unit volume is $\mathbf{J} \cdot \mathbf{E}$. Using Eq. 2.40, the expression of energy loss becomes

$$\mathbf{J} \cdot \mathbf{E} = \frac{1}{\mu_0} \mathbf{E} \cdot (\nabla \times \mathbf{B}) - \varepsilon \mathbf{E} \cdot \frac{\partial \mathbf{E}}{\partial t}. \tag{2.51}$$

Using the mathematical identity

$$\mathbf{E} \cdot (\nabla \times \mathbf{B}) = -\nabla \cdot (\mathbf{E} \times \mathbf{B}) + \mathbf{B} \cdot (\nabla \times \mathbf{E}), \tag{2.52}$$

Eq. 2.51 becomes

$$\mathbf{J} \cdot \mathbf{E} = -\nabla \cdot \left(\frac{1}{\mu_0} \mathbf{E} \times \mathbf{B} \right) + \frac{1}{\mu_0} \mathbf{B} \cdot (\nabla \times \mathbf{E}) - \varepsilon \mathbf{E} \cdot \frac{\partial \mathbf{E}}{\partial t}. \tag{2.53}$$

Using Eq. 2.39, Eq. 2.53 becomes

$$\mathbf{J} \cdot \mathbf{E} = -\nabla \cdot \left(\frac{1}{\mu_0} \mathbf{E} \times \mathbf{B} \right) - \frac{\partial}{\partial t} \left(\frac{\varepsilon}{2} E^2 + \frac{1}{2\mu_0} B^2 \right). \tag{2.54}$$

The right-hand side of Eq. 2.54 has a straightforward explanation. The energy density of the electromagnetic fields is

$$W = \frac{\varepsilon}{2} E^2 + \frac{1}{2\mu_0} B^2, \tag{2.55}$$

and the power density of the electromagnetic field per unit area is

$$\mathbf{S} = \frac{1}{\mu_0} \mathbf{E} \times \mathbf{B}.$$ (2.56)

The vector $\mathbf{S}$ is called the *Poynting vector* after its discoverer.

For an electromagnetic wave, according to Eq. 2.50, $cB_y = nE_x$. The magnitude of the Poynting vector along the direction of propagation is

$$S_z = \frac{n}{\mu_0 c} E_x^2.$$ (2.57)

2.2.3 Fresnel Formulas

Consider two media of refractive indices n_1 and n_2 with an interface at $z = 0$, as shown in Fig. 2.3. The incident light is moving in the z direction with wavevector k_I,

$$k_I = \frac{\omega n_1}{c}.$$ (2.58)

The field intensities of the incident light are

$$E_I = E_{I0}\, e^{i(k_I z - \omega t)},$$ (2.59)

$$B_I = \frac{n_1}{c} E_{I0}\, e^{i(k_I z - \omega t)},$$ (2.60)

where E_{I0} characterizes the intensity of incident light.

For transmitted light, the wavevector is determined by the refractive index of medium 2,

$$k_T = \frac{\omega n_2}{c}.$$ (2.61)

The field intensities of the transmitted light are

$$E_T = E_{T0}\, e^{i(k_T z - \omega t)},$$ (2.62)

$$B_T = \frac{n_2}{c} E_{T0}\, e^{i(k_T z - \omega t)},$$ (2.63)

The constant E_{T0} characterizing the intensity of the transmitted light is to be determined by the boundary conditions required by Maxwell's equations.

For reflected light, because it is in the same medium as the incident light, the absolute value of the wavevector is identical to that of the incident light. However, the direction of z is reversed. By using the same notation k_I, the field intensities of the reflected light are

$$E_R = E_{R0}\, e^{i(-k_I z - \omega t)},$$ (2.64)

$$B_R = -\frac{n_1}{c} E_{R0}\, e^{i(-k_I z - \omega t)}.$$ (2.65)

Notice the negative sign of the magnetic field intensity B_R. Again, the constant R characterizes the intensity of the reflected light.

Figure 2.3 Derivation of Fresnel formulas. Two media with indices of refraction n_1 and n_2 share an interface at $z = 0$. The incident light has a wavevector k_I. The wavevector of transmitted light is k_T. The wavevector of reflected light is identical to that of incident light but with opposite sign. By applying Maxwell's equations at the interface, the relations between the three components of light can be derived.

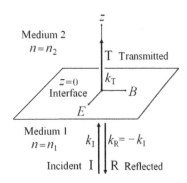

On the interface, $z = 0$, following Eqs. 2.1 and 2.2, both electric field intensity and magnetic field intensity should be continuous. In other words,

$$E_I + E_R = E_T, \tag{2.66}$$

$$B_I + B_R = B_T. \tag{2.67}$$

Using Eqs. 2.59–2.67, we find

$$E_{I0} + E_{R0} = E_{T0}, \tag{2.68}$$

$$n_1(E_{I0} - E_{R0}) = n_2 E_{T0}. \tag{2.69}$$

The solutions of Eqs.2.68 and 2.69 are

$$E_{R0} = \frac{n_1 - n_2}{n_1 + n_2} E_{I0}, \tag{2.70}$$

$$E_{T0} = \frac{2n_1}{n_1 + n_2} E_{I0}. \tag{2.71}$$

Equations 2.70 and 2.71 are the *Fresnel formulas* for the case of normal incidence. Obviously, if $n_1 = n_2$, there is no reflected light, and 100% of incident light is transmitted through the interface.

The power densities of the incident, transmitted, and reflected light can be evaluated using Eqs. 2.70 and 2.71 and the expression of the Poynting vector, Eq. 2.56. For incident light, the magnitude is

$$S_I = \frac{1}{\mu_0} E_I B_I = \frac{n_1}{\mu_0 c} E_{I0}^2. \tag{2.72}$$

For transmitted light,

$$S_T = \frac{1}{\mu_0} E_T B_T = \frac{n_2}{\mu_0 c} E_{T0}^2. \tag{2.73}$$

Using Eq. 2.70,

$$S_T = \frac{4n_1 n_2}{(n_1 + n_2)^2} \frac{n_1}{\mu_0 c} E_{I0}^2 = \frac{4n_1 n_2}{(n_1 + n_2)^2} S_I. \tag{2.74}$$

A dimensionless coefficient of transmission is defined as

$$\mathcal{T} \equiv \frac{S_{\mathrm{T}}}{S_{\mathrm{I}}} = \frac{4 n_1 n_2}{(n_1 + n_2)^2}. \tag{2.75}$$

Following Eqs. 2.73 and 2.74, the intensity of reflected light can be determined, and a dimensionless coefficient of reflection is defined as

$$\mathcal{R} \equiv \frac{S_{\mathrm{R}}}{S_{\mathrm{I}}} = \left(\frac{n_1 - n_2}{n_1 + n_2} \right)^2. \tag{2.76}$$

For semiconductors, the reflection loss can be significant. For example, for silicon, $n = 3.49$. The reflection coefficient is

$$\mathcal{R} = \frac{(1 - 3.49)^2}{(1 + 3.49)^2} \approx 0.3076. \tag{2.77}$$

More than 30% of light is lost by reflection. To build high-efficiency solar cells, an *antireflection coating* is essential. We will discuss this in Section 9.4.

2.2.4   Optics of metals

In the study of selective absorption surface, we encounter the optical phenomenon in metals. To study the optics of metals, we need to add Ohm's law to Maxwell's equations, Eq. 2.5:

$$\mathbf{J} = \sigma \mathbf{E}. \tag{2.78}$$

For simplicity, we consider the case of plane waves propagating in the z-direction, with one polarization, the electric field in the x-direction. Also, we assume that the metal is non-magnetic, such that dielectric constant and magnetic constant can be represented by the values in vacuum. Equation 2.23 becomes

$$-\frac{\partial B_y}{\partial z} = \frac{1}{c^2} \frac{\partial E_x}{\partial t} + \mu_0 \sigma E_x. \tag{2.79}$$

Differentiate with respect to time, Eq. 2.79 becomes

$$-\frac{\partial^2 B_y}{\partial z \partial t} = \frac{1}{c^2} \frac{\partial^2 E_x}{\partial t^2} + \mu_0 \sigma \frac{\partial E_x}{\partial t}. \tag{2.80}$$

Combining with Eq. 2.22, we find a differential equation for E_x:

$$\frac{\partial^2 E_x}{\partial z^2} = \frac{1}{c^2} \frac{\partial^2 E_x}{\partial t^2} + \mu_0 \sigma \frac{\partial E_x}{\partial t}. \tag{2.81}$$

For a electromagnetic wave of a fixed frequency ν, Eq. 2.81 becomes

$$\frac{d^2 E_x(z)}{dz^2} = -\left(\frac{\omega^2}{c^2} + i \omega \mu_0 \sigma \right) E_x(z), \tag{2.82}$$

here we explicitly indicate that E_x is a function of z. For good conductors, the second term in the bracket of Eq. 2.82 is much greater than the first term,

$$\frac{\omega \mu_0 \sigma c^2}{\omega^2} = \frac{\sigma}{\omega \varepsilon_0} \gg 1, \tag{2.83}$$

where Eq. 2.6 is applied. Equation 2.82 is reduced to

$$\frac{d^2 E_x(z)}{dz^2} = -i\omega\mu_0\sigma E_x(z). \tag{2.84}$$

To resolve the above differential equation, we assume that the form of solution is similar to that of non-metals,

$$E_x(z) = E_{x0}\, e^{kz}. \tag{2.85}$$

Insert into Eq. 2.84, we find

$$k = \sqrt{-i\omega\mu_0\sigma} = \pm\sqrt{\frac{\omega\mu_0\sigma}{2}}(1-i). \tag{2.86}$$

If the z-direction is pointing from vacuum to the metal, a negative sign should be chosen. Introducing a *skin depth* δ

$$\delta = \sqrt{\frac{2}{\omega\mu_0\sigma}}, \tag{2.87}$$

the wave vector k is of the form

$$k = \frac{-1+i}{\delta}. \tag{2.88}$$

The electric field intensity is

$$E_x(z,t) = E_{x0} \exp\left(-\frac{z}{\delta} + i\frac{z}{\delta} - i\omega t\right). \tag{2.89}$$

The skin depths of electromagnetic waves of interest in solar energy applications for several metals of interest are listed in Table 2.3. As shown, the skin depths are of the order of about 10 nanometers. The meaning is as follows. For a uniform metal film of the thickness of skin depth, the field intensity decays by a factor of e. The intensity

Table 2.3: Skin depths of several metals in nanometers

Wavelength (μm)	0.5	1.0	2.0	5.0	10.0
Al, $\sigma = 3.54 \times 10^7 \mho/\text{m}$	3.45	4.88	6.91	10.92	15.44
Cu, $\sigma = 5.80 \times 10^7 \mho/\text{m}$	2.70	3.81	5.39	8.53	12.06
Ag, $\sigma = 6.15 \times 10^7 \mho/\text{m}$	2.62	3.70	5.24	8.28	11.71

decays by a factor of e^2, which is about an order of magnitude. If the thickness of the metal film is many times the skin depth, it becomes dead opaque. Actually, silver has the smallest skin depth of all metals. A silver film of several tens of nanometers thick could block the visible and infrared radiation altogether.

To study the optical phenomenon at the vacuum–metal interface, we follow the derivation of the Fresnel formula for the transparent media, Eqs. 2.75 and 2.76. Now the transmitted wave is in the metal. In other words, Eq. 2.89 is

$$E_{(t)} = E_{(t)0} \exp\left(-\frac{z}{\delta} + i\frac{z}{\delta} - i\omega t\right). \tag{2.90}$$

The Faraday induction law, Eq. 2.22, is still valid,

$$\frac{\partial E_{(t)}}{\partial z} = -\frac{\partial B_{(t)}}{\partial t}. \tag{2.91}$$

and the magnetic field should follow the same temporal dependence as the electric field. Using Eq. 2.90, we have

$$i\omega B_{(t)} = \left(-\frac{1}{\delta} + \frac{i}{\delta}\right) E_{(t)}. \tag{2.92}$$

In other words,

$$B_{(t)} = \frac{1+i}{\omega\delta} E_{(t)}. \tag{2.93}$$

The magnetic field intensities of the incident and reflected light still follows the rules of transparent media. take $n_1 = 1$, Eq. 2.67 becomes

$$\frac{1}{c}\left(E_{I0} - E_{R0}\right) = \frac{1+i}{\omega\delta} E_{(t)0}. \tag{2.94}$$

Combining with Eq. 2.68, we find

$$2E_{I0} = \left(1 + \frac{c(1+i)}{\omega\delta}\right) E_{(t)0}, \tag{2.95}$$

and

$$2E_{R0} = \left(1 - \frac{c(1+i)}{\omega\delta}\right) E_{(t)0}. \tag{2.96}$$

The ratio between the electric field intensity of the reflected light and the electric field intensity of the incident light is

$$\frac{E_{R0}}{E_{I0}} = \frac{1 - \dfrac{c(1+i)}{\omega\delta}}{1 + \dfrac{c(1+i)}{\omega\delta}} \approx -\left(1 - \frac{2\pi\delta(1-i)}{\lambda}\right). \tag{2.97}$$

Here the relation $\omega/c = 2\pi/\lambda$ is used. Notice that $\delta \ll \lambda$, by taking the absolute value, the imaginary term is a second-order infinitesimal. We have

$$\left|\frac{E_{R0}}{E_{I0}}\right| \approx 1 - \frac{2\pi\delta}{\lambda}. \tag{2.98}$$

Table 2.4: Absorptivity of several metals in %

Wavelength (μm)	0.5	1.0	2.0	5.0	10.0
Al, $\sigma = 3.54 \times 10^7 \mho/\text{m}$	8.68	6.14	4.34	2.74	1.94
Cu, $\sigma = 5.80 \times 10^7 \mho/\text{m}$	6.78	4.79	3.39	2.14	1.52
Ag, $\sigma = 6.15 \times 10^7 \mho/\text{m}$	6.58	4.66	3.29	2.08	1.47

Again, because $\delta \ll \lambda$, the reflectivity of a metal mirror is

$$\mathcal{R} = \frac{S_{\mathbf{R}}}{S_{\mathbf{I}}} = \left| \frac{E_{\mathbf{R0}}}{E_{\mathbf{I0}}} \right|^2 \approx 1 - \frac{4\pi\delta}{\lambda}. \tag{2.99}$$

Conservation of energy requires that the tramnsmission coefficent is

$$\mathcal{T} = 1 - \mathcal{R} \approx \frac{4\pi\delta}{\lambda}. \tag{2.100}$$

Obviously, since all the radiation transmitted into the metal is absorbed, here the transmission coefficient is the absorption coefficient, or *absorptivity*. As shown in Table 2.4, for good metals, the absorptivity of far-infrared radiation can be smaller than 2%. Silver has the smallest absorptivity. Note that the above analysis is an approximation based on the direct-current conductivity. At optical frequencies, resonance of electrons makes the conductivity frequency dependent. More realistic theoretical models and accurate measurements show that the far-infrared absorptivity of silver can be much smaller than 1%. As shown, the reflectivity of a metal mirror can never be 100%. There is always a small percentage of power loss due to ohmic effect inside the skin depth near the metal surface. Among all materials, silver has the highest reflectivity. We will come to this point in Chapter 11.

2.3 Blackbody Radiation

It was known for centuries that a hot body emits radiation. At around 700°C, a body becomes red hot. At even higher temperatures, a body emits much more radiation, and the color changes to orange, yellow, white, and even blue. In the late 19th century, in order to understand phenomena related to industry technology such as steel making and incandescent light bulbs, it became a hot research subject.

Although all hot bodies emit radiation, blackbodies emit the maximum amount of radiation at a given temperature. At equilibrium, radiation emitted must equal radiation absorbed. Therefore, the body that emits the maximum amount also absorbs the maximum amount—which should look black. Practically, a blackbody is constructed by opening a small hole on a large cavity, as shown in Fig. 2.4. Any light ray passing through the hole with area A experiences multiple reflections on the internal surface of

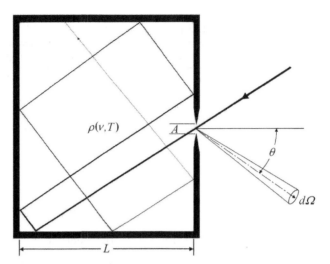

Figure 2.4   Blackbody radiation. A large cavity with a small hole is a good blackbody. The light enters the hole will experience multiple reflections, and all be absorbed and thus looks black. A blackbody emits maximum amount of radiation when heated.

the cavity. Because no material is not absolutely reflective, after several impingements, the light will eventually be completely absorbed by the cavity. Therefore, the small hole on the large cavity always looks black, which represents a blackbody.

2.3.1   Rayleigh–Jeans Law

The energy spectrum of radiation was studied in the late 19th century by Lord Rayleigh and then by Sir James Jeans using classical statistical physics. They treated standing electromagnetic waves in a cavity as individual modes, and the modes follow the equal-partition law of Maxwell–Boltzmann statistics.

Because the vector potential and the scalar potential satisfy the same wave equation, see Eqs. 2.16 and 2.17, we start with the study of standing waves of the scalar potential in a cubic cavity with metal surfaces of side L. Following the standard procedure for partial differential equations, we write the potential field as a product of a space-dependent function $\varphi(\mathbf{r})$ and a time-dependent function $f(t)$,

$$\phi(\mathbf{r}, t) = \varphi(\mathbf{r}) f(t). \tag{2.101}$$

Insert Eq. 2.101 into the wave equation of the scalar potential, Eq. 2.17, we find

$$\nabla^2 \left[\varphi(\mathbf{r}) f(t) \right] = \frac{1}{c^2} \frac{\partial^2}{\partial t^2} \left[\varphi(\mathbf{r}) f(t) \right]. \tag{2.102}$$

Divide both sides by $\varphi(\mathbf{r}) f(t)$, we find

$$\frac{1}{\varphi(\mathbf{r})} \nabla^2 \varphi(\mathbf{r}) = \frac{1}{c^2 f(t)} \frac{\partial^2 f(t)}{\partial t^2}. \tag{2.103}$$

Because the left-hand side only depends on $\mathbf{r}$ and the right-hand side only depends on t, it must be a constant independent of $\mathbf{r}$ and t. Denote that constant as $-k^2$, we find

$$\nabla^2 \varphi(\mathbf{r}) + k^2 \varphi(\mathbf{r}) = 0. \tag{2.104}$$

Equation 2.104 is the *Helmholz equation*, and the constant k, with the dimension of L^{-1}, is the absolute value of the *wave vector*. Denote

$$kc = \omega, \quad \text{that is,} \quad k = \frac{\omega}{c}, \tag{2.105}$$

we find a differential equation in time t,

$$\frac{\partial^2 f(t)}{\partial t^2} + \omega^2 f(t) = 0. \tag{2.106}$$

The general solution of the time-dependent factor $f(t)$ is

$$f(t) = a \cos \omega t + b \sin \omega t. \tag{2.107}$$

Here a and b are constants. In terms of frequency ν, using

$$\omega = 2\pi\nu, \tag{2.108}$$

the general solution Eq. 2.107 becomes

$$f(t) = a \cos 2\pi\nu t + b \sin 2\pi\nu t. \tag{2.109}$$

Now we focus our attention on the space-dependent factor $\varphi(\mathbf{r})$ through Eq. 2.104. Assuming that the cavity is made of metal. On the walls of the cavity, scalar potential must be zero. By looking for a solution as the product of a function of x, a function of y, and a function of z, the boundary condition is, at $x = 0$, $x = L$, $x = y$, $y = L$, $z = 0$, and $z = L$, $\varphi(\mathbf{r})$ must be zero. A simple argument leads to

$$\varphi(\mathbf{r}) = \varphi_0 \sin(k_x x) \sin(k_y y) \sin(k_z z). \tag{2.110}$$

Where φ_0 is a constant. The components of the wavevectors should be

$$k_x = \frac{\pi n_x}{L}, \quad k_y = \frac{\pi n_y}{L}, \quad k_z = \frac{\pi n_z}{L}, \tag{2.111}$$

where n_x, n_y, and n_z are positive integers. By direct substitution one finds that the solution, Eq. 2.110, satisfies differential equation 2.101 and the boundary conditions at the walls. Each set of the integers, n_x, n_y, n_z, represents a pattern of electromagnetic wave in the cavity. Inserting Eq. 2.105 into Eq. 2.101 yields

$$k_x^2 + k_y^2 + k_z^2 = \frac{4\pi^2 \nu^2}{c^2}, \tag{2.112}$$

and in terms of the numbers n_x, n_y, and n_z, Eq. 2.112 becomes

$$n_x^2 + n_y^2 + n_z^2 = \frac{4\nu^2 L^2}{c^2}. \tag{2.113}$$

Now, we count the number of standing waves with frequencies ν by considering a sphere of radius $\sqrt{n_x^2 + n_y^2 + n_z^2} = 2\nu L/c$. The number N of modes with positive $n_x, n_y,$ and n_z up to ν is

$$N = \frac{1}{8}\frac{4}{3}\pi \left(\frac{2\nu L}{c}\right)^3 = \frac{4\pi\nu^3 L^3}{3c^3}, \tag{2.114}$$

For each type of standing wave, there are two polarizations. Therefore, the number of modes of standing electromagnetic waves is

$$N = \frac{8\pi\nu^3 L^3}{3c^3}, \tag{2.115}$$

where L^3 is the volume, and the *density of states* at frequency ν is

$$\frac{d}{d\nu}\left(\frac{N}{L^3}\right) = \frac{8\pi\nu^2}{c^3}. \tag{2.116}$$

According to Maxwell–Boltzmann statistics, at absolute temperature T, each degree of freedom contributes energy $k_B T$, where k_B is the Boltzmann constant, and the energy density is

$$\rho(\nu, T) = \frac{d}{d\nu}\left(\frac{N}{L^3}\right) k_B T = \frac{8\pi\nu^2}{c^3} k_B T. \tag{2.117}$$

Equation 2.127 is the energy density of radiation per unit frequency interval in a cavity of temperature T. It is not directly observable. The directly observable quantity is the spectral radiance $u(\nu, T)$, that is, the energy radiating from a unit area of the hole per unit frequency range. To calculate $u(\nu, T)$ from $\rho(\nu, T)$, first we consider a simplified situation: If the field has a well-defined direction of radiation with velocity c, we have

$$u(\nu, T) = c\,\rho(\nu, T). \tag{2.118}$$

Because the hole is small, the radiation field in a cavity is isotropic. As the radiation only comes through a hole of a well-defined direction, $u(\nu, T)$ should be a fraction of $c\rho(\nu, T)$. The value of the fraction can be determined using the following argument. Consider a sphere of radius R. The surface area of the sphere is $4\pi R^2$. If the radiation inside the sphere is allowed to emit over all directions, the area is $4\pi R^2$. If the radiation is allowed to emit in only one direction, the area is a disc with radius R, that is, πR^2. Consequently, the factor is $1/4$. Equation 2.128 becomes

$$u(\nu, T) = \frac{1}{4}c\rho(\nu, T). \tag{2.119}$$

A detailed proof of the factor $1/4$ is as follows. Consider the radiation from a small hole of area A on the cavity; see Fig. 2.4. Because the electromagnetic wave is isotropic and the speed of light is c, the energy radiated through a solid angle $d\Omega$ at an angle θ is

$$\frac{dE}{dt\,d\Omega} = \frac{c}{4\pi}\rho(\nu, T)\,A\,\cos\theta \qquad (2.120)$$

because the area of the hole observed from an angle θ is $A\cos\theta$. Integrating over the hemisphere, the total irradiation per unit area is

$$u(\nu, T) = \frac{c}{4\pi}\int_0^{\pi/2} 2\pi\cos\theta\,\sin\theta\,d\theta\,\rho(\nu, T) = \frac{c}{4}\,\rho(\nu, T), \qquad (2.121)$$

confirming Eq. 2.119. Using Eq. 2.127, we finally obtain the Rayleigh–Jeans distribution of blackbody radiation,

$$u(\nu, T) = \frac{2\pi\nu^2}{c^2}\,k_{\mathrm{B}}T. \qquad (2.122)$$

The Rayleigh–Jeans distribution fits the experimental data at low frequencies. At higher frequencies, the spectral irradiance increases, and the sum becomes infinite. This contradicts the experimental fact that the total blackbody radiation is finite, and the spectral density decreases at higher frequencies; see Fig. 2.5.

2.3.2 Planck Formula and Stefan–Boltzmann's Law

In 1900, Max Planck found an empirical formula for the experimental data,

$$u(\nu, T) = \frac{2\pi\nu^2}{c^2}\,\frac{h\nu}{e^{h\nu/k_{\mathrm{B}}T} - 1}. \qquad (2.123)$$

The constant h in the formula, Planck's constant, was initially obtained by fitting with experimental blackbody radiation data. Later, Planck found a mathematical explanation of his formula by assuming that the energy of radiation can only take discrete values. Specifically, he assumed that the energy of radiation with frequency ν can only take integer multiples of a basic value $h\nu$, the *energy quantum*,

$$\epsilon = 0, \quad h\nu, \quad 2h\nu, \quad 3h\nu, \quad .... \qquad (2.124)$$

According to Maxwell–Boltzmann statistics, the probability of finding a state with energy $nh\nu$ is $\exp(-nh\nu/k_{\mathrm{B}}T)$. The average value of energy of a given component of radiation with frequency ν is

$$\bar{\epsilon} = \frac{\displaystyle\sum_{n=0}^{\infty} nh\nu\,e^{-nh\nu/k_{\mathrm{B}}T}}{\displaystyle\sum_{n=0}^{\infty} e^{-nh\nu/k_{\mathrm{B}}T}} = \frac{h\nu}{e^{h\nu/k_{\mathrm{B}}T} - 1}. \qquad (2.125)$$

instead of $k_B T$. By replacing the expression $k_B T$ in Eq. 2.122 with Eq. 2.125, we recovered Eq. 2.123.

By integrating the spectral radiance over frequency, the total radiation is

$$U(T) = \int_0^\infty \frac{2\pi h \nu^3}{c^2} \frac{d\nu}{e^{h\nu/k_B T} - 1}$$

$$= \frac{2\pi h}{c^2} \left(\frac{k_B T}{h}\right)^4 \int_0^\infty \frac{x^3 \, dx}{e^x - 1} \tag{2.126}$$

$$= \frac{2}{15} \frac{\pi^5 k_B^4}{c^2 h^3} T^4.$$

Here a mathematical identity is applied,

$$\int_0^\infty \frac{x^3 \, dx}{e^x - 1} = \frac{\pi^4}{15}. \tag{2.127}$$

Equation 2.126 is *Stefan–Boltzmann's law*, discovered experimentally before the Planck formula and backed by an argument using thermodynamics. The constant in Eq. 2.126,

$$\sigma \equiv \frac{2}{15} \frac{\pi^5 k_B^4}{c^2 h^3} = \frac{\pi^2 k_B^4}{60 \, c^2 \hbar^3} = 5.67 \times 10^{-8} \frac{W}{m^2 \cdot K^4}, \tag{2.128}$$

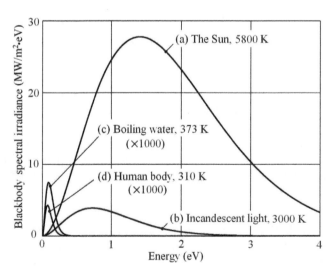

Figure 2.5 Blackbody spectral irradiance. The blackbody spectral irradiance, or the radiation power emitted per square meter per unit energy interval (here in electron volts) at an energy value (also in electron volts) at four different temperatures is shown. The maximum of solar irradiance is at 1.4 eV, with a value of 27.77 MW/m²·eV. The temperature of the filament of an incandescent light is about 3000 K. The radiation power density at the filament surface is only about 7% that on the Sun. The spectral irradiance from a blackbody at the boiling points of water and the human body are also shown, in units of kW/m²·eV.

Table 2.5: Blackbody radiation at different temperatures

Radiator	Temperature (K)	Power (W/m^2)	Peak ϵ (eV)	Peak λ (μm)	Peak u (W/m^2·eV)
The Sun	5800	6.31×10^7	1.410	0.88	2.81×10^7
Light bulb	3000	4.59×10^6	0.728	1.70	3.88×10^6
Boiling water	373	1.10×10^3	0.091	13.6	7.46×10^3
Human body	310	5.24×10^2	0.075	16.5	4.28×10^3

is called the *Stefan–Boltzmann's constant*. It can be memorized using the mnemonic: 45678. The total radiance is proportional to the fourth power of absolute temperature, and the coefficient is 5.67 times the inverse eighth power of 10.

For applications in solar cells, the electron volt is the most convenient unit of photon energy; see Fig. 2.5. The Planck formula for blackbody spectral irradiance in terms of photon energy ϵ in units of electron volts is

$$u(\epsilon, T) = \frac{2\pi q^4}{c^2 h^3} \frac{\epsilon^3}{e^{\epsilon/\epsilon_T} - 1} = 1.587 \times 10^8 \frac{\epsilon^3}{e^{\epsilon/\epsilon_T} - 1} \frac{\text{W}}{\text{m}^2 \cdot \text{eV}}, \qquad (2.129)$$

where $\epsilon_T = k_B T$ is the value of $k_B T$ in electron volts. Numerically, it equals $\epsilon_T = T/11,600$. For the Sun, $T_\odot = 5800$ K; thus $\epsilon_\odot = 0.5$ eV. At the location of Earth, the radiation is diluted by the distance from the Sun to Earth, the astronomical constant $A_\odot = 1.5 \times 10^{11}$ m. Introducing a geometric factor f representing the solid angle of the Sun with radius $r_\odot = 6.96 \times 10^8$ m as observed from Earth

$$f = \left(\frac{r_\odot}{A_\odot} \right)^2 = \frac{[6.96 \times 10^8]^2}{[1.5 \times 10^{11}]^2} = 2.15 \times 10^{-5}, \qquad (2.130)$$

the spectrum of the AM0 solar radiation (outside the atmosphere at the location of Earth) is

$$u_\oplus(\epsilon, T) = f u_\odot(\epsilon, T) = 3.41 \times 10^3 \frac{\epsilon^3}{e^{\epsilon/\epsilon_\odot} - 1} \frac{\text{W}}{\text{m}^2 \cdot \text{eV}}. \qquad (2.131)$$

The position of the peak in blackbody spectral irradiance can be of a transcendental equation

$$\frac{d}{dx} [3 \log x - \log (e^x - 1)] = 0, \qquad (2.132)$$

and can be obtained by numerical computation,

$$x = 2.82. \qquad (2.133)$$

In other words, the peak of blackbody spectral irradiance is at

$$\epsilon_{\text{MAX}} = 2.82 \, \epsilon_T = 2.43 \times 10^{-4} \, T \, (\text{eV}). \qquad (2.134)$$

The peak value for the function $x^3/(e^x - 1)$ is 1.42. Therefore, the peak value of the spectral irradiance is

$$u_{\text{MAX}} = 1.42 \, \frac{2\pi e k_{\text{B}}^3}{c^2 \, h^3} \, T^3 \cong 1.44 \times 10^{-4} \, T^3 \, \frac{\text{W}}{\text{m}^2 \cdot \text{eV}}. \tag{2.135}$$

Table 2.5 lists the data for some frequently encountered cases.

The profound significance of the concept of quantization of radiation and the meaning of Planck's constant were discovered by Albert Einstein in his interpretation of the photoelectric effect, that became the conceptual foundation of quantum physics.

2.4 Photoelectric Effect and Concept of Photons

The photoelectric effect was discovered accidentally by Heinrich Hertz in 1887 during experiments to generate electromagnetic waves. Since then, a number of studies have been conducted in an attempt to understand the phenomena. Around 1900, Phillip Lenard did a series of critical studies on the relation of the kinetic energy of ejected electrons with the intensity and wavelength of the impinging light [56]. His results were in direct conflict with the wave theory of light and inspired Albert Einstein to develop his theory of photons.

Figure 2.6 shows schematically the experimental apparatus of Phillip Lenard. The entire setup was enclosed in a vacuum chamber. An electric arc lamp, using carbon rods or zinc rods as the electrodes, generates strong UV light. A quartz window allows

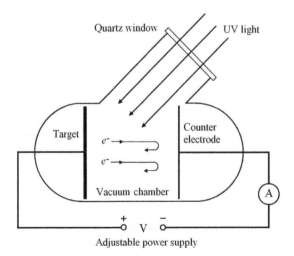

Figure 2.6 Lenard's apparatus for studying photoelectric effect. A quartz window allows the UV light from an electric arc lamp to shine on a target. The voltage between the target and the counter electrode is controlled by an adjustable power supply. An ammeter is used to measure the photocurrent. The voltage with which the photocurrent becomes zero is the *stopping voltage* [56].

Table 2.6: Stopping voltage for photocurrent

Rod material	Driving current (A)	Distance to target (cm)	Photocurrent (pA)	Stopping voltage (V)
Carbon	28	33.6	276	−1.07
Carbon	20	33.6	174	−1.12
Carbon	28	68	31.7	−1.10
Carbon	8	33.6	4.1	−1.06
Zinc	27	33.6	2180	−0.85
Zinc	27	87.9	319	−0.86

such UV light to shine on a target made of different metals. The target and a counter electrode are connected to an adjustable power supply. An ammeter is used to measure the electric current generated by the UV light, the *photocurrent*, especially when the voltages between the two electrodes are very small. By gradually increasing the voltage, which tends to reflect the electrons back to the target, the photocurrent is reduced. The voltage with which the photocurrent becomes zero is recorded as the *stopping voltage*.

The stopping voltage is apparently related to the kinetic energy of the electrons ejected from the target:

$$qV = \frac{1}{2}mv^2. \tag{2.136}$$

Understandably, the photocurrent varies with the intensity of light. By changing the magnitude of the current that drives the arc or the distance from the arc lamp to the target, the photocurrent could change by two orders of magnitude: for example, from 4.1 to 276 pA. An unexpected and dramatic effect Lenard observed was that no matter how strong or how weak the light is, and no matter how large or how small the photocurrent is, the stopping voltage does not change; see Table 2.6. The stopping voltage changes only when the material for the electric arc lamp changes. However, for a given type of arc, the stopping voltage stays unchanged.

The effect Lenard observed has no explanation in the framework of the wave theory of light. According to the wave theory of light, the more intense the light is, the more kinetic energy the electrons acquire.

2.4.1 Einstein's Theory of Photons

In 1905, while employed as a patent examiner at the Swiss Patent Office, Albert Einstein wrote five papers, published in *Annalen der Physik*, that initiated the twentieth century revolution in science. For general public, Einstein is mostly known for his theory of relativity. Therefore, when the Swedish Academy announced in 1922 that Einstein had won the Nobel Prize "for services to theoretical physics and especially for the discovery of the law of the photoelectric effect," referring to his paper *On a Heuristic Viewpoint*

Concerning the Production and Transformation of Light [27], the public was surprised. In hindsight, the Nobel Committee was correct: His paper on photoelectric effect is the most revolutionary and the most controversial.

The controversy originated from a misunderstanding of Einstein's concept of *energy quantization*. In the title of his prize-winning paper, Einstein only stated that during its generation and conversion into other forms of energy, the energy of the radiation appears in integer multiples of an elementary unit ϵ, depending on its frequency,

$$\epsilon = h\nu, \tag{2.137}$$

where $h = 6.63 \times 10^{-34}\,\mathrm{J \cdot s}$ is Planck's constant and ν is the frequency of light. For example, for green light, $\lambda = 0.53\,\mu\mathrm{m}$, and the frequency is $5.6 \times 10^{14}\,\mathrm{s}^{-1}$. The elementary amount of energy is $3.7 \times 10^{-19}\mathrm{J}$, or 2.3 eV.

The elementary unit of radiation energy is later named a *photon* by chemist Gilbert N. Lewis in 1926. Nevertheless, in popular science, the word photon is often misunderstood to be a geometrical point moving in space, having a well-defined coordinates $\mathbf{r}(t)$ in three-dimensional space at any given time. In a cavity such as Fig. 2.4, the electromagnetic wave always spread out in the entire cavity. The energy of such electromagnetic wave can only take an integer multiple of an elementary unit $\epsilon = h\nu$, called photon. It is never a geometrical point localized in space.

When radiation interacts with a piece of metal, it transfers the energy of a quantum $\epsilon = h\nu$ a time to an electron. The electron could escape from the metal by overcoming the *work function* ϕ of the metal, typically a few electron volts. If the energy of the photon is smaller than the work function of the metal, the electron would stay in the metal. If the energy of the photon is greater than the work function of the metal, then the electron can escape from the metal surface *with an excess kinetic energy,*

$$\frac{1}{2}mv^2 = h\nu - \phi. \tag{2.138}$$

The kinetic energy of an escaping electron can be measured by an external voltage, or electric field, to turn it back onto the target. Voltage that just is enough to cancel the kinetic energy is called the *stopping voltage,*

$$e\,V_{\text{stop}} = \frac{1}{2}mv^2 = h\nu - \phi, \tag{2.139}$$

where q is the electron charge, 1.60×10^{-19} C. According to Einstein's quantum theory of light, the stopping voltage is linearly dependent on the frequency of the photon and *independent of the intensity of light*. The slope should be a universal constant, which provides a direct method to determine the value of Planck's constant,

$$\frac{\Delta V_{\text{stop}}}{\Delta\nu} = \frac{h}{e}. \tag{2.140}$$

2.4.2   Millikan's Experimental Verification

Einstein's theory of photons was rejected by a number of prominent physicists for many years, including Max Planck, Niels Bohr, and notably Robert Millikan. Starting in 1905, for 10 years Millikan tried to disprove Einstein's theory experimentally. Finally, in 1916, Millikan published a long paper on *Physical Review*, entitled *A Direct Photoelectric Determination of Planck's h* [70]. The conclusion reads as follows:

1. Einstein's photoelectric equation has been subject to very searching tests and it appears in every case to predict exactly the observed results.

2. Planck's *h* has been photoelectrically determined with a precision of about .5 percent.

In 1923, Millikan received a Nobel Prize "for his work on the elementary charge of electricity and on the photoelectric effect."

An interesting fact in the history of science is that in the same paper Millikan emphatically rejected the theory of photons. He said that Einstein's photon hypothesis "may well be called reckless first because an electromagnetic disturbance which remains localized in space seems a violation of the very conception of an electromagnetic disturbance, and second because it flies in the face of the thoroughly established facts of interference." Millikan wrote that Einstein's photoelectric equation, although accurately representing the experimental data, "cannot in my judgment be looked upon at present as resting upon any sort of a satisfactory theoretical foundation [70]." Here, again, Millikan objected to the popular misunderstanding that photons are point particles moving in three-dimensional space. After the establishment of relativistic quantum field theory in the second half of the 20th century, especially quantum electrodynamics, the puzzle was resolved: the universe consists of only continuous fields. Although energy, mass, and electrical charge are quantized to integer numbers, there is no point-like particles whatsoever. Both Einstein and Millikan were correct.

2.4.3   Electron as a Field

Einstein's theory on the quantization of radiation means that although electromagnetic fields are continuous physical reality, the energy is quantized into indivisible units, the photons. Such a phenomenon could be more universal, applicable to other objects in Nature, such as electrons and other elementary particles.

In 1924, French physicist Louis de Broglie extended Einstein's theory of quantization of electromagnetic fields to electrons in his Ph.D. thesis entitled *Recherches sur la théorie des quanta*. Here is his argument that leads to the concept that electron is also a field, similar to electromagnetic fields, not a geometrical point. See Fig. 2.7.

In modern times, Einstein's quantization relation Eq. 2.1 is written in terms of circular frequency $\omega = 2\pi\nu$ as

$$\epsilon = h\nu = \hbar\omega, \tag{2.141}$$

where $\hbar$ is a reduced Planck's constant often called the *Dirac constant*

$$\hbar \equiv \frac{h}{2\pi} = 1.054 \times 10^{-34} \, \text{J} \cdot \text{s}. \tag{2.142}$$

Following special relativity, the general relation between energy E and momentum p is

$$\epsilon^2 = m^2 c^4 + p^2 c^2. \tag{2.143}$$

For photons, because the rest mass m is zero,

$$\epsilon = pc. \tag{2.144}$$

Using Eq. 2.141, Eq. 2.144 becomes

$$p = \frac{\hbar \omega}{c}. \tag{2.145}$$

According to Eq. 2.28, the relation between angular frequency and wave vector is

$$k = \frac{\omega}{c}. \tag{2.146}$$

Combining Eq. 2.145 and Eq. 2.146, we obtain a relation between wave vector and momentum for photons

$$p = \hbar k. \tag{2.147}$$

Louis de Broglie proposed that Eq. 2.147 should be applicable also to electrons. Together with the quantization relation for energy, Eq. 2.141, we have two quantization relations for electrons and other elementary particles.

As proposed by de Broglie, the electron is also a wave. For light, as an electromagnetic wave, the energy is quantized. The elementary unit, the energy quantum, is

Figure 2.7 Louis de Broglie. French physicist (1892–1987). In his 1924 Ph.D. thesis, he postulated the wave nature of electrons and suggested that all matter have wave properties, then won the 1929 Nobel Prize in Physics. He was born to a noble family in France and became the 7th duc de Broglie in 1960. In 1942, he was elected as the Perpetual Secretary of the French Academy of Sciences. After the Second World War, he proposed the establishment of multi-national research laboratories, leading to the establishment of the European Organization for Nuclear Research (CERN).

proportional to the frequency. Electron is also a wave. The electrical charge is quantized. The unit is a universal constant, the elementary charge e. Because the electron has a well-defined charge–mass ratio, the mass of the electron is also quantized. The relation between the wave vector k and the momentum p of a massive particle is the same for photons, Eq. 2.147.

According to the historical records, in 1925, Einstein received a preprint of de Broglie's thesis. Einstein highly appreciated the idea, then immediately sent a letter to Schrödinger for his attention. In just a few months, based on the idea of de Broglie, Schrödinger formulated his wave equation of electrons and derived the Rydberg formula for the hydrogen atom. Quantum mechanics, in its most productive form, was born. We will present a quantum mechanics primer in Chapter 7.

2.5 Einstein's Derivation of Blackbody Formula

Based on the concept of photons and the interaction of photons with matter, Einstein made a very simple derivation of the blackbody radiation formula. The key of his derivation is the introduction of *stimulated emission of radiation*, which gave birth to the laser, an acronym for *light amplification by stimulated emission of radiation*, and provides a better understanding of the interaction of radiation with matter.

Einstein studied a simple two-state atomic system; see Fig. 2.8. The radiation field is represented by an energy density $\sigma(\nu)$, where ν is the frequency. The atomic system has two states with energy levels E_1 and E_2. The photons with energy $h\nu$ are associated with a transition between the two states. The energy relation is

$$h\nu = E_2 - E_1. \tag{2.148}$$

According to Maxwell–Boltzmann statistics, the populations of the two states are

$$N_1 \propto e^{-E_1/k_\mathrm{B}T} \tag{2.149}$$

and

$$N_2 \propto e^{-E_2/k_\mathrm{B}T}. \tag{2.150}$$

Using Eq. 2.148, we have

$$\frac{N_1}{N_2} = e^{h\nu/k_\mathrm{B}T}. \tag{2.151}$$

Einstein assumed three transition coefficients: the absorption coefficient B_{12}, the spontaneous emission coefficient A, and the stimulated emission coefficient B_{21}. The rate equations are

$$\frac{dN_2}{dt} = B_{12}N_1\sigma(\nu) - B_{21}N_2\rho(\nu) - AN_2, \tag{2.152}$$

$$\frac{dN_1}{dt} = -B_{12}N_1\rho(\nu) + B_{21}N_2\rho(\nu) + AN_2. \tag{2.153}$$

Figure 2.8 Einstein's derivation of blackbody radiation formula. The radiation field $\rho(\nu)$ interacts with a two-level atomic system. Three interaction modes are assumed: absorption, to lift the atomic system from state 1 to state 2; spontaneous emission and stimulated emission, the atomic system decays from state 2 to state 1, giving out energy to the radiation field.

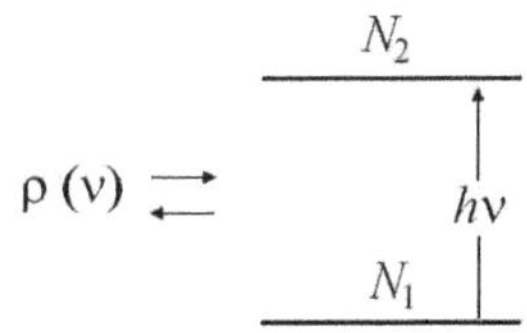

$$\rho(\nu) \underset{\longleftarrow}{\longrightarrow}$$

At equilibrium, both dN_1/dt and dN_2/dt should vanish. Therefore,

$$\frac{N_1}{N_2} = \frac{A + B_{21}\rho(\nu)}{B_{12}\rho(\nu)} = e^{h\nu/k_\mathrm{B}T}. \tag{2.154}$$

The coefficients are independent on temperature and frequency. At a high temperature, the power density should be high, and the right-hand side of Eq. 2.154 should approach unity. Therefore, the absorption coefficient B_{12} should be equal to the stimulated emission coefficient B_{21}. Both can be represented by one coefficient B:

$$B_{12} = B_{21} = B. \tag{2.155}$$

Equation 2.154 becomes

$$\frac{A}{B}\frac{1}{\rho(\nu)} + 1 = e^{h\nu/k_\mathrm{B}T}. \tag{2.156}$$

The power density distribution of radiation is then

$$\rho(\nu) = \frac{A}{B}\frac{1}{e^{h\nu/k_\mathrm{B}T} - 1}. \tag{2.157}$$

For radiations of low photon energy, Eq. 2.157 reduces to

$$\rho(\nu) \to \frac{A}{B}\frac{k_\mathrm{B}T}{h\nu}. \tag{2.158}$$

It should be identical to the Rayleigh–Jeans formula. Comparing with Eq. 2.122, we find the ratio of coefficients A and B,

$$\frac{A}{B} = \frac{8\pi h\nu^3}{c^3}. \tag{2.159}$$

Finally, Planck's formula is recovered,

$$\rho(\nu) = \frac{8\pi h\nu^3}{c^3}\frac{1}{e^{h\nu/k_\mathrm{B}T} - 1}. \tag{2.160}$$

Problems

2.1. Show that the capacitance C of a parallel-plate capacitor with vacuum as the dielectric is

$$C = \frac{\varepsilon_0 A}{d} \text{ [F]},\qquad(2.161)$$

where A is the area and d is the distance between the electrodes.

2.2. Show that the capacitance C of a parallel-plate capacitor with a medium of relative dielectric constant ε_r is

$$C = \frac{\varepsilon_0 \varepsilon_r A}{d} \text{ [F]}.\qquad(2.162)$$

Calculate the capacitance of a capacitor with $A = 1\,\mathrm{m}^2$ and $s = 1\,\mathrm{mm}$ for glass and silicon.

2.3. Show that the inductance L of an inductor made of a long solenoid of N loops with cross-sectional area A and length l is

$$L = \frac{\mu_0 N^2 A}{l} \text{ [H]}.\qquad(2.163)$$

2.4. Show that the speed of light v in a medium of relative dielectric constant ε_r is

$$v = \frac{c}{\sqrt{\varepsilon_r}}.\qquad(2.164)$$

Calculate the speed of light v in glass and silicon (the relative dielectric constants ε_r for glass and silicon are 2.25 and 11.7, respectively).

2.5. The refractive index of window glass is $n = 1.50$. How much light power is lost when going through a sheet of glass at normal incidence? (*Hint*: there are two glass–air interfaces.)

2.6. The radius of the Sun is $R = 6.96 \times 10^8$ m, and the distance between the Sun and Earth is $D = 1.5 \times 10^{11}$ m. The solar constant is 1366 W/m^2. Estimate the surface temperature of the Sun. (*Hint*: use the Stefan–Boltzmann law.)

2.7. What is the magnitude of the electric field intensity of the sunlight just outside the atmosphere of Earth?

2.8. What is the electric field intensity of the electron in a hydrogen atom at the distance of one Bohr radius from the proton?

2.9. Derive the blackbody radiation spectral density per unit wavelength in unit of micrometers.

2.10. Using the blackbody radiation formula per unit wavelength, derive the Wien displacement law in micrometers.

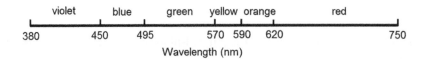

Figure 2.9 Wavelengths of visible lights.

2.11. The wavelengths of visible light with different colors in nanometers are shown in Fig. 2.9. Compute the frequencies and energy values of the photons, in both joules and electron volts.

2.12. What is the solar constant of Venus? Assume that the Sun is a blackbody emitter at 5800 K and the mean Venus-Sun distance is 1.08×10^{11} m.

2.13. To compute the blackbody irradiation for photon energy from ϵ_0 to infinity, an easy-to-use formula can be obtained by introducing $x_0 = \epsilon_0 / k_B T$ and expanding the denominator of Eq. 2.126 into

$$
\begin{aligned}
U(T, \epsilon_0) &= \frac{2\pi (k_B T)^4}{c^2 h^3} \int_{x_0}^{\infty} \frac{e^{-x} x^3 \, dx}{1 - e^{-x}} \\
&= \frac{2\pi (k_B T)^4}{c^2 h^3} \sum_{n=1}^{\infty} \int_{x_0}^{\infty} e^{-nx} x^3 \, dx.
\end{aligned}
\tag{2.165}
$$

Prove that

$$
U(T, \epsilon_0) = \frac{2\pi (k_B T)^4}{c^2 h^3} \sum_{n=1}^{\infty} e^{-nx_0} \left[\frac{x_0^3}{n} + \frac{3x_0^2}{n^2} + \frac{6x_0}{n^3} + \frac{6}{n^4} \right],
\tag{2.166}
$$

with

$$
x_0 = \frac{\epsilon_0}{k_B T}.
\tag{2.167}
$$

2.14. Assuming that the Sun is a blackbody emitter at 5800 K, what fraction of solar radiation is green (wavelength between 495 and 570 nm)?

2.15. Assuming that the Sun is a blackbody emitter at 5800 K, what fraction of solar radiation has photon energy greater than 1.1 eV?

Chapter 3

Origin of Solar Energy

Since the late 19th century, measurement of solar radiation power density has been diligently pursued. After the advancement of artificial satellites, solar radiation data outside the atmosphere became available. From the data accumulated to date, variation in the solar radiation power density at the average position of Earth outside the atmosphere has not exceeded 0.1% over a century. Although not a physical constant, this quantity, S, is often called the *solar constant*,

$$S = 1366 \pm 3 \, \text{W/m}^2. \tag{3.1}$$

The total radiation power of the Sun, $L_\odot$, or the *solar luminosity*, can be evaluated using the solar constant and the average distance between the Sun and Earth, $A_\odot$, the *astronomical unit of length*,

$$A_\odot = 1.5 \times 10^{11} \, \text{m}, \tag{3.2}$$

which gives (see Fig. 3.1)

$$L_\odot = 4\pi A_\odot^2 S = 3.84 \times 10^{26} \, \text{W}. \tag{3.3}$$

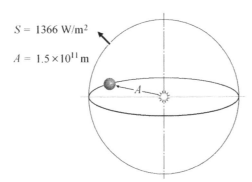

$S = 1366 \, \text{W/m}^2$

$A = 1.5 \times 10^{11} \, \text{m}$

Figure 3.1 Luminosity of the Sun. The average radiation power density of sunlight outside the atmosphere of Earth, the *solar constant*, is $S = 1366 \, \text{W/m}^2$. The average distance between the Sun and Earth, the *astronomical constant*, is $A_\odot = 1.5 \times 10^{11}$ m. The luminosity of the Sun, or the total power of solar radiation, is then $L_\odot = 3.84 \times 10^{26}$ W.

The total power of solar radiation is about twenty trillion times the energy consumption of the entire world, 1.6×10^{13} W. For the purpose of utilizing solar power, we should ask some basic questions: What is the origin of solar energy? How long mankind can enjoy the sunlight? How much variation of solar radiation can one expect?

3.1   Basic Parameters of the Sun

For centuries, the Sun has been the subject of intense study by astronomers. The basic parameters and the basic structure of the Sun are well understood [8, 72, 85, 103]. Here is a brief summary.

3.1.1   Distance

The distance between the Sun and Earth has been measured using triangulation and *radar echoes*. Because the orbit of Earth around the Sun is an ellipse, the distance is not constant. Around January 3, at *perihelion*, the distance is at a minimum, 1.471×10^{11} m. Around July 3, at *aphelion*, the distance is at a maximum, 1.521×10^{11} m. The average distance, $A_\odot = 1.5 \times 10^{11}$ m (see Eq. 3.2) is treated as a basic parameter in astronomy.

Besides the distance in meters, the average time for light to travel between the Sun and Earth, the *light time for the astronomical unit of distance* $\tau_\odot$, has been determined with high accuracy. For all practical applications in solar energy, the approximate value, accurate to 0.2%, is worth memorizing:

$$\tau_\odot = 500 \, \text{s}. \tag{3.4}$$

3.1.2   Mass

The mass of the Sun is measured using Kepler's law and the orbital parameters of the planets. Again, for all practical applications, an approximate value, accurate to 1%, is worth memorizing:

$$m_\odot = 2 \times 10^{30} \, \text{kg}. \tag{3.5}$$

It is 333,000 times the mass of Earth. Due to radiation and solar wind, the Sun loses 10^{17} kg of its mass per year, which is negligible. Practically, the mass of the Sun is constant over its lifetime.

3.1.3   Radius

The radius of the Sun is measured by the angular diameter of the visible disc. Because the distance between the Sun and Earth is not constant, the angular diameter of the Sun is not constant. It ranges from 31.6′ to 32.7′. The average value is 32′, or 0.533°. Using the average distance $A_\odot$, the radius of the Sun is then

$$r_\odot = \frac{1.5 \times 10^{11} \times 0.533}{2 \times 57.3} \text{ m} \cong 6.96 \times 10^8 \text{ m}. \qquad (3.6)$$

Calculated from the radius, the volume of the Sun is $1.412 \times 10^{27} \text{m}^3$, and its average density is 1.408 g/cm^3.

3.1.4   Emission Power

The emission power of the surface of the Sun, in watts per square meter, can be calculated from Eqs. 3.3 and 3.6:

$$U_\odot = \frac{L_\odot}{4\pi r_\odot^2} = \frac{3.84 \times 10^{26}}{4\pi (6.96 \times 10^8)^2} \cong 63.1 \text{ MW/m}^2. \qquad (3.7)$$

3.1.5   Surface Temperature

Considering the Sun as a blackbody, the temperature can be calculated from the Stefan–Boltzmann law. Using Eqs. 3.7, 2.126, and 2.128,

$$T_\odot = \left[\frac{U_\odot}{\sigma}\right]^{1/4} = \left[\frac{63.1 \times 10^6}{5.67 \times 10^{-8}}\right]^{1/4} \cong 5800 \text{ K}. \qquad (3.8)$$

In the literature, the surface temperature of the Sun varies from one source to another, from 5600 to 6000 K. The difference is insignificant. Throughout this book, we will use 5800 K as the nominal surface temperature of the Sun. Not only is this easy to remember, but it is exactly equal to 0.5 eV, which is convenient for the treatment of solar cells.

Table 3.1: Chemical composition of the Sun

Element	Z	Molecular Weight	Abundance (% of Number of Atoms)	Abundance (% of Mass)
Hydrogen	1	1.008	91.2	71.0
Helium	2	4.003	8.7	27.1
Oxygen	8	16.000	0.078	0.97
Carbon	6	12.011	0.043	0.40
Nitrogen	7	14.007	0.0088	0.096
Silicon	14	28.086	0.0045	0.099
Magnesium	12	24.312	0.0038	0.076
Neon	10	20.183	0.0035	0.058
Iron	26	55.847	0.0030	0.14

Source: Ref. [103].

3.1.6 Composition

The spectrum of solar radiation does not precisely match the spectrum of blackbody radiation. The fine structure of the solar spectrum provides evidence of its chemical composition (Table 3.1). Actually, the second most abundant element in the Sun, helium, which was discovered from the spectrum of the solar radiation, comes from the Greek word *helios*.

3.2 Kelvin–Helmholtz Time Scale

For centuries, the origin of sunlight has been a fundamental inquiry of the scientific community. In the middle of the 19th century, the field of thermodynamics matured. The first law of thermodynamics states that energy can only change its form and can neither be created nor be destroyed. The immense solar radiation is apparently draining enormous amounts of energy from the Sun. In the 1850s, Sir William Thomson (Fig. 3.2), one of the founders of thermodynamics, made a study on the origin of the Sun's energy based on the knowledge of physics at that time [106]. He argued that chemical energy could not be the answer, because even if the Sun is made of coal and burns completely, it can only supply a few thousand years of sunlight. The only explanation based on the knowledge of that time was the gravitational contraction of the Sun, also independently proposed by German physicist Hermann von Helmholtz; see Fig. 3.3(a). When two meteorites with masses m_1 and m_2 approach from infinity

Figure 3.2 Sir William Thomson. Irish-born Scottish physicist (1824-1907), aka Lord Kelvin, one of the founders of thermodynamics. He attributed the origin of the Sun's energy to gravitational interaction, and asserted that the Sun could not shine for more than 30 million years. He used his theory to refuse Sir Charles Lyell's geology, and especially Charles Darwin's evolutionary biology; both require that the lifetime of the Sun must be at least one billion years. The discrepancy was resolved by Albert Einstein and Hans Bethe in the 20th century. Artwork by Hubert Herkomer, courtesy of Smithsonian Museum.

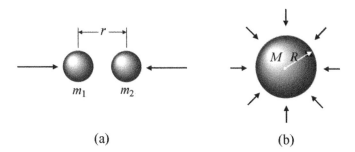

(a)   (b)

Figure 3.3   The Kelvin–Helmholtz model. A model of the origin of solar energy based on gravitational contraction. (a) When two meteorites approach each other from faraway, the gravitational potential energy is reduced and the kinetic energy is increased. (b) If the Sun is formed by coalition of a large number of meteorites, the gravitational energy can be converted into kinetic energy, and then radiated as heat.

to a distance r, the potential energy is decreased by

$$U = -\frac{Gm_1m_2}{r},\tag{3.9}$$

where $G = 6.67 \times 10^{-11} \text{N} \cdot \text{m}^2/\text{kg}^2$ is the gravitational constant. Because of gravitational attraction, the system gains a kinetic energy

$$E_\text{K} = \frac{G\, m_1\, m_2}{r}.\tag{3.10}$$

Assuming that the Sun is formed by the coalescence of large number of meteorites into a sphere of radius $r_\odot$, as shown in Fig. 3.3(b), kinetic energy can be gained by gravitational attraction. The value of the energy gain depends on the distribution of the mass. By analogy to Eq. 3.10, we can make a good order-of-magnitude estimate of the energy gain,

$$E_\text{K} \cong \frac{G\, M_\odot^2}{r_\odot}.\tag{3.11}$$

Obviously, it is an overestimate. For example, by calculating the energy gain with a uniform mass distribution, a numerical factor of 0.6 appears. However, Eq. 3.11 gives an upper limit. Based on the Kelvin–Helmholtz assumption, the lifetime of the Sun can be estimated by dividing that energy gain by the rate of energy loss, or the solar luminosity $L_\odot$:

$$\tau_\text{KH} \cong \frac{G\, M_\odot^2}{r_\odot\, L_\odot} = \frac{6.67 \times 10^{-11} \times \left(2 \times 10^{30}\right)^2}{6.96 \times 10^8 \times 3.84 \times 10^{26}} \approx 10^{15}\,\text{s}.\tag{3.12}$$

The quantity τ_KH, about 30 million years, is called the *Kelvin–Helmholtz time scale*. Based on this calculation, Lord Kelvin asserted that the Sun's life span must not exceed 30 million years [106].

At that time, geology was already well established based on two century's accumulation of discoveries and studies of fossils as well as geological strata, summarized by Sir Charles Lyell in the 1830s in a monograph *Elements of Geology* [60]. Geological evidence showed that Earth existed for more than one billion years under basically uniform sunlight conditions. Charles Darwin explained the sequence of discoveries in paleontology by evolution through natural selection in his monumental 1859 monograph *The Origin of Species* [21]. According to Lyell and Darwin [60, 21], countless evidence in paleontology has shown that the Sun has shined consistently for at least 1 billion years. The assertion of Lord Kelvin that the Sun's life expectancy is about 20 million years is in direct conflict with the findings in geology and biology.

The discrepancy is due to the limitation in the theory of physics in the Victorian period. On April 27, 1900, Lord Kelvin gave a lecture at the Royal Institution of Great Britain entitled *Nineteenth-Century Clouds over the Dynamical Theory of Heat and Light* [107]. Kelvin mentioned that the "beauty and clearness of theory" was overshadowed by "two clouds": the null result of the Michelson–Morley experiment and the difficulties in explaining the Stefan–Boltzmann law of blackbody radiation based on classical statistical mechanics. Kelvin believed that these two problems were minor and could be resolved within the framework of classical physics. Nevertheless, two years before his death, in 1905, these "two clouds" evolved into a perfect storm in theoretical physics with the emergence of relativity and quantum theory, which completely overturned the Kelvin–Helmholtz theory on the origin of solar energy.

The mystery of the origin of solar energy was resolved after Albert Einstein established the relation between energy and mass [26], and Hans Bethe's account of nuclear fusion as the origin of stellar energy [9]. The time scale of Bethe perfectly matches with the estimate of the Sun's age according to Lyell and Darwin.

3.3   Energy Source of the Sun

The answer to the source of stellar energy resides in the last of five 1905 papers by Einstein, entitled *Does the Inertia of a Body Depends Upon its Energy Content?* In contemporary notations, Einstein's statement is as follows [26]:

> If a body gives off energy ΔE in the form of radiation, its mass diminishes by $\Delta E/c^2$. Because whether the energy withdrawn from the body becomes radiation or else makes no difference, we might make a more general conclusion that the mass of a body is a measure of its energy content; if the energy changes by ΔE, the mass changes accordingly by $\Delta E/c^2$.

Accordingly, if the initial mass of the body is m_0 and its final mass is m_1, with $\Delta m = m_0 - m_1$, radiation can be emitted with energy

$$\Delta E = \Delta m c^2. \tag{3.13}$$

Using the vast experimental data on nuclear reaction already available at that time, in 1938, Hans Bethe (Fig. 3.4) made a thorough study of all possible nuclear reactions which could generate stellar energy [9]. He concluded that, for stars with mass similar to or less than the Sun, the proton–proton chain dominates. For stars with mass much larger than the Sun, the process catalyzed by carbon, nitrogen, and oxygen (the carbon chain) dominates.

3.3.1 The $p - p$ Chain

The mass of proton is 1.672623×10^{-27} kg, and the mass of a helium nucleus (the alpha particle) is 6.644656×10^{-27} kg. Every time four protons fuse into one alpha particle, there is an excess of mass,

$$\Delta m = (4 \times 1.672623 - 6.644656) \times 10^{-27}\text{kg} = 4.5836 \times 10^{-29}\text{kg}. \tag{3.14}$$

Using Einstein's equation, the excess energy is

$$\Delta E = 4.5836 \times 10^{-29} \times \left(2.99792 \times 10^8\right)^2 = 4.11952 \times 10^{-12}\text{J}, \tag{3.15}$$

or 25.7148 MeV. In other words, every hydrogen atom participating in the reaction generates 6.4287 MeV of energy.

However, as shown below, part of the energy is radiated in the form of neutrinos, for which the Sun and Earth are transparent. Therefore, the radiation energy generated

Figure 3.4 Hans Albrecht Bethe. German-born American physicist (1906–2005) and Nobel laureate for his 1938 theory on nuclear fusion as the origin of stellar energy. A versatile theoretical physicist, Bethe also made important contributions to quantum electrodynamics, nuclear physics, solid-state physics, and astrophysics. Bethe left Germany in 1933 when the Nazis came to power and he lost his job at the University of Tübingen. He moved first to England, then to the United States in 1935, and joined the faculty at Cornell University, a position which he occupied for the rest of his career. During World War II, he was appointed by John Oppenheimer as the Director of Theoretical Division of the Manhattan Project at Los Alamos laboratory. From 1948 to 1949 he was a visiting professor at Columbia University. Photograph courtesy of Mickael Okoniewski. Taken at Cornell University on December 19, 1996.

in the nuclear reaction is reduced:

$$^1\text{H} + {}^1\text{H} \longrightarrow {}^2\text{D} + \text{e}^+ + \nu + 0.164\,\text{MeV}, \tag{3.16}$$

$$^2\text{D} + {}^1\text{H} \longrightarrow {}^3\text{He} + \gamma + 5.49\,\text{MeV}, \tag{3.17}$$

$$^3\text{He} + {}^3\text{He} \longrightarrow {}^4\text{He} + 2\,{}^1\text{H} + 12.85\,\text{MeV}. \tag{3.18}$$

The overall reaction is

$$4\,{}^1\text{H} \longrightarrow {}^4\text{He} + 2\text{e}^+ + 2\nu + 2\gamma + 24.16\,\text{MeV}. \tag{3.19}$$

3.3.2 Carbon Chain

For stars with mass greater than the Sun, the carbon chain, or the CNO chain, dominates the energy generation. Note that the carbon and nitrogen nuclei only act as catalysts. The net result is again combining four protons to form a helium nucleus, or an alpha particle,

$$^{12}\text{C} + {}^1\text{H} \longrightarrow {}^{13}\text{N} + \gamma + 1.95\,\text{MeV}, \tag{3.20}$$

$$^{13}\text{N} \longrightarrow {}^{13}\text{C} + \text{e}^+ + \nu + 1.50\,\text{MeV}, \tag{3.21}$$

$$^{13}\text{C} + {}^1\text{H} \longrightarrow {}^{14}\text{N} + \gamma + 7.54\,\text{MeV}, \tag{3.22}$$

$$^{14}\text{N} + {}^1\text{H} \longrightarrow {}^{15}\text{O} + \gamma + 7.35\,\text{MeV}, \tag{3.23}$$

$$^{15}\text{O} \longrightarrow {}^{15}\text{N} + \text{e}^+ + \nu + 1.73\,\text{MeV}, \tag{3.24}$$

$$^{15}\text{N} + {}^1\text{H} \longrightarrow {}^{12}\text{C} + {}^4\text{He} + 4.96\,\text{MeV}. \tag{3.25}$$

The overall reaction is

$$4\,{}^1\text{H} \longrightarrow {}^4\text{He} + 2\text{e}^+ + 2\nu + 3\gamma + 25.03\,\text{MeV}. \tag{3.26}$$

3.3.3 Internal Structure of the Sun

Observing from Earth, the Sun is a sphere of roughly equal temperature, 5800 K. The internal structure of the Sun is not observable. However, it can be inferred using the laws of physics. A simple model of the structure of the Sun is based on an approximation that the Sun is made of spherical layers of uniform materials. The only spatial parameter is the radial distance from the center. A balance between the gravitational force and radiation pressure determines its density and temperature distribution. The source of the energy is a nuclear reaction as presented in Section 3.3. The parameters of the model are determined by comparing with observations.

The current standard model is shown in Fig. 3.5 [8, 85]. The core of the Sun, with a density on the order of 100 g/cm^3 and a temperature of $10 \times 10^6 - 15 \times 10^6$ K, is the engine of energy creation. Here the nuclei and electrons combine to form a dense and superhot plasma. The radiation energy generated in the core, mostly gamma rays, travels to the surface through the radiation zone. The density of the radiation

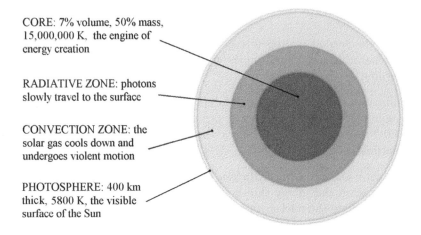

CORE: 7% volume, 50% mass, 15,000,000 K, the engine of energy creation

RADIATIVE ZONE: photons slowly travel to the surface

CONVECTION ZONE: the solar gas cools down and undergoes violent motion

PHOTOSPHERE: 400 km thick, 5800 K, the visible surface of the Sun

Figure 3.5 Internal structure of the Sun. The core of the Sun, with a density on the order of 100 g/cm^3 and temperature of $10 \times 10^6 - 15 \times 10^6$ K, is the engine of energy creation. The radiation energy generated in the core travels to the surface through the radiative zone. Through the convection zone, the radiation energy is emitted by the photosphere. The boundaries between adjacent spheres are gradual rather than well defined.

zone decreases from about 5 to about 1 g/cm^3. It takes a photon about 170,000 years to travel through that region. The convection zone, with a density less than that of water, is a zone of violent motion. The photosphere, a plasma of hydrogen and helium with a density much lower than that of Earth's atmosphere but totally opaque, is the blackbody radiator that emits sunlight.

Problems

3.1. Based on the current hydrogen reserve in the Sun and the energy output, if the efficiency of generating radiation from the $p - p$ chain is 90%, how many years can the Sun keep burning?

3.2. Assuming that the nuclear radius R follows the simple formula $R = r_0 A^{1/3}$, where A is the atomic mass number (number of protons Z plus number of neutrons N) and $r_0 = 1.25 \times 10^{15}$ m, how much kinetic energy is required for a proton (hydrogen nucleus) to touch a carbon nucleus?

3.3. If the kinetic energy of the hydrogen nucleus is $E_k = 1/2 k_B T$, to have 0.1% of hydrogen nuclei reach the energy threshold for a CNO reaction, what is the minimum temperature to initiate the nuclear fusion?

Chapter 4

Tracking Sunlight

Because of the rotation and the orbital motion of Earth around the Sun, the apparent position of the Sun in the sky changes over time. To utilize the solar energy efficiently, we must understand the apparent motion of the Sun. The accurate theory of the solar system in astronomy and the data in *The Astronomical Almanac* can be overwhelmingly complicated. In this chapter, we present a simple model which results in formulas easily programmable on a microcomputer and accurate enough for solar energy utilization. Analytic formulas are derived and presented.

4.1   Rotation of Earth: Latitude and Longitude

Figure 4.1 shows the apparent motion of the stars in the night sky. This motion is due to the rotation of Earth on its axis. For solar energy applications, we can consider

Figure 4.1   The night sky. By orienting a camera toward the sky in the night and exposing for some time, the stars seem to rotate around the celestial North Pole. Photo taken by Robert Knapp, Portland, Oregon. See www.modernartphotograph.com. *Source*: Courtesy of Robert Knapp.

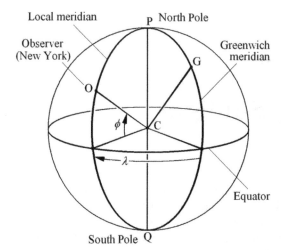

Figure 4.2 Latitude and longitude. The zero point of latitude, the prime meridian, is defined as the meridian passing through the Royal Greenwich Observatory. East of the prime meridian is the Eastern Hemisphere, and in the west is the Western Hemisphere. Similarly, north of the equator is the Northern Hemisphere, and south of the equator is the Southern Hemisphere. The position of the observer is identified by its *latitude* ϕ and *longitude* λ, as marked on the map. The convention of sign is eastward is positive and westward is negative. For example, the latitude of New York City is $\phi = 40°47'$ N, or $+0.712$ rad, and its longitude is $\lambda = 73°58'$ W, or -1.29 rad.

Earth as a perfect sphere rotating with a constant angular velocity on a fixed axis. The axis of rotation of Earth crosses the surface of Earth at two points: the *North Pole* and the *South Pole*. The great circle perpendicular to the axis is the *equator*. A location on Earth can be specified by two coordinates, the *latitude* ϕ and the *longitude* λ, as marked on the map and can be determined using GPS (the Global Positioning System). The longitude specifies a *meridian* (a half great circle passing through the two poles and the location). While the latitude is uniquely defined by the poles and the equator, the longitude requires an origin as the zero point, the *prime meridian*. The prime meridian was chosen by the International Meridian Conference held in October 1884 in Washington, DC, as the meridian passing through a marked point in the Royal Greenwich Observatory near London. Therefore, the prime meridian is often called the *Greenwich meridian*. Figure 4.2 shows the definition of longitude and latitude.

4.2   Celestial Sphere

From the point of view of an observer on Earth, the Sun, as well as any star, is located on a sphere of a large but undefined radius, called the *celestial sphere*. There are two commonly used coordinate systems to describe the position of an astronomical object on the celestial sphere, the *horizon system* and the *equatorial system*.

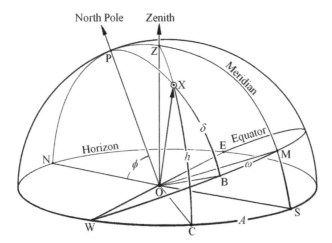

Figure 4.3   Celestial sphere and coordinate transformation. The horizon system defines the position of a celestial body X as directly perceived by the observer. The angular distance of a celestial body X to the horizon is its *height h*, also called *altitude* or *elevation*. The other coordinate is *azimuth A*. The zero point of the azimuth is defined as the south point of the horizon. In the equatorial coordinate system, the North Pole is the reference point. The celestial equator is the basic plane. The distance of a celestial body to the equator is its *declination δ*. The other coordinate is *hour angle ω*. The zero point of the hour angle is defined as the meridian. The half great circle passes through the north celestial pole and the zenith. To convert the coordinates from the horizon system to the equatorial system and *vice versa*, identities in spherical trigonometry are applied on a spherical triangle formed by vertices Z (zenith), P (north celestial pole), and celestial body X.

The extension of the center of Earth and an observer O into the sky is pointing to the *zenith Z* (see Fig.4.2), and the plane perpendicular to that line or the corresponding great circle is the *horizon*. The horizon divides the sphere into two hemispheres. The upper hemisphere is visible to the observer, and the lower hemisphere is hidden under the horizon. The angular distance of a celestial body above the horizon is its *height h*. In the astronomy literature, the term *altitude* or *elevation* is also used. Obviously, the height of the North Pole P equals the geographical latitude of the observer, ϕ.

To completely identify the position of a celestial body, X, we need another reference point. The great circle connecting the zenith with the North Pole is called the *meridian*. It intersects the horizon at point S, the *south point of the horizon*. To identify the position of a celestial body with respect to the south point, we draw a great circle through the zenith and the star, which intersects the horizon at point C. The angle $\widehat{SC}$ is defined as the *azimuth A*, or the horizontal direction of the celestial body. Regarding the utilization of solar energy, we take the definition of azimuth to be westward. Therefore, the azimuth of the Sun always increases over time.

The horizon system defines the position of a celestial body as directly perceived by the observer. However, because Earth is rotating on its axis, those coordinates depend on the location of the observer and vary over time. In the *equatorial coordinate system*,

Table 4.1: Notations in positional astronomy

Quantity	Notation	Definition
Latitude	ϕ	Geographical coordinate
Longitude	λ	Geographical coordinate
Height	h	Also called altitude or elevation
Azimuth	A	Horizontal direction or bearing
Declination	δ	Angular distance to the equator
Hour angle	ω	In radians, westward
Sunset hour angle	ω_s	In radians, always positive
East–west hour angle	ω_{ew}	In radians, always positive
Right ascension	α	Absolute celestial coordinate
Mean ecliptic longitude	l	On ecliptic plane
True ecliptic longitude	θ	On ecliptic plane
Eccentricity of orbit	e	Currently ≈ 0.0167
Obliquity of ecliptic	ε	Currently $\approx 23.44°$

the position of the Sun is independent of the location of the observer. The coordinates of the Sun in the horizon system can be obtained using a coordinate transformation from its coordinates in the equatorial coordinate system.

In the equatorial system, the coordinate equivalent to the latitude of Earth is the *declination*, δ, and the coordinate equivalent to the longitude of Earth for a fixed observer is the *hour angle*, ω; see Fig. 4.3.

As shown, the declination is the angular distance of the celestial body to the celestial equator. It is positive for the stars to the north of the celestial equator and negative for the stars to the south of the celestial equator.

As we mentioned previously, the great circle connecting the zenith and the celestial pole is the *meridian*, which is the reference point equivalent to the Greenwich meridian in geography, which intersects the equator at point M. The great circle connecting the celestial pole and the star is called the *hour circle*, which intersects the equator at point B. The angle $\widehat{MB}$ is the *hour angle* ω, equivalent to longitude on Earth. The convention is, if the celestial body is to the west of the meridian, the hour angle is positive. This is natural and convenient for the position of the Sun, as its hour angle defines the *solar time*.

Another coordinate frequently used in astronomy in place of the hour angle is the *right ascension* α, which takes the *vernal equinox* as the reference point. For a list of notations in positional astronomy, see Table 4.1.

4.2.1 Coordinate Transformation: Cartesian Coordinates

The standard method of coordinate transformation in positional astronomy is using *spherical trigonometry*, a brief summary of which is presented in Appendix B. Here we

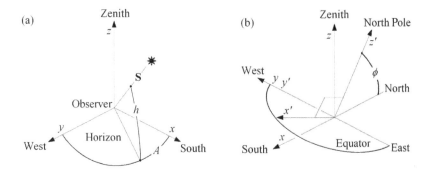

Figure 4.4 Coordinate transformation in Cartesian coordinates. (a) Cartesian coordinates for the horizon system use the conventions that south is x, west is y, and the zenith is z. The position of a celestial body is determined by two angles, height h and azimuth angle A. (b) Cartesian coordinates for the equatorial system use the convention that z'-axis points to the North Pole, while the east-west axis y'-axis is identical to that in the horizon system. The x'-axis is perpendicular to both. The position of a celestial body is determined by declination δ and hour angle ω, see Fig. 4.3.

present a treatment based on three dimensional Cartesian coordinates and spherical polar coordinates. Treating a two-dimensional problem with three-dimensional methods is overkill. However, because most physicists and engineers are familiar with such an approach, it may be easier to understand.

The Cartesian coordinates for the horizon system is shown in Fig. 4.4(a). Because all celestial bodies move from east to west, the convention is, south is x, west is y, and the zenith is z. A unit vector pointing to a celestial body is determined by two angles. The polar angle is the height h. The azimuth angle A takes south as the zero point. As shown in Fig. 4.4(a), using the three unit vectors $\mathbf{i}$, $\mathbf{j}$, and $\mathbf{k}$ pointing to x, y, and z, respectively, a unit vector $\mathbf{S}$ pointing to the Sun is given as

$$\mathbf{S} = \mathbf{i} \cos h \, \cos A + \mathbf{j} \cos h \, \sin A + \mathbf{k} \sin h. \tag{4.1}$$

The Cartesian coordinates for the equatorial system and its relation to the horizon system are shown in Fig. 4.4(b). The y'-axis is identical to that in the horizon system. The z'-axis points to the North Pole, and the x'-axis is perpendicular to both. The unit vectors in the equatorial system pointing to the x', y', and z'-axes are $\mathbf{i}'$, $\mathbf{j}'$, and $\mathbf{k}'$, respectively. Using declination δ and hour angle ω defined in Fig. 4.3, a unit vector $\mathbf{S}$ pointing to the Sun in terms of $\mathbf{i}'$, $\mathbf{j}'$, and $\mathbf{k}'$ is given as

$$\mathbf{S} = \mathbf{i}' \cos \delta \, \cos \omega + \mathbf{j}' \cos \delta \, \sin \omega + \mathbf{k}' \sin \delta. \tag{4.2}$$

From Fig. 4.4(b), the transformations between the two sets of unit vectors, $\mathbf{i}$, $\mathbf{j}$, $\mathbf{k}$ and $\mathbf{i}'$, $\mathbf{j}'$, $\mathbf{k}'$ are

$$\mathbf{i}' = \mathbf{i} \sin \phi + \mathbf{k} \cos \phi,$$

$$\mathbf{j}' = \mathbf{j}, \tag{4.3}$$

$$\mathbf{k}' = -\mathbf{i} \cos \phi + \mathbf{k} \sin \phi,$$

and

$$\mathbf{i} = \mathbf{i}' \sin \phi - \mathbf{k}' \cos \phi,$$

$$\mathbf{j} = \mathbf{j}',$$

$$\mathbf{k} = \mathbf{i}' \cos \phi + \mathbf{k}' \sin \phi,$$

(4.4)

where ϕ is the (geographical) latitude of the observer; see Figs 4.2 and 4.4. Using Eqs. 7.119 and 4.2, we obtain

$$\mathbf{S} = \mathbf{i} \left(\cos \delta \cos \omega \sin \phi - \sin \delta \cos \phi \right)$$

$$+ \mathbf{j} \cos \delta \sin \omega$$

$$+ \mathbf{k} \left(\cos \delta \cos \omega \cos \phi + \sin \delta \sin \phi \right),$$

(4.5)

and

$$\mathbf{S} = \mathbf{i}' \left(\cos h \cos A \sin \phi + \sin h \cos \phi \right)$$

$$+ \mathbf{j}' \cos h \sin A$$

$$+ \mathbf{k}' \left(- \cos h \cos A \cos \phi + \sin h \sin \phi \right).$$

(4.6)

Comparing Eqs. 4.5 and 4.6 with Eqs. 7.119 and 4.2, we obtain the transformation formulas for the two sets of angles:

$$\cos h \cos A = \cos \delta \cos \omega \sin \phi - \sin \delta \cos \phi,$$

(4.7)

$$\cos h \sin A = \cos \delta \sin \omega,$$

(4.8)

$$\sin h = \cos \delta \cos \omega \cos \phi + \sin \delta \sin \phi;$$

(4.9)

and

$$\cos \delta \cos \omega = \cos h \cos A \sin \phi + \sin h \cos \phi,$$

(4.10)

$$\cos \delta \sin \omega = \cos h \sin A,$$

(4.11)

$$\sin \delta = - \cos h \cos A \cos \phi + \sin h \sin \phi.$$

(4.12)

4.2.2 Coordinate Transformation: Spherical Trigonometry

The coordinate transformation formulas can be easily obtained using formulas in spherical trigonometry; see Fig. 4.3. We should focus our attention on the spherical triangle PZX, with three arcs $p = \widehat{ZX}$, $z = \widehat{XP}$, and $x = \widehat{PZ}$. As seen from Fig. 4.3, the relations between the elements of the spherical triangle and the quantities of interest are

$$P = \omega,$$

$$Z = 180° - A,$$

$$p = 90° - h,$$

$$z = 90° - \delta,$$

$$x = 90° - \phi.$$

(4.13)

First, consider the case of given declination δ and hour angle ω in the equatorial system to find height h and azimuth A in the horizon coordinate system. The latitude of the observer's location ϕ is obviously a necessary parameter. Using the cosine formula

$$\cos p = \cos x \cos z + \sin x \sin z \cos P, \tag{4.14}$$

with Eqs. 4.13, we obtain

$$\sin h = \sin \delta \sin \phi + \cos \delta \cos \omega \cos \phi. \tag{4.15}$$

Further, using the sine formula

$$\frac{\sin Z}{\sin z} = \frac{\sin P}{\sin p}, \tag{4.16}$$

we find

$$\cos h \sin A = \sin \omega \cos \delta. \tag{4.17}$$

Finally, applying Formula C in Appendix B,

$$\sin p \cos Z = \cos z \sin x - \sin z \cos x \cos P, \tag{4.18}$$

and using Eqs. 4.13, we find

$$\cos h \cos A = \sin \phi \cos \delta \cos \omega - \cos \phi \sin \delta. \tag{4.19}$$

Equations 4.15, 4.18, and 4.19 are identical to Eqs. 4.7, 4.8, and 4.9.

Next, we consider the case of given height h and azimuth A in the horizon coordinate system to find declination δ and hour angle ω in the equatorial system. Again, the latitude of the observer's location ϕ is a necessary parameter.

Using the cosine formula

$$\cos z = \cos p \cos x + \sin p \sin x \cos Z, \tag{4.20}$$

we have

$$\sin \delta = -\cos h \cos A \cos \phi + \sin h \sin \phi. \tag{4.21}$$

By rearranging Eq. 4.18, we find

$$\cos \delta \sin \omega = \cos h \sin A. \tag{4.22}$$

Similar to the derivation of Eq. 4.19, using formula C,

$$\sin z \cos P = \cos p \sin x - \sin p \cos x \cos Z, \tag{4.23}$$

we obtain

$$\cos \delta \cos \omega = \cos h \cos A \sin \phi + \sin h \cos \phi. \tag{4.24}$$

Those equations are identical to Eqs. 4.10–4.12.

4.3 Treatment in Solar Time

Since the prehistory era, human activities have been revolving around the apparent motion of the Sun across the sky. The *solar time* $t_\odot$, which is based on the hour angle of the Sun, is an intuitive measure of time and used for thousands of years in all cultures of the world. As we will discuss in Section 4.4, because the apparent motion of the Sun is nonuniform and depends on location, it is not an accurate measure of time. The difference between solar time and standard time as used in everyday life, even with proper alignment, can be more than ± 15 min.

To make an estimate of solar radiation, sometimes high accuracy is not required. Thus, the simple and intuitive solar time is widely used in the solar energy literature. For example, to compute the integrated values of total insolation over a day or a year, the time shift is irrelevant. The concepts become simple. For example, at solar noon, when the Sun is passing the meridian, solar time is zero. Sunrise time, which is always negative, equals sunset time in magnitude. In this case, the hour angle of the Sun is a measure of time. In other words, if $t_\odot$ is the solar time on an 24-h scale, the hour angle of the Sun ω_s in radians is

$$\omega_s = \pi \frac{t_\odot - 12}{12},\tag{4.25}$$

and

$$t_\odot = 12 + 12\frac{\omega_s}{\pi}.\tag{4.26}$$

4.3.1 Obliquity and Declination of the Sun

The orbital plane of Earth around the Sun, the *ecliptic*, is at an angle called the *obliquity* ϵ from the equator. From the point of view of an observer on Earth, the Sun is moving in the *ecliptic plane*; see Fig. 4.5. On a time scale of centuries, the obliquity angle varies over time. Currently, $\epsilon = 23.44°$. This is what causes the seasons.

Over a calendar year, the motion of the Sun is characterized by four cardinal points. At the *vernal equinox*, the trajectory of the Sun intersects the celestial equator, heading north. At the *summer solstice*, the trajectory of the Sun reaches its northern most point, about 23.44° above the celestial equator. At the *autumnal equinox*, the trajectory of the Sun intersects the celestial equator, heading south. At the *winter solstice*, the trajectory of the Sun reaches its lowest point, about 23.44° below the celestial equator. The dates and times of these four cardinal points vary year by year.

In line with the concept of solar time, the motion of the Sun along its orbital can be described by the *mean ecliptic longitude l*. At the vernal equinox, $l = 0$. At the summer solstice, $l = \pi/2$, or 90°. At the autumnal equinox, $l = \pi$, or 180°. At the winter solstice, $l = 3\pi/2$, or 270°.

An accurate formula for the declination of the Sun is presented in Section 4.4.7. Here we present a simple approximation by assuming that the declination varies sinusoidally with the mean ecliptic longitude l, which is linear to the number of the day in a year.

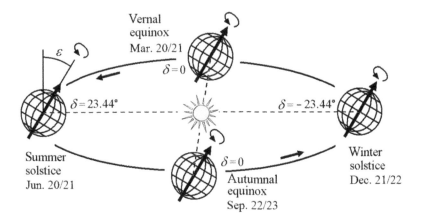

Figure 4.5 Obliquity and the seasons. The rotational axis and the orbital plane of Earth has a tilt angle ϵ, the *obliquity*, which is the origin of the seasons. It causes a periodic difference between solar time and civil time.

The error could be as large as $1.60°$, but is acceptable in many applications:

$$\delta \approx \varepsilon \sin l = \varepsilon \sin \left(\frac{2\pi\,(N - 80)}{365.2422} \right), \tag{4.27}$$

where ε is the obliquity of the ecliptic, currently $\varepsilon = 23.44°$; and N is the number of the day counting from 1 January, which can be computed using the formula

$$N = \mathrm{INT}\left(\frac{275 \times M}{9} \right) - K \times \mathrm{INT}\left(\frac{M + 9}{12} \right) + D - 30, \tag{4.28}$$

where M is the month number, D is the day of the month, and $K = 1$ for a leap year, $K = 2$ for a common year. A leap year is defined as divisible by 4, but not by 100, except if divisible by 400. INT means taking the integer part of the number. This formula can be verified directly. The number 80 in Eq. 4.27 is the number of the day of the vernal equinox, March 20 or 21. The actual date varies from year to year. It also differs between leap year and common year. Using Eq. 4.28, it can be shown that this date varies from 79 to 81. The most common number of days of vernal equinox is 80.

The apparent motion of the Sun from an observer on Earth is shown in Fig. 4.6. Earth is rotating on its axis OP eastward because of its spin. Therefore, apparently, the Sun is moving westward. Due to obliquity, on different days of the year, the declination of the Sun varies. At the winter solstice, the declination of the Sun reaches its minimum, $-\varepsilon$. At the summer solstice, the declination of the Sun reaches its maximum, $+\varepsilon$. At the vernal equinox or autumnal equinox, the declination of the Sun is zero, and the Sun is moving on the celestial equator.

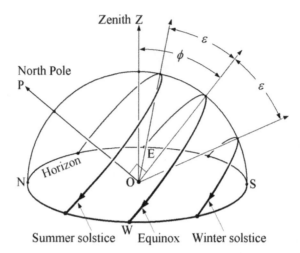

Figure 4.6 Apparent motion of the Sun. Earth is rotating on its axis OP eastward. The Sun is moving westward apparently. Because of obliquity, on different days of a year, the declination of the Sun varies. At the winter solstice, the declination of the Sun reaches its minimum, $-\varepsilon$. At the summer solstice, the declination of the Sun reaches its maximum, $+\varepsilon$. At the vernal equinox or autumnal equinox, the declination of the Sun is zero, and the Sun is moving on the celestial equator.

4.3.2 Sunrise and Sunset Time

Using the treatments in the previous section, we will give an example using the time of sunrise (or sunset) in solar time. The condition of sunrise is the time when the height of the Sun h is zero. According to Eq. 4.15, the condition is

$$\sin \delta \, \sin \phi + \cos \delta \, \cos \phi \, \cos \omega = 0, \tag{4.29}$$

or

$$\cos \omega_s = -\tan \delta \, \tan \phi. \tag{4.30}$$

For each value of the cosine, there are multiple values of the angle ω. For example, at the equator, $\phi = 0$, the times of sunrise and sunset are always at 6:00 am and 6:00 pm solar time. The duration of the day is always 12 hours. At the North Pole or South Pole, $\tan \phi = \infty$, there is never a sunrise or a sunset. In the Temperate Zones and the Torrid Zone, the sunrise and sunset times in terms of the 24-h solar time of the Nth day of the year is determined by

$$t_s = 12 \mp \frac{12}{\pi} \arccos\left[-\tan \delta \, \tan \phi\right]. \tag{4.31}$$

In the frigid zones, where

$$\phi > 90° - \varepsilon, \tag{4.32}$$

there is a period of time in a year when the Sun never rises or never sets.

Another time of the day which is useful for the computation of solar radiation is the solar time when the Sun crosses the East–West great circle. The condition is where the azimuth $A = \pm\pi/2$, or $\cos A = 0$. From Eq. 4.7, one finds

$$\cos\omega_{\mathrm{ew}} = \tan\delta\,\cot\phi, \tag{4.33}$$

or, in terms of solar time in hours,

$$t_{\mathrm{ew}} = 12 \pm \frac{12}{\pi}\arccos\left[\tan\delta\,\cot\phi\right]. \tag{4.34}$$

4.3.3 Direct Solar Radiation on an Arbitrary Surface

For a surface of arbitrary orientation, with polar angle β and azimuth angle γ, the unit vector of its norm $\mathbf{N}$ is given by

$$\mathbf{N} = \mathbf{i}\,\sin\beta\,\cos\gamma + \mathbf{j}\,\sin\beta\,\sin\gamma + \mathbf{k}\,\cos\beta. \tag{4.35}$$

The convention of the sign of the angle is the same as the hour angle: The origin of the azimuth is south and westward it is positive. Combining Eq. 4.35 with Eq. 4.5, the cosine between the norm of the surface and the solar radiation is

$$\begin{aligned}
\cos\theta = \mathbf{N}\cdot\mathbf{S} &= \sin\beta\,\cos\gamma\,(\cos\delta\,\cos\omega\,\sin\phi - \sin\delta\,\cos\phi)\\
&+ \sin\beta\,\sin\gamma\,\cos\delta\,\sin\omega\\
&+ \cos\beta\,(\cos\delta\,\cos\omega\,\cos\phi + \sin\delta\,\sin\phi),
\end{aligned} \tag{4.36}$$

or, rearranging

$$\begin{aligned}
\cos\theta &= \sin\delta\,(\sin\phi\,\cos\beta - \cos\phi\,\sin\beta\,\cos\gamma)\\
&+ \cos\delta\,(\cos\phi\,\cos\beta\,\cos\omega + \sin\phi\,\sin\beta\,\cos\gamma\,\cos\omega\\
&+ \sin\beta\,\sin\gamma\,\sin\omega).
\end{aligned} \tag{4.37}$$

For a surface facing south with $\gamma = 0$, Eq. 4.37 simplifies to

$$\cos\theta = \sin(\phi - \beta)\,\sin\delta + \cos(\phi - \beta)\,\cos\delta\,\cos\omega. \tag{4.38}$$

Consider special cases as follows: For a horizontal surface, $\beta = 0$,

$$\cos\theta = \sin\phi\,\sin\delta + \cos\phi\,\cos\delta\,\cos\omega. \tag{4.39}$$

At the North Pole, where $\phi = \pi/2$,

$$\cos\theta = \sin\delta. \tag{4.40}$$

And at the equator, where $\phi = 0$,

$$\cos\theta = \cos\delta\,\cos\omega. \tag{4.41}$$

For a vertical surface facing south, $\beta = \pi/2$ and $\gamma = 0$,

$$\cos\theta = -\cos\phi \sin\delta + \sin\phi \cos\delta \cos\omega. \tag{4.42}$$

At the North Pole, where $\phi = \pi/2$,

$$\cos\theta = \cos\delta \cos\omega. \tag{4.43}$$

And at the equator, where $\phi = 0$,

$$\cos\theta = \sin\delta. \tag{4.44}$$

Of particular importance is a surface with *latitude tilt*, or $\beta = \phi$. Equation 4.38 is greatly simplified:

$$\cos\theta = \cos\delta \cos\omega. \tag{4.45}$$

The surface can get high radiation energy over the entire year, because $\cos\delta$ is always greater than 0.93.

4.3.4   Direct Daily Solar Radiation Energy

An important application in terms of solar time is the computation of direct solar radiation energy H_D on a surface on a clear day. The effect of clouds and scattered sunlight will be treated in Chapter 5. In a clear day, on a surface perpendicular to the sunlight, the power is $1\,\text{kW/m}^2$, and the total radiation energy in an hour is $1\,\text{kWh/m}^2$. When the sunlight is tilted with an angle θ, the radiation energy is reduced to $\cos\theta \times 1\,\text{kWh/m}^2$. Therefore, the daily direct solar radiation energy in units of kilowatt-hours per square meter is the integration of $\cos\theta$ over 24 h.

Consider first a vertical surface facing south in the northern hemisphere, namely, $\beta = \pi/2$ and $\gamma = 0$. From Eq. 4.36, one obtains

$$\cos\theta = \cos\delta \cos\omega \sin\phi - \sin\delta \cos\phi. \tag{4.46}$$

During days between the vernal equinox and the autumnal equinox, sunlight can shine on the south surface only when the Sun locates in the southern half of the sky. The direct daily solar radiation energy H_D in kilowatt-hours per square meter is

$$\begin{aligned}
H_D &= \frac{12}{\pi} \cos\delta \sin\phi \int_{-\omega_{ew}}^{\omega_{ew}} \cos\omega \, d\omega - \frac{24}{\pi} \omega_{ew} \sin\delta \cos\phi \\
&= \frac{24}{\pi} (\cos\delta \sin\phi \sin\omega_{ew} - \omega_{ew} \sin\delta \cos\phi).
\end{aligned} \tag{4.47}$$

During the days between the autumnal equinox and the vernal equinox of the next year, the available sunlight is limited from sunrise to sunset. The daily solar radiation energy in kilowatt-hours per square meter is

$$H_{\mathrm{D}} = \frac{12}{\pi} \cos \delta \, \sin \phi \int_{-\omega_s}^{\omega_s} \cos \omega \, d\omega - \frac{24}{\pi} \omega_s \sin \delta \, \cos \phi$$

$$= \frac{24}{\pi} \left(\cos \delta \, \sin \phi \, \sin \omega_s - \omega_s \sin \delta \, \cos \phi \right). \tag{4.48}$$

The direct daily solar radiation energy on a surface facing south for various latitudes in the northern hemisphere of the entire year is shown in Fig. 4.7. As shown, in the temperate zone, especially where $\phi > 30°$ and $\phi < 50°$, in the winter, the surface enjoys almost full sunlight. In the summer, the solar radiation is much weaker because of tilting. From the point of view of passive solar buildings, south-facing windows are highly preferred. For solar photovoltaics applications, south-facing panels are much more efficient in the winter. In the northern frigid zone, solar radiation energy in the winter is reduced because of late sunrise and early sunset, or no sunlight at all.

For rooftop applications in large cities, to avoid structural damages from wind, horizontal placement of solar panels surface is often used. Since $\beta = 0$, using Eq. 4.36,

$$\cos \theta = \cos \delta \, \cos \omega \, \cos \phi + \sin \delta \, \sin \phi. \tag{4.49}$$

By integrating the cosine over time from sunrise to sunset, the daily radiation energy

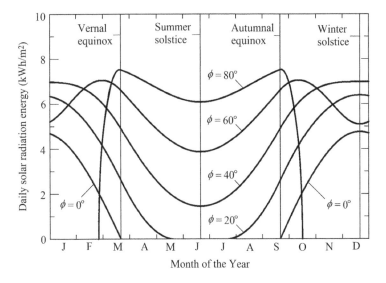

Figure 4.7 Daily solar radiation energy on a vertical surface facing south. In the winter, in the temperate zone, the surface enjoys almost full sunlight, except for places with very high latitude. In the summer, the solar radiation is much weaker. From the point of view of passive solar buildings, south-facing windows are highly preferred. In the northern frigid zone, solar radiation energy in the winter is reduced because of late sunrise and early sunset, or no sunlight at all.

is

$$H_D = \frac{12}{\pi} \cos\delta \, \cos\phi \int_{-\omega_s}^{\omega_s} \cos\omega \, d\omega + \frac{24}{\pi} \, \omega_s \sin\delta \, \sin\phi$$

$$= \frac{24}{\pi} \left(\cos\delta \, \cos\phi \, \sin\omega_s + \omega_s \sin\delta \, \sin\phi \right). \tag{4.50}$$

The variation of the radiation energy over a year is shown in Fig. 4.8. As shown, in the summer, especially in locations with low latitude, the radiation energy is strong. However, in winter, especially in locations of higher latitude, the daily radiation energy is weak.

A much better choice for solar panel placement is *latitude tilt*. From Eq. 4.45, between the vernal equinox and the autumnal equinox, the daily radiation energy is only limited by the angle of incidence,

$$H_D = \frac{12}{\pi} \cos\delta \int_{-\pi/2}^{\pi/2} \cos\omega \, d\omega$$

$$= \frac{24}{\pi} \cos\delta. \tag{4.51}$$

As shown in Fig. 4.9, in this period of the year, the daily radiation energy is independent of latitude. However, between the autumnal equinox and the vernal equinox of the next year, sunrise is later than 6:00 am and sunset is earlier than 6:00 pm. The daily

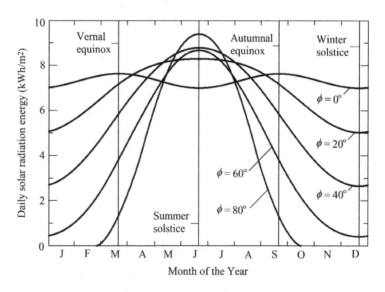

Figure 4.8 Daily solar radiation energy on a horizontal surface. In the summer, especially in locations with low latitude, the radiation energy is strong. However, in the winter, especially in locations of higher latitude, the daily radiation energy is weak.

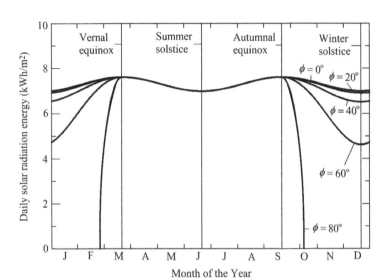

Figure 4.9 Daily solar radiation energy on a latitude-tilt surface. Solar panels placed on a latitude-tilt surface enjoy almost the maximum solar radiation over the entire year.

radiation energy is

$$H_D = \frac{12}{\pi} \cos\delta \int_{-\omega_s}^{\omega_s} \cos\omega \, d\omega$$
$$= \frac{24}{\pi} \cos\delta \sin\omega_s.$$

(4.52)

Also shown in Fig 4.9, the radiation energy is reduced and depends on the latitude of the location, but not by much. Over the entire year, maximum solar radiation energy is obtained.

If the surface is allowed to follow the apparent motion of the Sun, the daily radiation energy can be further enhanced. Consider a solar panel mounted on an axis parallel to the axis of Earth and that rotates uniformly one turn per day. The daily solar radiation energy is

$$H_D = \frac{24}{\pi} \cos\delta \, \omega_s.$$

(4.53)

As shown in Fig. 4.10, the daily solar radiation on a surface with single-axis tracking is substantially higher than for fixed surfaces. The advantage is more apparent in the average direct daily solar radiation $\overline{H}_D$ over a year, as shown in Table 4.2. The solar radiation on a surface with tracking is more than 50% higher than all fixed surfaces. Nevertheless, because of the effect of shadows, it does not save horizontal area.

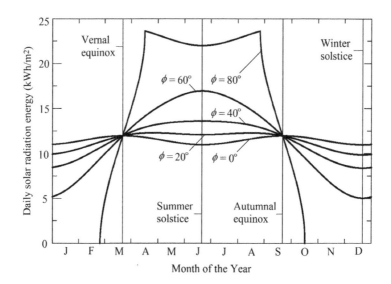

Figure 4.10 Daily solar radiation energy on a surface with tracking. Daily radiation received by a solar panel mounted on an axis parallel to the axis of Earth and the rotates uniformly one turn per day, similar to that on Plate 6. The average daily solar radiation over the entire year is independent of latitude, see Table 4.2. In the frigid zone, in the summer, sunlight could be available 24 hours a day.

4.3.5 The 24 Solar Terms

As shown previously, the motion of the Sun in a calendar year is determined by the Sun's *ecliptic longitude* l. The vernal equinox at $l = 0°$ or $360°$, the summer solstice at $l = 90°$, the autumnal equinox at $l = 180°$, and the winter solstice at $l = 270°$ are the four *cardinal points*. To study the position of the Sun over a year, more points are needed. In traditional Western calendars, including the Julian calendar and the Gregorian calendar, the definitions of the 12 months are not based on natural phenomena. The calendars are not synchronized with the motion of the Sun. Therefore, using calendar dates to specify the motion of the Sunit is inaccurate.

Table 4.2: Average daily solar radiation on various surfaces

$\overline{H}_{\mathrm{D}}$ in kWh/day at Latitude	30°	40°	50°	60°
Vertical, facing south	3.72	4.57	5.25	5.60
Vertical, facing west or east	3.31	2.93	2.46	1.91
Horizontal	6.25	5.43	4.39	3.11
Latitude tilt facing south	7.27	7.21	7.08	6.77
Optimum single-axis tracking	11.50	11.50	11.50	11.50

In East Asia, however, for over 2000 years, a purely solar-based calendar system has been used: the 24 *solar terms*. It is defined solely based on the orbital motion of Earth around the Sun. Similar to dividing a mean solar day into 24 h, the solar term system divides a solar year (the time between two consecutive vernal equinoxes) into 24 equal parts; see Table 4.3. Each year, the exact date and *time of day* of each of the 24 solar terms is published in the Almanac. The date and time for the four cardinal points in years 2024 through 2030 are shown in Table 4.4. Because of rough coordination of the date in the Gregorian calendar with vernal equinox, the dates differ for only 1 or 2 days from year to year. Since the *ecliptic longitude* l of the Sun at each solar term is well defined, by using the formula of the Sun's declination (Eq. 4.27), the declination of the Sun at the 24 solar terms can be calculated,

$$\delta \approx \varepsilon \sin l = \varepsilon \sin \left(\frac{(S - 6)\,\pi}{12} \right), \tag{4.54}$$

Table 4.3: The 24 solar terms

No.	Name	In Pinyin	l (deg)	Approx. date
0	Winter solstice	Dōngzhì	270°	December 22
1	Minor cold	Xiǎohán	285°	January 6
2	Major cold	Dàhán	300°	January 20
3	Spring commences	Lìchūn	315°	February 4
4	Rain water	Yǔshuǐ	330°	February 19
5	Insect awakes	Jīngzhé	345°	March 6
6	Vernal equinox	Chūnfēn	0°	March 21
7	Pure brightness	Qīngmíng	15°	April 5
8	Grain rain	Gǔyǔ	30°	April 20
9	Summer commences	Lìxià	45°	May 6
10	Grain forms	Xiǎomǎn	60°	May 21
11	Grain in ear	Mángzhòng	75°	June 6
12	Summer solstice	Xiàzhì	90°	June 21
13	Minor heat	Xiǎoshǔ	105°	July 7
14	Major heat	Dàshǔ	120°	July 23
15	Autumn commences	Lìqiū	135°	August 8
16	Heat recedes	Chǔshǔ	150°	August 23
17	White dew	Báilù	165°	September 8
18	Autumnal equinox	Qiūfēn	180°	September 23
19	Cold dew	Hánlù	195°	October 8
20	Frost descents	Shuāngjiàng	210°	October 23
21	Winter commences	Lìdōng	225°	November 7
22	Minor snow	Xiǎoxuě	240°	November 22
23	Major snow	Dàxuě	255°	December 7

where S is the order of the solar term, shown in Table 4.3. The number 6 is taken because in Table 4.3, vernal equinox is number 6.

Traditionally, in the Asian calendar, the first solar term is defined as *spring commences*. It is equivalent to defining 3 am early morning as the starting time of a day. This tradition was established probably because spring commences means the starting of agricultural activities of a year. The logical starting solar term is the *winter solstice*, which is equivalent to defining midnight as the starting time of a day.

4.4　Treatment in Standard Time

For several millennia, all over the world, humans have been using the motion of the Sun for time keeping, the *solar time*. However, because the motion of the Sun is not uniform, solar time shows significant deviation from the time defined by uniform motion, for example, the rotation of Earth, the pendulum, or an atomic clock, represented by the motion of the fictitious mean sun. The difference could be as much as 16 min or more. In order to make a sunlight tracking system based on standard time, the difference must be taken into account to reasonable accuracy. In this section, we will present a treatment which is simple to understand and program and yet accurate enough for solar energy utilization.

4.4.1　Sidereal Time and Solar Time

To very high accuracy, the angular velocity of Earth's rotation is a constant. Therefore, the time interval between two consecutive passages of a given fixed star over an observer's meridian is a constant, which is an accurate measure of time, the *sidereal day*. The word sidereal was derived from the Latin *sideus*, which means "star". However, because Earth also has an orbital motion around the Sun, the duration of the solar day is different from the sidereal day; see Fig. 4.11. For example, at midnight, an observer sees a distant fixed star at its meridian. At the time the same star passes the meridian again, Earth has rotated about the Sun by an angle

$$\Delta\phi = \frac{360°}{365.2422} = 59'08''. \tag{4.55}$$

Therefore, a solar day is 0.273% longer than the sidereal day. Because solar time is taken as time as we know, and remember that a 360° angle of rotation corresponds to 24 h, it equals

$$24^h \text{mean solar time} = \frac{24^h \times 366.2422}{365.2422} = 24^h 3^m 56^s \text{ sidereal time.} \tag{4.56}$$

the length of one sidereal day is

$$24^h \text{ sidereal time} = \frac{24^h \times 365.2422}{366.2422} = 23^h 56^m 04^s \text{ mean solar time.} \tag{4.57}$$

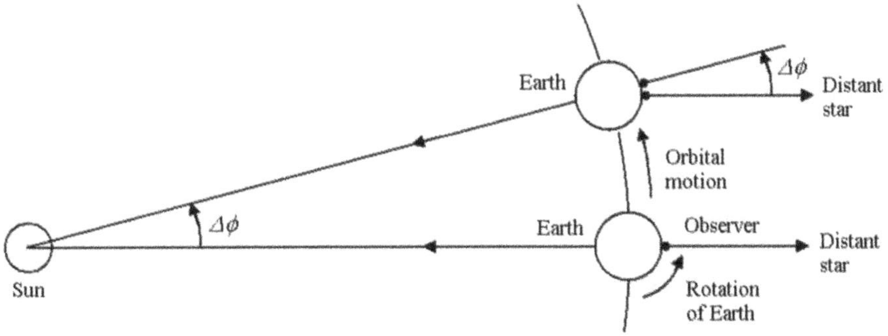

Figure 4.11   Sidereal time and solar time. Because Earth has a rotation on its axis and an orbital motion around the Sun, the duration of the solar day is different from the sidereal day. A solar day is 0.273% longer than the sidereal day.

In the above equations, we introduced the term *mean solar time*. Here we are using a simplified model for the orbital motion of Earth around the Sun, namely, a perfect circle on the sidereal equator with a uniform angular speed. From the point of view of an observer on Earth, such a fictitious Sun of uniform motion along the celestial equator is called the *mean sun*.

From Fig. 4.11, the actual value of time depends on the location of the observer. Specifically, it depends on the longitude of the observer. To make a universal time, a standard longitude must be selected. Similar to the case of the origin of longitude, the standard longitude for the universal time is selected as the *prime longitude* located at Greenwich. The time starting at midnight at Greenwich is called Greenwich mean time (GMT), or more often, *universal time* (UT).

The world is divided into *time zones*. Each zone has a definition of time which differs mostly an integer number of hours from the GMT or UT. For example, Eastern standard time (EST) is defined as UT−5h. In the summer, Eastern daylight saving time (EDT) is defined as UT−4h.

4.4.2   Right Ascension of the Sun

As discussed in section 4.2, in the equatorial coordinate system, the position of a star can be characterized by the declination and the hour angle. The declination of a star does not vary over time. However, because of the rotation of Earth, the hour angle of a star varies over time. Using the *right ascension* α instead of the hour angle, the coordinates of each fixed star are, to a high degree of accuracy, fixed.

Similar to the case of longitude on Earth, a fixed point on the celestial sphere should be chosen as the reference point. The point universally chosen is the *vernal equinox* ♈, the point where the Sun passes the celestial equator while heading north; see Fig. 4.12. The angle between the intersection of the meridian passing through the star S with the equator B and the vernal equinox ♈ is called the right ascension of the star. The sign convention is opposite to that of the hour angle: It is measured eastward. For the treatment of the motion of the Sun, this convention is natural, because it increases with the number of the day in the year.

4.4.3 Time Difference Originated from Obliquity

As mentioned previously, there are two origins of the difference between the motion of the true sun and the mean sun. The first is the obliquity ε. Time is measured by the angle of the mean sun on the equator, but the true sun is moving on the ecliptic plane. Assuming that the Sun is moving uniformly on the ecliptic circle, its *mean longitude l* is a linear function of time. The mean longitude of the Sun increases by 2π during a calendar year; see Fig. 4.12. Taking the number of the day N as the unit of time and the vernal equinox Υ as the origin, which is March 20 or 21, approximately the 80st day of the year, the mean longitude in radians is

$$l = \frac{2\pi (N - 80)}{365.2422}. \tag{4.58}$$

The number of the day N can be computed using Eq. 4.28. The exact date and time for the four cardinal points, vernal equinox, autumnal equinox, summer solstice, and winter solstice, from 2024 to 2030, are shown in Table 4.4. To ensure definitiveness, the time is in Greenwich mean time (GMT).

Because of obliquity, the projection of the longitude l on the celestial equator, the right ascension α, is not linear over time. Using a formula for a rectangular spherical triangle, Eq. B.33, we find the relation

$$\tan \alpha = \cos \varepsilon \tan l. \tag{4.59}$$

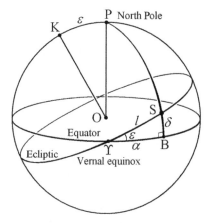

Figure 4.12   Obliquity and equation of time. Because of obliquity, even if Earth is orbiting the Sun with a uniform angular speed, the projection of the speed of the apparent motion of Sun on the equator is not uniform. It gives rise to a term of the difference between mean solar time and true solar time with a periodicity of half a year.

Table 4.4: Cardinal points in years 2024 through 2030, in GMT

Year	Perihelion	Aphelion	Equinoxes	Solstices
2024	Jan 3, 01 h	Jul 5, 5 h	Mar 20, 03:06	Jun 20, 20:51
			Sep 22, 12:44	Dec 21, 09:21
2025	Jan 4, 13 h	Jul 3, 20 h	Mar 20, 09:01	Jun 21, 02:42
			Sep 22, 18:19	Dec 21, 15:03
2026	Jan 3, 17 h	Jul 6, 18 h	Mar 20, 14:46	Jun 21, 08:24
			Sep 23, 00:05	Dec 21, 20:50
2027	Jan 3, 03 h	Jul 5, 5 h	Mar 20, 20:25	Jun 21, 14:11
			Sep 23, 06:02	Dec 22, 02:42
2028	Jan 5, 12 h	Jul 3, 22 h	Mar 20, 02:17	Jun 20, 20:02
			Sep 22, 11:45	Dec 21, 08:19
2029	Jan 2, 18 h	Jul 6, 05 h	Mar 20, 08:02	Jun 21, 01:48
			Sep 22, 17:38	Dec 21, 14:14
2030	Jan 3, 10 h	Jul 4, 13 h	Mar 20, 13:52	Jun 21, 07:31
			Sep 22, 23:27	Dec 21, 20:09

Currently, $\varepsilon \approx 23.44°$, and $\cos \varepsilon \approx 0.917$, very close to 1. Using a trigonometry identity

$$\cos \varepsilon = \frac{1 - \tan^2 \frac{\varepsilon}{2}}{1 + \tan^2 \frac{\varepsilon}{2}}, \tag{4.60}$$

Eq. 4.59 can be written as

$$\frac{\tan l - \tan \alpha}{\tan l + \tan \alpha} = \tan^2 \frac{\varepsilon}{2}. \tag{4.61}$$

Using the obvious relation

$$\frac{\tan l - \tan \alpha}{\tan l + \tan \alpha} = \frac{\sin l \cos \alpha - \cos l \sin \alpha}{\sin l \cos \alpha + \cos l \sin \alpha} = \frac{\sin(l - \alpha)}{\sin(l + \alpha)}, \tag{4.62}$$

we find

$$\frac{\sin(l - \alpha)}{\sin(l + \alpha)} = \tan^2 \frac{\varepsilon}{2}. \tag{4.63}$$

Because $l - \alpha$ is a small quantity, and hence $\sin(l - \alpha) \approx l - \alpha$ and $\sin(l + \alpha) \approx \sin(2l)$, we obtain a formula up to the first order of ε^2,

$$l - \alpha \approx \tan^2 \frac{\varepsilon}{2} \sin(2l) \approx 0.043 \sin(2l). \tag{4.64}$$

4.4.4 Aphelion and Perihelion

The second difference between the fictitious mean sun and the true sun is that the orbit of Earth is elliptical rather than circular. The distance between Earth and the Sun is farthest at *aphelion* and closest at *perihelion*. Aphelion is derived from the Greek words *apo* ("away from") and *helios* ("sun"), while perihelion includes the Greek word *peri* ("near"). The date and the hour in the day of each aphelion and perihelion for 2024 – 2030 are shown in Table 4.4. The distance from the Earth to the Sun is about 3% further at aphelion than it is at perihelion. Because radiation intensity is inversely proportional to the square of distance, the solar radiation power in early January is 6% stronger than that in early July.

4.4.5 Time Difference Originated from Eccentricity

The eccentricity of the orbit of Earth around the Sun gives rise to a second term in the equation of time, see Fig. 4.13. According to Kepler's first law, the orbit of Earth around the Sun is an ellipse, and the position of the Sun is at a focus of the ellipse. From the point of view of Earth, the Sun is orbiting Earth along an ellipse,

$$r = \frac{p}{1 + e\cos(\theta - \theta_0)}, \tag{4.65}$$

where r is the instantaneous distance between the Sun and Earth; e is the eccentricity of the ellipse, currently $e = 0.0167$; θ is the true longitude of the Sun along the ecliptic; θ_0 is the true longitude of the perihelion; and p is a constant.

According to Kepler's second law, the radius vector of the ellipse sweeps out equal areas in equal times:

$$\frac{1}{2}\int_0^t r^2\, d\theta = \frac{1}{2}\int_0^t \frac{p^2}{[1 + e\cos(\theta - \theta_0)]^2}\, d\theta \propto t. \tag{4.66}$$

Because the eccentricity e is small, the integrand of Eq. 4.66 can be expanded into a power series,

$$t \propto \int_0^t [1 - 2e\cos(\theta - \theta_0)]\, d\theta = \theta - 2e\sin(\theta - \theta_0). \tag{4.67}$$

For an entire year, the longitude increases by 2π, and the sine function returns to its original value.

Now, we relate the true longitude of the Sun, θ, with the mean longitude l discussed in the last section, which is proportional to time. For a calendar year, both quantities increase by 2π. Therefore,

$$l = \theta - 2e\sin(\theta - \theta_0). \tag{4.68}$$

If the eccentricity e is negligibly small, we have

$$\theta = l. \tag{4.69}$$

Figure 4.13 Eccentricity of Earth's orbit: Kepler's laws. According to Kepler's laws, Earth is moving around the Sun on an elliptical orbit. It gives rise to a term of the difference between mean solar time and true solar time with periodicity of a year.

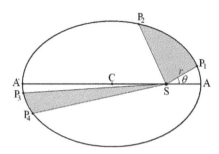

For a first-order approximation, we substitute θ in the second term of Eq. 4.68 with l, which yields

$$\theta = l + 2e \sin(l - l_0), \tag{4.70}$$

where l_0 is the longitude of perihelion. The second term in Eq. 4.70 arises from the eccentricity of the orbit.

4.4.6   Equation of Time

By combining Eq. 4.70 with Eq. 4.64, we obtain a complete expression of the equation of time (ET) up to the first order of obliquity and the first order of eccentricity, $e = 0.0167$,

$$\begin{aligned} \text{ET} = l - \alpha &= \tan^2 \frac{\varepsilon}{2} \sin(2l) - 2e \sin(l - l_0) \\ &\approx 0.043 \sin(2l) - 0.0334 \sin(l - l_0). \end{aligned} \tag{4.71}$$

The unit of angles, l and α, is in radians. A complete circle is 2π. To convert the unit to time as we know, we notice that the mean Sun revolves around Earth every 24 hours. The time it takes for the mean Sun to catch up the true Sun, or *vice versa*, following Eq. 4.71, is

$$\begin{aligned} \text{ET(min)} &= \frac{24 \times 60}{2\pi} \left[0.043 \sin(2l) - 0.0334 \sin(l - l_0) \right] \\ &= 9.85 \sin(2l) - 7.65 \sin(l - l_0). \end{aligned} \tag{4.72}$$

In Eq. 4.72, as usual, the time difference is expressed in minutes. Using the approximate formula for the mean longitude l, Eq. 4.27, an explicit expression of the equation of time can be obtained,

$$\text{ET} = \left[9.85 \sin\left(\frac{4\pi\,(N - 80)}{365.2422} \right) - 7.65 \sin\left(\frac{2\pi\,(N - 3)}{365.2422} \right) \right] \text{(min)}. \tag{4.73}$$

Here the date of perihelion is assumed to be January 3. The number of the day in a year N can be computed using Eq. 4.28. Equation 4.73 is sufficiently accurate to deal with problems in sunlight tracking. A chart is shown in Fig. 4.14. Remember that hour angle ω is to the negative of right ascension α, the equation of time ET should be added to the hour angle of the Sun, which means that if ET is positive, the real Sun is faster than the mean Sun.

Equation 4.73 can be used to convert standard time to solar time. Standard time is defined by an offset Δ from UT, almost always an integer number of hours. For example, Eastern Standard Time (EST) is defined by UT–5 h, or $\Delta = -5$; Eastern Daylight Saving Time (EDT) is defined by UT–4 h, or $\Delta = -4$. The solar time $t_\odot$ is

$$t_\odot = \text{UT} + \frac{1}{15}\lambda - \Delta + \text{ET},\tag{4.74}$$

where λ is the longitude of the observer, each hour corresponds to $15°$; and ET is the equation of time given by Eq. 4.71.

By setting $t_\odot = 0$, the standard time for solar noon, T_0, can be determined from Eq. 4.74,

$$T_0 = \Delta - \frac{1}{15}\lambda - \text{ET}.\tag{4.75}$$

As an example, New York City, where $\lambda = -73°58'$, corresponds to minus 4 h and 56 min. For EST, $\Delta = -5$. At EST noon, the mean solar time is 12:04 in the afternoon. In mid-November, Eq. 4.73 gives ET = 16 min, which means the real Sun is 16 min faster than the mean Sun. Therefore, at 12:00 EST, solar time is 12:20 in the afternoon. The Sun is $5°$ west of the meridian.

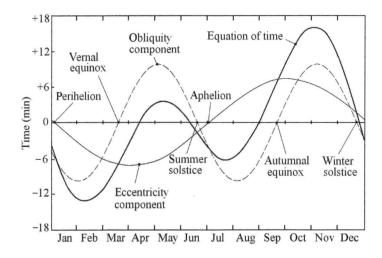

Figure 4.14 Equation of time. Thick solid curve, the difference between mean solar time and true solar time, the so-called equation of time has two terms. The first term, the thin solid curve, has a periodicity of a year, originated from the eccentricity of the orbit, which starts at aphelion. The second term, the dashed curve, originated from the obliquity of the ecliptic, has a periodicity of half a year, and starts at the vernal equinox.

Figure 4.15 The analemma: the apparent motion of the Sun. By fixing a camera toward the southern sky, taking one picture at the same time of the day for a year, then superimposing the pictures of the sunny days, an "8"-like pattern is recovered, which is called the *analemma*. It is a combined result of equation of time and the variation of declination, see Section 4.4. This photograph is compiled by Greek astronomer Anthony Ayiomamitis from 47 clear-day photos taken in 2003 near the Temple of Apollo, Corinth, Greece. Apollo is the God of the Sun in Greek legends. Apollo is also the God of light, truth, music, poetry, and the arts. *Source*: Courtesy of Anthony Ayiomamitis.

4.4.7 Declination of the Sun

In Section 4.3.1, we presented an approximate equation for the declination of the Sun (Eq. 4.27). Here, we derive a more accurate equation for the declination δ of the Sun. The reference point of the Sun's declination is the March equinox, where by definition the declination is zero and the ecliptic longitude is also zero. If the Sun's longitude runs linear over time, the sine formula gives

$$\sin \delta = \sin \varepsilon \sin l. \tag{4.76}$$

However, the ecliptic motion of Earth gives rise to an additional term,

$$\sin \delta = \sin \varepsilon \sin \left\{ l + 2e \left[\sin(l - l_0) - \sin l_0 \right] \right\}. \tag{4.77}$$

Consequently,

$$\delta = \arcsin \left(\sin \varepsilon \sin \left\{ l + 2e \left[\sin(l - l_0) - \sin l_0 \right] \right\} \right). \tag{4.78}$$

4.4.8 Analemma

As a combined result of Equation of Time and the variation of declination, the apparent position of the Sun at a given time of the day varies over the date in a year, which forms a well-defined trajectory on the sky, the *analemma*. The trajectory can be recorded by fixing a camera towards the Southern sky (in the Northern hemisphere), taking one picture at the same time of the day everyday, then superimpose the pictures of the sunny days. A "8"-like pattern is revealed, see Fig. 4.15.

Problems

4.1. For a place in the frigid zone with latitude ϕ that satisfies $\phi > \pi/2 - \varepsilon$, determine the starting day and ending day in a year that the Sun never sets.

4.2. For a place in the frigid zone with latitude ϕ that satisfies $\phi > \pi/2 - \varepsilon$, determine the starting day in a year and ending day in the next year that the Sun never rises.

4.3. For a south-facing window of area A, calculate the total solar radiation for any day in the year.

Hint: For days after the vernal equinox and before the autumnal equinox, the time of radiation is between the two points where the Sun crosses the E–W great circle. Otherwise, the Sun is at the north side of the building.

4.4. On gage C3 of the 2009 *Astronomical Almanac*, the leading terms of the equation of time in units of seconds are (using our notations)

$$\mathrm{ET} = -108.5 \sin l + 596.0 \sin 2l - 428.2 \cos l, \tag{4.79}$$

where

$$l = 279°.791 + 0.985647N, \tag{4.80}$$

where N is the number of days counted from January 1, and l is the mean longitude of the Sun.

Questions:
 1. What is the meaning of the number 0.985647?
 2. What is the meaning of the phase angle 279°.791?

4.5. Write the sine and cosine terms in Eq. 4.79 in the form of $\sin(l - l_0)$, where l_0 is a constant phase angle. Explain the meaning of the constant l_0.

4.6. Determine the sunset time (in solar time) and the length of daytime of New York City on New Years Day, Memorial Day, Labor Day, and Thanksgiving Day.

4.7. What is the ratio of solar power on a vertical surface facing south, a horizontal surface, and a latitude-tilt surface, at solar noon of an equinox (vernal equinox or autumnal equinox) for New York City?

4.8. In New York City, where the latitude is $40°47\prime$, on Memorial Day (May 25) and Thanksgiving Day (Nov 27) of 2009, at solar noon, determine the declination of the Sun, the height of the Sun, and the power density of direct sunlight in watts per square meter on a horizontal surface at that time.

4.9. In New York City, where the latitude is $40°47\prime$ and longitude is $-73°58\prime$, on Memorial Day (May 25) and Thanksgiving Day (Nov 27) of 2009, determine the civil time (EST, or EDT if necessary) of solar noon (the time the Sun passes the local meridian).

4.10. For a surface of arbitrary orientation β and γ, find out the starting time and ending time of solar radiation on the surface.

4.11. For a surface facing south, $\gamma = 0$ but $\beta \neq 0$, find out the starting time and ending time of solar radiation on the surface.

4.12. For a surface facing south, $\gamma = 0$ but $\beta \neq 0$, find out the direct daily radiation.

Chapter 5

Interaction of Sunlight with Earth

In this chapter, we study the interaction of sunlight with Earth. Due to the effect of the atmosphere, about one-half of sunlight is reflected, scattered, or absorbed before reaching the ground. The available sunlight varies with location and time. Furthermore, the sunlight absorbed by the ground penetrates into Earth, is stored as heat, and becomes *shallow geothermal energy*, a significant component of renewable energy.

5.1 Interaction of Radiation with Matter

In this section, the general physical phenomena of the interaction of radiation with matter are described.

5.1.1 Absorptivity, Reflectivity, and Transmittivity

When a ray of radiation falls on a piece of matter, in general, part of the radiation is *reflected*, another part is *absorbed*, and yet another part is *transmitted*. To describe and characterize the interaction of radiation with matter, the following three dimensionless coefficients are introduced:

$A(\lambda)$, *absorptivity*: fraction of incident radiation of wavelength λ absorbed

$R(\lambda)$, *reflectivity*: fraction of incident radiation of wavelength λ reflected

$T(\lambda)$, *transmittivity*: fraction of incident radiation of wavelength λ transmitted

Figure 5.1 Absorptivity, reflectivity, and transmittivity. Part of the radiation falling on a piece of matter is *absorbed*, another part is *reflected*, and the rest is *transmitted*. Three dimensionless coefficients are introduced: *absorptivity* $A(\lambda)$, *reflectivity* $R(\lambda)$, and *transmittivity* $T(\lambda)$. Conservation of energy requires that $A(\lambda) + R(\lambda) + T(\lambda) = 1$.

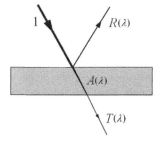

If the nature of the radiation does not change or its wavelength stays unchanged, conservation of energy requires that incident radiation be absorbed, reflected, or transmitted; see Fig. 5.1. Therefore,

$$A(\lambda) + R(\lambda) + T(\lambda) = 1. \tag{5.1}$$

For opaque surfaces, the transmittivity is zero. Conservation of energy requires that incident radiation be either absorbed or reflected,

$$A(\lambda) + R(\lambda) = 1. \tag{5.2}$$

5.1.2   Emissivity and Kirchhoff's Law

When matter is heated, it emits radiation (Fig. 5.2). The actual power density of radiation also depends on the nature of the surface. However, it never exceeds that of a blackbody. The actual radiation from a surface as a fraction of blackbody radiation at a given wavelength is called its *emissivity* $E(\lambda)$. It is less than 1, except it equals exactly 1 for a blackbody. Based on classical thermodynamics, Kirchhoff showed that *the emissivity of a surface at a given wavelength must equal its absorptivity*. At thermal equilibrium, the radiation energy emitted must equal the radiation energy absorbed. Otherwise, heat can transfer from a cold reservoir to a hot reservoir, which violates the second law of thermodynamics (see Chapter 6). Therefore, the absorptivity at a given wavelength must equal the emissivity at the same wavelength,

$$E(\lambda) = A(\lambda). \tag{5.3}$$

5.1.3   Bouguer–Lambert–Beer's Law

An empirical relationship between the absorption of light and the property of the absorbing medium was discovered by Pierre Bouguer before 1729, then was formulated by Johann Heinrich Lambert in 1760 in his monograph "Photometria." It states that the light intensity depends exponentially on the thickness of the optical path z,

Figure 5.2 Emissivity and absorptivity. When heated, matter emitts radiation. The maximum radiation power spectrum follows Planck's law. Ine general, the radiation from a surface as a fraction of blackbody radiation at a given wavelength is called its *emissivity*, $E(\lambda)$. At thermal equilibrium, the radiation energy emitted must equal the radiation energy absorbed. Consequently, $E(\lambda) = A(\lambda)$.

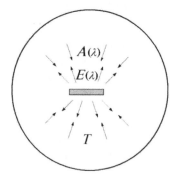

$$I_\lambda(z) = I_\lambda(0)\, e^{-A(\lambda)\, z}, \tag{5.4}$$

where $A(\lambda)$ is the absorption coefficient of the medium at wavelength λ, which has a dimension of inverse length and z is the optical path.

The empirical relation was further developed by August Beer in 1852 to correlate the absorption coefficient with the *concentration of absorbing particles*, known as the Beer's law or Bouguer–Lambert–Beer's law,

$$I_\lambda(z) = I_\lambda(0)\, e^{-N\sigma(\lambda)\, z}, \tag{5.5}$$

where N is the number of absorbing particles per unit volume and $\sigma(\lambda)$ is the *absorption cross section* of the absorbent at wavelength λ. An intuitive proof is shown in Fig. 5.3. For a thin slice in the absorption path of thickness dz and cross sectional area S, the fractional area dS occupied by the absorbing particles is

$$dS = N\sigma(\lambda)\, S\, dz. \tag{5.6}$$

Obviously, a proportion dS/S of radiation energy is blocked by the absorbing particles,

$$\frac{dI_\lambda(z)}{I_\lambda(z)} = -\frac{dS}{S} = -N\sigma(\lambda)\, dz. \tag{5.7}$$

Integrating over z and using the initial condition at $z = 0$, one obtains

$$I_\lambda(z) = I_\lambda(0)\, e^{-N\sigma(\lambda)\, z}. \tag{5.8}$$

Comparing with Eq. 5.4, one finds $A(\lambda) = N\sigma(\lambda)$. If the concentration of particles is not uniform over the optical path, which can be described by a concentration distribution $N(z)$, the above relation can be extended to

$$I_\lambda(z) = I_\lambda(0) \exp\left(-\int_0^z \sigma(\lambda)\, N(z)\, dz\right). \tag{5.9}$$

Now we look into a situation relevant to the attenuation of sunlight by the atmosphere; see Fig. 5.4. In the atmosphere, to a good approximation, the distribution of

Figure 5.3 Bouguer–Lambert–Beer's law. Variation of light intensity with the concentration of the absorbing particles and the length of the optical path plays an important role in the study of the effect of the atmosphere on sunlight.

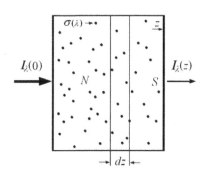

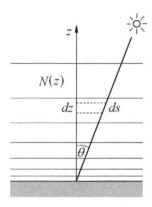

Figure 5.4 Attenuation of sunlight at azimuth θ. Variation of light intensity with the concentration of the absorbing particles and the length of the optical path plays an important role in the study of the effect of the atmosphere on sunlight.

molecules and particles is a function of height z. If the Sun is at the zenith, according to Eq. 5.9, the absorbance over the entire atmosphere is

$$\alpha(\lambda, 0) = \frac{I_\lambda(0)}{I_\lambda(\infty)} = \exp\left(-\int_0^\infty \sigma(\lambda)\, N(z)\, dz.\right). \tag{5.10}$$

If the azimuth of the Sun is θ — notice that $dz = ds \cos\theta$ (see Fig. 5.4) — the absorbance becomes

$$\alpha(\lambda, \theta) = \exp\left(-\int_0^\infty \sigma(\lambda)\, N(z)\, ds\right)$$
$$= \frac{1}{\cos\theta} \exp\left(-\int_0^\infty \sigma(\lambda)\, N(z)\, dz\right) \tag{5.11}$$
$$= \frac{\alpha(\lambda, 0)}{\cos\theta}.$$

5.2 Interaction of Sunlight with Atmosphere

The interaction of sunlight with the atmosphere has been studied extensively by climate scientists. Here is a summary. Approximately, 30% of solar radiation is reflected or scattered back to space, see Fig. 5.5. Six percent is scattered by air; 20% is reflected by clouds; 4% is reflected by the surface of Earth; and 20% is absorbed by the atmosphere: 16% is absorbed by water vapor, dust, and O_3. Another 4% is absorbed by clouds. The solar radiation thus absorbed heats up the atmosphere. Fifty percent is absorbed by the solid surface of Earth. The total radiation energy received by the atmosphere and the solid Earth is about 70%.

Earth should be in thermal equilibrium with the surroundings. Indeed, 70% of solar energy reaching the Earth is transferred back to space as heat radiation.

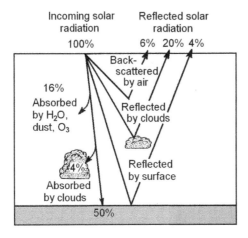

Figure 5.5 Interaction of sunlight with atmosphere. Approximately 30% of solar radiation is reflected or scattered back to space immediately, 20% is absorbed by the atmosphere and cloud, and 50% is absorbed by Earth. (See Ref. [82], p. 94).

5.2.1 AM1.5 Reference Solar Spectral Irradiance

According to several decades of careful measurements, the power density of solar radiation outside the atmosphere is 1366 W/m^2. On the surface of Earth, due to scattering and absorption, even under a perfectly clear sky, when the Sun is right at the zenith, solar radiation is reduced by about 22%. Because, on average, the Sun should have an azimuth angle with the horizon, the reduction should be on average more than 22%.

To standardize the measurement of solar energy applications, in 1982, the American Society for Testing and Materials (ASTM) started to promulgate *Standard Tables of Reference Solar Spectral Irradiance at Air Mass 1.5*. The standard was revised in 2003 as ASTM G173-03. A separate standard for zero air mass was promulgated in 2006 to become ASTM E490-06. An extension to different tilt angles was promulgated in 2008 as ASTM G197-08. In both ASTM G173-03 and G197-08, atmospheric and climatic conditions are identical. These standards represent the solar radiation under reasonable cloudless atmospheric conditions favorable for the computerized simulation, comparative rating, or experimental testing of fenestration systems. The U.S. standard was adapted by the International Organization for Standardization (ISO) to become ISO 9845-1, 1992.

The meaning of AM is as follows. When the Sun is at the zenith under reasonable cloudless atmospheric conditions, the absorption due to the atmosphere is defined as AM 1. In most cases, the zenith angle of the Sun is not zero. ASTM chose as the standard the condition when the absorption is 1.5 times the normal air mass, abbreviated as AM 1.5. According to Eq. 5.11, the standard zenith angle is

$$\theta = \arccos \frac{1}{1.5} = 48.19°. \tag{5.12}$$

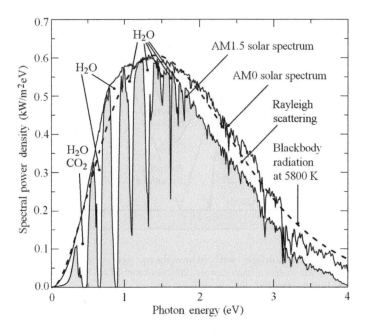

Figure 5.6 AM0 and AM1.5 solar radiation spectra. The AM0 spectrum is the solar radiation outside the atmosphere. It is approximately a blackbody radiation at 5800 K. The power density is 1.366 kW/m². The AM1.5 spectrum shows a number of atmospheric effects. On the blue side, there is a broad-band reduction of power density due to Rayleigh scattering from molecules and dust particles. On the infrared side, water vapor contributes the most absorption, followed by carbon dioxide.

Also see Fig. 5.4. The integrated power density of the AM 1.5 solar radiation is 1 kW/m². The quantity 1 kW/m² is defined as a unit of radiation, called *one sun*. We shall use this unit throughout the book.

The spectral irradiance or the radiation power spectrum for AM0 and AM1.5 solar radiation is shown in Fig. 5.6. The AM0 spectrum is the solar radiation outside the atmosphere. It is approximately a blackbody radiation at 5800 K. The integrated power density is 1.366 kW/m². The AM1.5 spectrum shows a number of atmospheric effects. On the blue side, there is a broad-band reduction of power density due to Rayleigh scattering from molecules and dust particles. The probability of Rayleigh scattering is proportional to the inverse fourth power of the wavelength of the radiation, and thus the short-wavelength radiation is reduced heavily. On the infrared side, water vapor contributes the most absorption followed by carbon dioxide. Plate 1 is a colored version of Fig. 5.6. A table of the data is shown in Appendix E.

5.2.2 Annual Insolation Map

In Chapter 4, we discussed the variation of direct solar radiation as a function of time (day in a year and time in a day) and location (latitude and longitude). Because of the

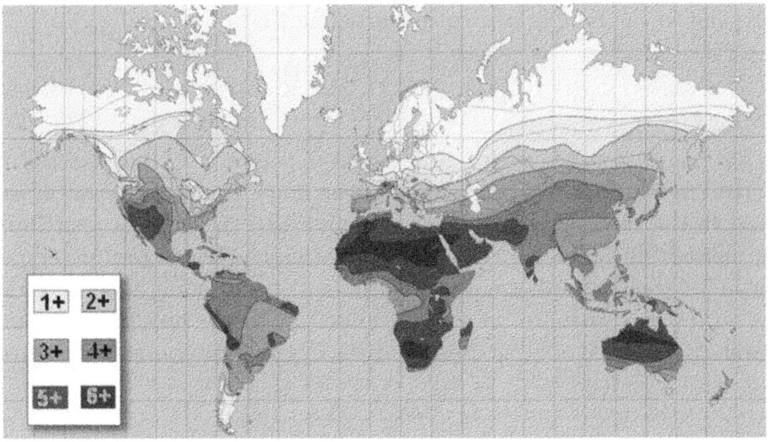

Figure 5.7 Insolation map of the world. Solar radiation per day on a surface of 1 m^2 in kilojoules averaged over a year. As shown, large areas in Northern Africa have the highest insolation. *Source*: www.bpsolar.com.

interaction of solar radiation with the atmosphere, the actual solar radiation received at the surface is always less, and the percentage of reduction depends on the location. A frequently used representation is the *annual insolation map*; see Plates 3–5. There are two conventions:

1. Annual radiation energy in kilowatt-hours per square meter. The standard solar radiation is defined as one sun, or 1 kW/m^2. Therefore, the insolation is often expressed in hours per year. The number ranges from more than 2000 h/year (Sahara desert, part of Outback of Australia, part of South Africa) to less than 600 h/year (Greenland, northern parts of Siberia, Finland, and Canada).

2. Average diurnal radiation energy in kilowatt-hours per square meter over a year. Similarly, it is often expressed in h/day. The number ranges from more than 6 h/day to less than 2 h/day.

There is an obvious relation between those two conventions:

$$\text{Annual insolation} = 365.2422 \times \text{average diurnal insolation.} \tag{5.13}$$

Figure 5.7 is an average diurnal insolation map of the world.

5.3 Penetration of Solar Energy into Earth

The theory of the storage of solar energy in the earth is based on Fourier's law of heat conduction, see Fig. 5.8. In a solid where the temperature T is a function of time t and depth z, it is an experimental fact that the heat flux q across an area A along direction

z is proportional to the temperature gradient along that direction,

$$q(z) = -kA\frac{\partial T(z,t)}{\partial z},\qquad(5.14)$$

where k is the thermal conductivity, a constant depending on the material nature of the solid; and the negative sign means the heat flows in the direction of decreasing temperature. In the International System of Units (SI), the unit for heat flux is watts per square meter. The unit for thermal conductivity is watts per meter per degree Kelvin. Table 5.1 shows the thermal conductivities of commonly used materials.

Look at a thin slab with area A and thickness Δz. If the density of the material is ρ and its heat capacitance is c_p, at temperature T, the heat content in that slab is

$$Q = \rho c_p T\, A\Delta z.\qquad(5.15)$$

The rate of change of the heat content Q in the slab is the difference of the incoming heat flow through the top of the slab, $q(z)$, and the outgoing heat flow through the bottom of the slab, $q(z + \Delta z)$. According to Eq. 5.14,

$$\frac{\partial Q}{\partial t} = -kA\frac{\partial T(z,t)}{\partial z}\bigg|_z + kA\frac{\partial T(z,t)}{\partial z}\bigg|_{z+\Delta z} = kA\frac{\partial^2 T(z,t)}{\partial z^2}\Delta z.\qquad(5.16)$$

On the other hand, from Eq. 5.15, the temperature in the slab changes according to

$$\frac{\partial Q}{\partial t} = \rho c_p A\Delta z\,\frac{\partial T}{\partial t}.\qquad(5.17)$$

Combining Eqs 5.16 and 5.17, we find the heat conduction equation,

$$\frac{\partial T}{\partial t} = \frac{k}{\rho c_p}\frac{\partial^2 T}{\partial z^2}.\qquad(5.18)$$

Let

$$\alpha = \frac{k}{\rho c_p},\qquad(5.19)$$

Equation 5.18 then becomes

$$\frac{\partial T}{\partial t} = \alpha\frac{\partial^2 T}{\partial z^2}.\qquad(5.20)$$

Figure 5.8 Derivation of the heat-conduction equation. The rate of change of the heat content Q in the thin slab is the difference of the heat flow from the top of the slab $q(z)$ and the heat flow through the bottom of the slab $q(z + \Delta z)$. The heat conduction equation Eq. 5.18 follows.

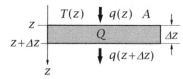

Table 5.1: Thermal property of earth

Material	Density, ρ	Heat capacity, c_p	Thermal conductivity, k	Coefficient, α
	(10^3kg/m^3)	(10^3J/kg·K)	(W/m·K)	$(10^{-7} \text{m}^2/\text{s})$
Limestone	2.18	0.91	1.5	6.26
Granite	3.0	0.79	3.5	14.7
Earth (wet)	1.7	2.1	2.5	7.0
Earth (dry)	1.26	0.795	0.25	2.5

Source: American Institute of Physics Handbook,
American Institute of Physics, New York, 3rd ed., 1972; and Ref. [30].

We seek a solution of Eq. 5.20 for a semi-infinite space $z \geq 0$ with boundary conditions

$$T = T_0 + \Delta T \, \cos \omega t, \qquad z = 0, \tag{5.21}$$
$$T = T_0, \qquad z = \infty, \tag{5.22}$$

where ΔT is the amplitude of the temperature variation at $z = 0$ and ω is its circular frequency. In the case we are considering, that is, the annual variation of temperature, the circular frequency is

$$\omega = \frac{2\pi}{(365.25 \times 86400)} \approx 2 \times 10^{-7} \text{s}^{-1}. \tag{5.23}$$

Introducing a dimensionless temperature,

$$\Theta = \frac{T - T_0}{\Delta T}, \tag{5.24}$$

Equation 5.20 becomes

$$\frac{\partial \Theta}{\partial t} = \alpha \frac{\partial^2 \Theta}{\partial z^2} \tag{5.25}$$

with boundary conditions

$$\Theta = \cos \omega t, \qquad z = 0, \tag{5.26}$$
$$\Theta = 0, \qquad z = \infty. \tag{5.27}$$

Equation 5.25 can be resolved much easier in complex numbers using the Euler relation $e^{ix} = \cos x + i \sin x$, or $\cos x = \text{Re}[e^{ix}]$. The boundary conditions now become

$$\Theta = \text{Re}\left[e^{-i\omega t}\right], \qquad z = 0, \tag{5.28}$$
$$\Theta = 0, \qquad z = \infty, \tag{5.29}$$

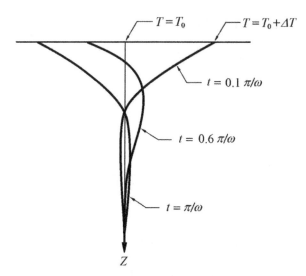

Figure 5.9 Penetration of solar energy into Earth. On the ground, the average temperature is T_0, and the amplitude of annual temperature variation is ΔT. The heat penetrates into the ground with a time delay. At a certain depth, the profile of annual temperature variation is *reversed*.

and Eq. 5.25 can be resolved using the *Ansatz*,

$$\Theta = \mathrm{Re}\left[e^{\lambda z - i\omega t}\right]. \tag{5.30}$$

Obviously, the suggested solution (Eq. 5.30), satisfies the boundary condition at $z = 0$ (Eq. 5.28). The constant λ can be determined by the differential equation Eq. 5.25. In fact, it gives

$$\alpha\,\lambda^2 = -i\omega. \tag{5.31}$$

There are two solutions to Eq. 5.31,

$$\lambda = \pm\left[\sqrt{\frac{\omega}{2\alpha}} - i\sqrt{\frac{\omega}{2\alpha}}\right]. \tag{5.32}$$

Because of the boundary condition at $z = \infty$, only the negative sign is admissible. Finally,

$$\Theta = \exp\left\{-\sqrt{\frac{\omega}{2\alpha}}z\right\}\cos\left(\sqrt{\frac{\omega}{2\alpha}}z - \omega t\right). \tag{5.33}$$

Or, using Eq. 5.24, the solution is

$$T = T_0 + \Delta T\,\exp\left\{-\sqrt{\frac{\omega}{2\alpha}}z\right\}\cos\left(\sqrt{\frac{\omega}{2\alpha}}z - \omega t\right). \tag{5.34}$$

For a numerical example, we use $T_0 = 15°C$, $\Delta T = 12°C$, and limestone. From Eq. 5.23 and Table 5.1, one finds $\sqrt{\omega/2\alpha} = \sqrt{2 \times 10^{-7}/2 \times 0.626 \times 10^{-6}} \approx 0.33\,\mathrm{m}^{-1}$. Taking the number of the month as a parameter,

$$T = 15° + 12° \, e^{-0.33 \, z} \, \cos\left(0.33 \, z - \frac{2\pi \times (\text{month} - 7)}{12}\right). \tag{5.35}$$

At a distance z where $\sqrt{\omega/2\alpha} = \pi$, the value of the cosine function becomes its negative. For example, according to Eq. 5.35, when $0.33z = \pi$, or $z = 9.5$ m, the temperature in the summer is the lowest and the temperature in the winter is the highest. The profile of the annual temperature variation is *reversed*; see Fig. 5.9. Therefore, the stored solar energy can be effectively utilized. We will discuss the details of its implementation in Chapter 6.

Problems

5.1. Modeling the Sun as a blackbody radiator of $T_\odot = 5800$ K, calculate Earth's surface temperature $T_\oplus$, assuming that the temperature is uniform over its entire surface which is also a blackbody radiator.

5.2. As in the previous problem, assume the Sun as a blackbody radiator of $T_\odot = 5800$ K. If the absorptivity of Earth is 0.65 over the entire spectral range and its entire surface, what is Earth's surface temperature $T_\oplus$?

5.3. By direct substitution, prove that

$$u(z,t) = \frac{A}{\sqrt{t}} \exp\left\{\frac{-z^2}{4\alpha t}\right\} \tag{5.36}$$

is a solution of the one-dimensional heat conduction equation

$$\frac{\partial u}{\partial t} = \alpha \frac{\partial^2 u}{\partial z^2}. \tag{5.37}$$

5.4. Because the heat conduction equation is linear, if $u(z,t)$ is a solution, then the infinite integral of $u(z,t)$ is also a solution. Prove that

$$u(z,t) = A\,\text{erf}\,\frac{z}{2\sqrt{\alpha t}} \tag{5.38}$$

is a solution of Eq. 5.37, where the error function $\text{erf}(x)$ is defined as

$$\text{erf}(x) = \frac{2}{\sqrt{(\pi)}} \int_0^x e^{-\xi^2}\,d\xi. \tag{5.39}$$

5.5. If the surface temperature of Earth suddenly changes from $0°C$ to T_0, how long does it take for the interior of Earth to be in equilibrium with its surface, which means it reaches 99% of the surface temperature? Make estimates for depths of 10 m, 100 m, and 1000 m, where the ground is made of granite.

5.6. Using the theory of solar energy penetration into Earth, determine the effect of the average daily variation of earth temperature. Assume that the amplitude of daily temperature is $\Delta T = 5°C$ and the ground is made of limestone:
   1. At what depth the phase of temperature profile is reversed (i.e., cooler at about 3 pm and warmer at about 3 am)?
   2. What is the ratio of the amplitude of temperature variation at the depth of temperature profile reversal versus the temperature amplitude at the surface?

5.7. The average temperature of Oklahoma City is $28°C$ in July and $4°C$ in January. The temperature profile is approximately a sinusoidal curve during the entire year. If the ground is made of granite, what is the monthly temperature variation (list the values for each month) 10 m beneath the ground?

Chapter 6

Thermodynamics of Solar Energy

Thermodynamics is a branch of physics devoted to the study of energy and its transformation. It was established in the first half of the 19th century for a better understanding of the underlying principles of heat engine, which converts heat energy to mechanical energy, such as the steam engine and the internal combustion engine. In the middle of 19th century, thermodynamics evolved into a logically consistent system starting with a few axioms, or laws, from which the entire theory could be deduced. In the 20th century, the theory was extended to refrigeration and heat pumps as well as other forms of energy, such as electric, magnetic, elastic, chemical, electrochemical, and nuclear. At the core is the first law of thermodynamics regarding the conservation of energy and the second law of thermodynamics regarding the conversion of thermal energy to other forms of energy. In some textbooks, there is a zeroth law and a third law. The zeroth law is a self-evident definition of temperature. The applications of the third law are unrelated to the study of solar energy. The theory of thermodynamics is macroscopic in nature, dealing directly with measurable physical quantities. The corresponding microscopic theory is statistical physics, where the laws of thermodynamics are derived from an atomic point of view.

In this chapter, we present concepts of thermodynamics that are essential for the understanding of solar energy. Complete presentations of thermodynamics, including the third law of thermodynamics, can be found in standard textbooks.

6.1 Definitions

In this section, we introduce several basic definitions. The object of thermodynamics is called a system. A typical system is a uniform body of substance with a well-defined boundary, such as a piece of solid, a volume of liquid, a package of gas. The physical objects outside the system boundary are called the surroundings. A state of a system represents the totality of its macroscopic properties. Two types of physical quantities are present in thermodynamics: the extensive quantity and the intensive quantity. An extensive quantity is proportional to the volume or mass of the system and is additive, such as mass, volume, energy, and entropy. Intensive quantities are independent of the volume or mass of the system and are not additive, such as temperature, density, pressure, and the specific values of extensive quantities such as energy density. A thermodynamic system can be connected to the surroundings or isolated. There is no heat or work exchange between an isolated system and its surroundings. The state of a system can go through a process where at least one of the physical quantities is changing over time. Two processes are of particular interest: the isothermal process, where the temperature of the system does not change, and the adiabatic process, where there is no heat change between the system and its surroundings.

An infinitesimal quantity of *work*, δW, is defined as the product of the force F acting on the boundary of the system and the length dL it moves in the direction of the force. Because the force F is a product of pressure P and area A, it is convenient to write infinitesimal work acting on the system as

$$\delta W = F\, dL = PA\, dL = -P\, dV, \tag{6.1}$$

where dV is the infinitesimal change of the volume of the system. The occurrence of the negative sign is because, when the pressure is positive, the force is pointing to the inside of the system; when the system expands, or when $dV > 0$, the displacement dL is pointing to the outside of the system. The total work during a process from state 1 to state 2 is then

$$W = \int_1^2 F\, dL = -\int_1^2 P\, dV. \tag{6.2}$$

Heat is defined as the energy transferred to the system across the boundary without moving the surface. If the temperature of the surroundings is higher than that of the system, the heat is transferred to the system, which is denoted as positive. Similarly, the total heat transferred to the system during a process from state 1 to state 2 is

$$Q = \int_1^2 dQ. \tag{6.3}$$

Temperature is defined by the zeroth law of thermodynamics:

> *When two systems are in thermodynamic equilibrium with a third system, the two systems are in thermodynamic equilibrium with each other, and all three systems have the same temperature.*

The zeroth law tells us how to compare temperature, but it does not define the scale of temperature. In thermodynamics, there are two independent definitions of temperature. The first one is defined by Lord Kelvin based on the Carnot cycle, which bears his name; (see Section 2.3.2). The second definition is based on the properties of ideal gas; (see Section 2.5.1). The two definitions are equivalent within a constant multiplier.

6.2   First Law of Thermodynamics

The first law of thermodynamics asserts that energy can be converted from one form to another but can never be created or annihilated. A succinct presentation is as follows:

It is impossible to build a perpetual motion that generates energy from nothing.

The first law of thermodynamics is by no means trivial, regarding the fact that each year, numerous patent applications on perpetual motion are still received by the patent offices of countries all over the world, even in the era of high technology. Searching on Google for "perpetual motion," you would be surprised by the exotic new designs of perpetual motion proposed by overambitious inventors.

In his autobiography, Max Planck described how his physics teacher taught him about the concept of energy: A construction worker lifted a brick and put it at the top of the building. His work increased the energy of the brick. But the increase of energy was in the form of potential energy and not explicit. One day, the brick fell down from the top. As it almost reached the ground, the brick moved fast, which had an explicit kinetic energy. Finally, the brick hit the ground and converted the energy to heat. In the first step, the construction worker did the work as

$$W = fh = mgh, \tag{6.4}$$

where m is the mass of the brick and $g = 9.81 m/s^2$ is the gravitational acceleration. The gravitational force $f = mg$. And h is the height, or the distance along the direction of force the brick moves. The increase of potential energy ΔE equals the work performed on the system,

$$\Delta E = W = mgh. \tag{6.5}$$

Just before the brick hit the ground, its velocity v is given as

$$v = \sqrt{2gh}, \tag{6.6}$$

which satisfies the law of conservation of energy,

$$\Delta E = \frac{1}{2}mv^2 = mgh. \tag{6.7}$$

The unit of energy is the product of the unit of force, the newton, and the unit of length, the meter. The unit of energy, newton-meter, or joule, is named after English physicist James Prescott Joule, who did the first experiment to demonstrate the equivalence of mechanical work and heat in 1844. A schematic of Joule's experiment is shown in Fig. 6.1.

In Joule's experiment, at the beginning, the weight is positioned at a predetermined distance from its equilibrium position. The initial temperature of the water, T_0, is measured. Then, by setting the weight to move and waiting to the end of motion, the final temperature of the water barrel, T_1, is measured. The heat generated by mechanical disturbance is

$$Q = (T_1 - T_0)M. \tag{6.8}$$

If M is the mass of water in grams, then the heat Q is in calories. The mechanical equivalence of one calorie of heat found by Joule is 4.159 J/cal, very close to the result

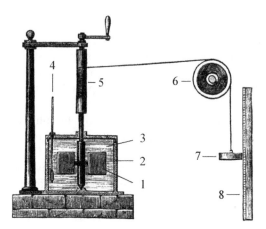

Figure 6.1 Joule's experiment. A paddle-wheel 1 is placed inside an insulated barrel 2 filled with water 3. The temperature is measured by a thermometer 4. The wheel is driven through a spindle 5 and a pulley 6 by a falling weight 7. The height of the weight is measured by the ruler 8. After setting the paddle-wheel to move, the mechanical energy is transformed into heat, which is measured by the thermometer.

of modern measurements. As a result of the equivalence of mechanical work and heat, the increment of the energy of a system is the sum of mechanical work and heat,

$$\Delta E = W + Q. \tag{6.9}$$

Energy could transfer as heat as well. By pushing two systems with heat capacity C_1 and C_2 at temperatures T_1 and T_2 into contact, as shown in Fig. 6.5, heat transfers from the hotter system to the cooler system. Assuming that the thermal capacities of the two systems are constant, that is, independent of temperature within the temperature range of interest, eventually, the temperature becomes a single value T_0,

$$T_0 = \frac{1}{C_1 + C_2}(C_1 T_1 + C_2 T_2). \tag{6.10}$$

The heat transferred from one system to another is

$$Q = \pm \frac{C_1 C_2}{C_1 + C_2}(T_1 + T_2). \tag{6.11}$$

In Joule's experiment, mechanical work transforms into heat. However, there no simple way to transform heat back to mechanical work. In the case of heat transfer, heat can spontaneously transfer from a system at a higher temperature to a system at a lower temperature. But heat can never spontaneously transfer from a system at a lower temperature to a system at a higher temperature without expending mechanical work. Such observations lead to the second law of thermodynamics.

6.3 Second Law of Thermodynamics

There are many ways to state the second law of thermodynamics. It can be shown that all those incarnations are equivalent. A succinct formulation, similar to that of Kelvin and Planck, is as follows:

> *It is impossible to build a machine that converts heat to mechanical work from a single source of heat.*

Because the heat in the ocean is unlimited, if *perpetual motion of the second type* could be built, mankind would never have to worry about having energy. Another formulation of the second law of thermodynamics, due to Clausius, is as follows:

> *It is impossible to transfer heat from a reservoir at a lower temperature to a reservoir at a higher temperature without spending mechanical work.*

In fact, if a machine to transfer heat from a cold reservoir to a hot reservoir without expending external mechanical energy could be built, everybody on Earth would be able to enjoy free heating and free air conditioning.

6.3.1 Carnot Cycle

The spirit of the second law of thermodynamics can be best understood using the Carnot cycle, proposed by Sadi Carnot in 1824 in the quest for the ultimate efficiency of heat engines [1]. A schematic is shown in Fig. 6.2. The engine consists of two heat reservoirs and a cylinder with a piston filled with a volume of working gas as the thermodynamic system.

The Carnot cycle is an idealization of a heat engine that generates mechanical work by transferring heat from a hot reservoir at temperature T_H to a cold reservoir at temperature T_L. A complete cycle consists of four processes. First, the system is in contact with the hot reservoir, undergoing an isothermal expansion. The system, always at temperature T_H, gains heat Q_H from the hot reservoir. Second, the system is isolated from the reservoir and is undergoing an adiabatic expansion. With no heat transfer, the temperature of the system is reduced to T_L. Third, the system is in contact with the cold reservoir, undergoing an isothermal compression. The system, always at temperature T_L, releases heat Q_L to the cold reservoir. Fourth, the system is isolated from the reservoir and is undergoing an adiabatic compression. With no heat transfer, the temperature of the system is raised to T_H. A net work W is performed to the surroundings. The first law of thermodynamics requires that

$$Q_H = Q_L + W. \tag{6.12}$$

The efficiency η of a heat engine is defined as the ratio of mechanical work W over the heat energy from the hot reservoir, Q_H:

$$\eta \equiv \frac{W}{Q_H} = 1 - \frac{Q_L}{Q_H}. \tag{6.13}$$

An essential assumption of the Carnot cycle is that the processes are *reversible*. The Carnot cycle in Fig. 6.2 can be operated as a refrigerator or a heat pump; see Fig. 6.3. The thermodynamic system – a body of gas confined in a cylinder with a piston – transfers heat from a cold reservoir to a hot reservoir with a cost of mechanical work. The four processes are as follows: First, the system is in contact with the cold reservoir, undergoing an *isothermal expansion* process. The system, always at temperature T_L, gains heat Q_L from the cold reservoir. Second, the system is isolated from the reservoir and is undergoing an *adiabatic compression* process. With no heat transfer, the temperature of the system is raised to T_H. Third, the system is in contact with the hot reservoir, undergoing an *isothermal compression* process. The system, always at temperature T_H, releases heat Q_H to the hot reservoir. Fourth, the system is isolated from the reservoir and is undergoing an *adiabatic expansion* process. With no heat transfer, the temperature of the system is reduced to T_L.

Because the Carnot cycle is reversible, the same Carnot machine can function as a heat engine or a heat pump (or equivalently, refrigerator). That fact has a far-reaching consequence: The efficiency of all Carnot cycles depends only on the temperatures of the two heat reservoirs,

$$\eta = 1 - \frac{Q_L}{Q_H} = f(T_H, T_L). \tag{6.14}$$

The ingenious proof, given by Sadi Carnot in 1824, is based on a logical argument as follows: If two Carnot machines have different efficiencies, one can use the Carnot

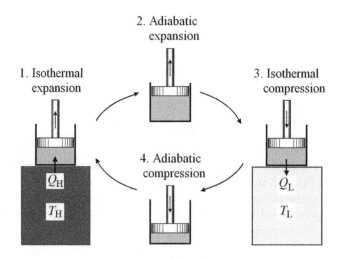

Figure 6.2 Carnot cycle. The thermodynamic system is a quantity of gas confined in a cylinder with a piston. It consists of four reversible processes: (1) an isothermal expansion process at temperature T_H, acquiring a heat Q_H from the hot reservoir; (2) an adiabatic expansion process to reduce the temperature to T_L; (3) an isothermal compression process at temperature T_L, releasing a heat Q_L to the cold reservoir; (4) an adiabatic compression process to raise the temperature to T_H. A net mechanical work W is done to the surroundings.

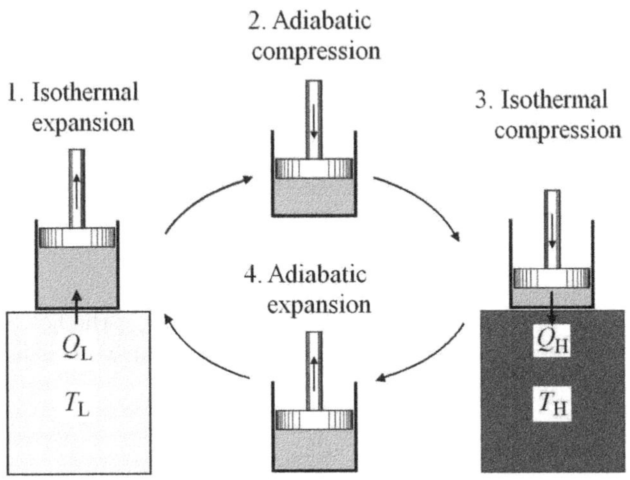

Figure 6.3 Reverse Carnot cycle. An idealized representation of a refrigerator or a heat pump, it consists of four processes: (1) an isothermal expansion process at temperature T_L, acquiring a heat Q_L from the cold reservoir; (2) an adiabatic compression process to raise the temperature to T_H;. (3) an isothermal compression process at temperature T_H, releasing a heat Q_H to the hot reservoir; (4) an adiabatic expansion process to reduce the temperature to T_L. A net mechanical work W is required to make the transfer.

machine with a higher efficiency as the heat engine and the one with a lower efficiency as the heat pump. The combined machine would contradict the second law of thermodynamics. There are two alternative proofs with regard to the two formulations of the second law of thermodynamics.

The first proof is as follows: With the same amount of heat from the hot reservoir, that heat engine can generate more work than is required to pump heat from the cold reservoir to recover the input heat back to the hot reservoir. Therefore, the combined machine can convert heat into mechanical work from a single heat reservoir, which is a *perpetual motion of the second type*. It contradicts the Kelvin–Planck formulation of the second law of thermodynamics.

An alternative proof is as follows: By using all the mechanical work generated with the heat Q_H from the hot reservoir using the more efficient machine, a heat greater than Q_H can be generated to put back into the hot reservoir. The combined machine is capable of transferring heat from a cold reservoir to a hot reservoir without requiring any external mechanical work. It contradicts the Clausius formulation of the second law of thermodynamics. The conclusions of the analysis based on the Carnot cycle is simple but far-reaching: The efficiency of a reversible Carnot cycle, regardless of the nature of the process and the nature of the substance, is uniquely determined by the two temperatures – the temperature of the hot reservoir and the temperature of the cold reservoir. The universality of the Carnot cycle motivated William Thompson (Lord Kelvin) to define the scale of thermodynamic temperature.

6.3.2 Thermodynamic Temperature

In Section 2.1, the condition of equality of temperature was defined. But the *scale* of temperature is yet to be defined. Based on the theory of the Carnot cycle, William Thomson (Lord Kelvin) defined the thermodynamic temperature, also known as the absolute temperature or the Kelvin temperature scale, abbreviated K.

For reversible Carnot cycles, the efficiency is defined as

$$\eta \equiv \frac{W}{Q_H} = 1 - \frac{Q_L}{Q_H}. \tag{6.15}$$

Now look at the ratio of Q_L and Q_H. If the temperature of the hot reservoir equals the temperature of the cold reservoir, then no mechanical work can be generated. Therefore, $Q_L = Q_H$. In other words,

$$\text{If} \quad \frac{Q_L}{Q_H} = 1, \quad \text{then} \quad \frac{T_L}{T_H} = 1. \tag{6.16}$$

Obviously, the temperature scale should also satisfy the following:

$$\text{If} \quad \frac{Q_L}{Q_H} < 1, \quad \text{then} \quad \frac{T_L}{T_H} < 1. \tag{6.17}$$

Lord Kelvin proposed a temperature scale satisfying these criteria [2]

$$\frac{T_L}{T_H} = \frac{Q_L}{Q_H}. \tag{6.18}$$

Lord Kelvin showed that this absolute temperature scale is equivalent to the temperature scale based on ideal gas properties. The Kelvin temperature is also identical to the temperature in statistical physics created by Maxwell and Boltzmann; see Appendix D.

In terms of the Kelvin temperature scale, the efficiency of a reversible Carnot cycle (Eq. 6.15), is

$$\eta_c = 1 - \frac{T_L}{T_H}, \tag{6.19}$$

which is the maximum efficiency any heat engine can achieve, often referred to as the Carnot efficiency. As shown, the lower the temperature T_L, the greater the efficiency. The efficiency can be artificially close to 1 if T_L is sufficiently low. However, the efficiency of any heat engine cannot equal to 1; otherwise the second law of thermodynamics is violated. Although 0 K could never be reached, we can formally define absolute zero temperature by

$$\text{If} \quad Q_L = 0, \quad \text{or} \quad W = Q_H, \quad \text{then} \quad T_L = 0. \tag{6.20}$$

The Celsius temperature scale is defined as a shifted Kelvin scale, with the triple point of water (the state in which the solid, liquid, and vapor phases exist together in equilibrium) defined as 0.01°C. On the Celsius scale, the boiling point of water is found experimentally to be 100.00°C. This definition also fixes the constant factor in the Kelvin scale. The relation between the Kelvin scale and the Celsius scale is

$$K = {}^\circ C + 273.15. \tag{6.21}$$

6.3.3 Entropy

Equation 6.18 can be rewritten as

$$\frac{Q_H}{T_H} = \frac{Q_L}{T_L}, \tag{6.22}$$

which has a significant consequence. Actually, Eq. 2.22 can be generalized to any number of steps in a cycle, where in each step an infinitesimal heat δQ is transferred. For a reversible cycle, the cyclic integral is zero,

$$\oint \frac{\delta Q}{T} = 0. \tag{6.23}$$

Therefore, a single-valued *function of the state* can be defined,

$$S_2 - S_1 \equiv \int_1^2 \frac{\delta Q}{T}. \tag{6.24}$$

This function, which plays a central role in thermodynamics, is called *entropy*. However, heat Q is not a function of the state, because the cyclic integral of δQ is, in general, not zero. Each state of the system can have multiple values of the same quantity. However, an infinitesimal amount of heat can be expressed in terms of the state function entropy as

$$\delta Q = T \, dS. \tag{6.25}$$

6.4 Thermodynamic Functions

As discussed in Section 2.1, for the fixed state of a system, the state functions are fixed. Examples of state functions include volume V, mass m, pressure P, temperature T, etc. Heat and mechanical work, on the other hand, are not state functions. We denote the differentials of heat and mechanical work as δQ and δW.

According to the first law of thermodynamics, the total energy of a system is a function of the state. A mathematical representation of the first law of thermodynamics is that the cyclic integral of the sum of mechanical work and heat is zero,

$$\oint (\delta Q + \delta W) = 0. \tag{6.26}$$

The difference of energy between state 1 and state 2 is defined as

$$U_2 - U_1 = \int_1^2 (\delta Q + \delta W).$$ (6.27)

Because mechanical work is related to pressure P and volume V as $\delta W = -P\,dV$, combining Eq. 6.27 with Eq. 6.25, the differential of energy is

$$dU = T\,dS - P\,dV.$$ (6.28)

6.4.1   Free Energy

The total energy of a system U can be defined as the capability of doing work to the surroundings or the capability of transferring heat to the surroundings. In the analysis of actual problems, sometimes it is necessary to emphasize the portion of energy related to the capability of doing mechanical work. The definition of *free energy* satisfies this purpose:

$$F = U - TS.$$ (6.29)

Intuitively speaking, the definition of free energy seems to eliminate the heat component of the total energy and retain the mechanical part. Similar to Eq. 6.25, the differential of free energy is

$$dF = -S\,dT - P\,dV.$$ (6.30)

Therefore, more precisely speaking, free energy is a measure of the capability of a state of the system to do mechanical work at a constant temperature.

6.4.2   Enthalpy

Sometimes it is useful to know the portion of the energy of a system related to the capability of delivering heat. The definition of *enthalpy* satisfies that purpose,

$$H = U + PV.$$ (6.31)

Intuitively speaking, the definition of enthalpy seems to eliminate the mechanical work component of the total energy and retain the heat content. Similar to Eq. 6.25, the differential of enthalpy is

$$dH = T\,dS + V\,dP.$$ (6.32)

More precisely speaking, enthalpy is a measure of the capability of a state of the system to deliver heat at a constant pressure.

When a liquid evaporates under constant pressure, the volume of the gas increases. The heat required for evaporation is the sum of the difference of internal energy U and the work needed to expand the volume, PV. The heat transferred during the phase change is the sum of U and PV, that is, the *enthalpy*,

$$\Delta H = \Delta (U + PV).$$ (6.33)

6.4.3 Gibbs Free Energy

The fourth state function, introduced by Willard Gibbs, is

$$G = U + PV - TS. \tag{6.34}$$

Similarly, the differential of the Gibbs free energy is

$$dG = -S\,dT + V\,dP. \tag{6.35}$$

From Eq. 6.35, it is clear that if the temperature and pressure of a system are kept constant and the quantity of the substance is constant, then the Gibbs free energy is constant. This property is important in the analysis of chemical reactions, because most chemical reactions occur under conditions of constant temperature and constant pressure.

First, consider a single-component system in which the quantity of the substance is a variable which by convention is expressed as the number of moles of the substance, N. When the temperature and the pressure of the system are kept constant, the Gibbs free energy can be expressed as

$$dG = -S\,dT + V\,dP + \mu\,dN. \tag{6.36}$$

The quantity μ is the Gibbs free energy per mole of the substance.

6.4.4 Chemical Potential

For a system with more than one component, the expression of the Gibbs free energy (Eq. 6.36), can be generalized to

$$dG = -S\,dT + V\,dP + \sum_i \mu_i\,dN_i, \tag{6.37}$$

where i is the index for the ith component. If there is a chemical reaction in the system, then the composition of the system, represented by the set of molar values N_i, will change. Under constant temperature and constant pressure, the equilibrium condition of the system is

$$dG = \sum_i \mu_i\,dN_i = 0. \tag{6.38}$$

The quantity μ_i is known as the *chemical potential* of the ith component of the system.

6.5 Ideal Gas

Experimentally, it is found that within a large range of temperature and pressure many commonly encountered gases satisfy a universal relation

$$PV = NRT, \tag{6.39}$$

where P is the pressure in pascals, V is the volume in cubic meters, N is the number of moles of the gas, T is the absolute temperature in kelvins; and R is a universal gas constant,

$$R = 8.3144 \, \frac{\text{J}}{\text{mol} \cdot \text{K}}. \tag{6.40}$$

In solar energy storage systems, for the gases commonly used, such as nitrogen, oxygen, argon, and methane, at temperatures of interest the ideal gas relation is accurate up to a pressure of 10 MPa, or 100 standard atmosphere pressure.

In the following, we will derive all the thermodynamic functions for the ideal gases. First, we will show that the energy U of an ideal gas depends only on temperature but not on volume. In fact, assuming that $U = U(T, V)$, using Eq. 6.25, the internal energy varies with volume as

$$\left(\frac{\partial U}{\partial V} \right)_T = T \left(\frac{\partial S}{\partial V} \right)_T - P. \tag{6.41}$$

Using the equation of state 6.39, Eq. 6.41 becomes

$$\left(\frac{\partial U}{\partial V} \right)_T = T \left(\frac{\partial P}{\partial T} \right)_V - P = T \frac{P}{T} - P = 0. \tag{6.42}$$

Defining the constant-volume specific heat per mole as

$$C_v \equiv \frac{1}{N} \left(\frac{\partial U}{\partial V} \right)_T, \tag{6.43}$$

the internal energy as a function of temperature is

$$U_2 - U_1 = \int_1^2 N C_v dT. \tag{6.44}$$

If in the temperature interval of interest C_v is a constant,

$$U_2 - U_1 = N C_v (T_2 - T_1). \tag{6.45}$$

To obtain an explicit expression of entropy, we use Eqs. 6.25, 6.39, and 6.43,

$$dS = \frac{dU}{T} + \frac{P \, dV}{T} = N C_v \frac{dT}{T} + N R \frac{dV}{V}. \tag{6.46}$$

Again, assuming that the specific heat C_v is a constant,

$$S_2 - S_1 = N \int_1^2 \left[\frac{C_v \, dT}{T} + R \frac{dV}{V} \right] = N \left[C_v \log \frac{T_2}{T_1} + R \log \frac{V_2}{V_1} \right]. \tag{6.47}$$

Another important quantity is the *constant-pressure specific heat* C_p. For ideal gases, there is a simple relation between C_v and C_p. To heat N moles of gas while keeping the pressure constant, work $P \Delta V$ is done to the surroundings. According to the first law of thermodynamics, an additional energy $P\Delta V$ must be supplied. Because for an ideal gas $P \Delta V = NR \Delta T$, the total energy required to raise the temperature is

$$\Delta Q = NC_v \Delta T + NR \Delta T = N(C_v + R) \Delta T. \tag{6.48}$$

Therefore,

$$C_p = C_v + R. \tag{6.49}$$

The ratio of constant-pressure specific heat and constant-volume specific heat is an important parameter in the study of the adiabatic process, often denoted as γ:

$$\gamma \equiv \frac{C_p}{C_v} = 1 + \frac{R}{C_v}. \tag{6.50}$$

During an adiabatic process, the heat transfer is zero. Therefore,

$$NC_v \, dT = -P \, dV. \tag{6.51}$$

On the other hand, from the equation of state for the ideal gas,

$$NR \, dT = V \, dP + P \, dV. \tag{6.52}$$

Combining Eqs 6.51 and 6.52, we obtain

$$\left(1 + \frac{R}{C_v}\right) P \, dV + V \, dP = 0. \tag{6.53}$$

The solution of the above differential equation is

$$PV^\gamma = \text{const}, \tag{6.54}$$

which is the state equation of an adiabatic process.

In the following, we show that the efficiency of a reversible Carnot cycle of ideal gas is $\eta = 1 - (T_L/T_H)$, and thus the thermodynamic temperature scale is identical to the temperature scale based on the ideal gas; see Fig. 6.4. In the first process, an isothermal expansion process at temperature T_H from state 1 to state 2, the gas medium receives heat from the hot reservoir,

$$Q_H = \int_1^2 P dV = NRT_H \int_1^2 \frac{dV}{V} NRT_H \log \frac{V_2}{V_1}, \tag{6.55}$$

where V_1 and V_2 are the volume of states 1 and 2, respectively. In the third process, an adiabatic contraction process from state 3 to state 4, the gas medium releases heat to the cold reservoir,

$$Q_L = \int_3^4 P dV = NRT_L \int_3^4 \frac{dV}{V} NRT_L \log \frac{V_3}{V_4}, \tag{6.56}$$

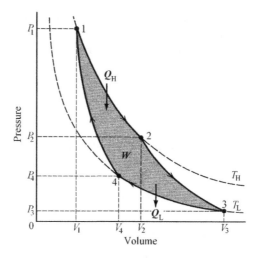

Figure 6.4 Carnot cycle with ideal gas as the system. Curve 1–2 is an isothermal expansion process at T_H. Curve 2–3 is an adiabatic expansion process. Curve 3–4 is an isothermal contraction process at T_L. Curve 4–1 is an adiabatic contraction process. The gas does work W to the surroundings.

where V_3 and V_4 are the volumes of state 3 and state 4, respectively.

Because processes 1–2 and 3–4 are isothermal, according to the equation of state,

$$\begin{aligned}
P_1V_1 &= P_2V_2, \\
P_3V_3 &= P_4V_4.
\end{aligned} \tag{6.57}$$

On the other hand, processes 2–3 and 4–1 are adiabatic. According to Eq. 6.54,

$$\begin{aligned}
P_2V_2^\gamma &= P_3V_3^\gamma, \\
P_4V_4^\gamma &= P_1V_1^\gamma.
\end{aligned} \tag{6.58}$$

Combining Eqs. 6.57 and 6.58, we have

$$\frac{V_1}{V_2} = \frac{V_3}{V_4}. \tag{6.59}$$

Applying this relation to Eqs. 6.55 and 6.56, we conclude that the thermodynamic temperature scale and the ideal gas temperature scale are identical,

$$\frac{T_L}{T_H} = \frac{Q_L}{Q_H}. \tag{6.60}$$

6.6   Ground Source Heat Pump and Air Conditioning

In Section 5.3, we discussed solar energy stored in the ground: At a depth of about 10 m, the temperature of Earth is cooler than the average in the summer and warmer than the average in the winter. Tapping into that stored solar energy could greatly reduce the use of energy for space heating and cooling as well as water heating. Although direct use of the stored energy for space cooling in the summer has been practiced for centuries in some places on the world, it allows very little control. The modern method of tapping into that stored solar energy is to use vapor compression heat pump and refrigeration systems [6, 58, 92]. This enables full control to the desired temperature. The energy saving can be as great as 50%.

Table 6.1 shows the accumulated wattage and number of ground source heat-pump installations in several countries up to 2004. Although the United States has the largest installation wattage and number of systems, Sweden has the largest percentage of heat pump installations. This is simply due to economics. In Sweden, the weather is cold, and there are no fossil fuel resources. Nevertheless, most of the electricity in Sweden is from hydropower, which is relatively inexpensive. The combination of cheap electricity and the high efficiency of geothermal systems makes perfect economic sense.

6.6.1   Theory

A schematic of a ground source heat pump is shown in Fig. 6.5. Several meters underground, the temperature is basically constant over the entire year, for example, at $T_0 = 15°C$. In winter, the ambient temperature can be as cold as $0°C$. To maintain a comfortable temperature, heat should be supplied into the house. In summer, the ambient temperature can be as hot as $30°C$. Heat should be extracted from the house. In Section 6.3.1, we showed that, by supplying mechanical work, heat can be transferred from a cold reservoir to a hot reservoir. Therefore, using a type of machine called *heat*

Table 6.1: Ground source heat pumps in selected countries

Country	Capacity (MWt)	Energy Use (GWh/year)	Number of Units Installed
Austria	275	370	23,000
Canada	435	600	36,000
Germany	640	930	46,400
Sweden	2,300	9,200	230,000
Switzerland	525	780	30,000
Uunited States	6,300	6,300	600,000

Source: Geothermal (Ground-Source) Heat Pumps:
A World Overview, GHC Bulletin, September 2004 [58] / Penwell Group.

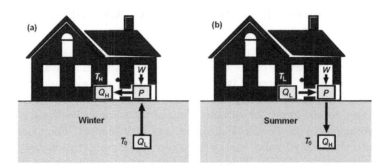

Figure 6.5 Ground source heat pump. The temperature in the ground about 10-m deep, T_0, is basically constant over the year. (a) In the winter, by applying a mechanical work W, a heat pump P transfers heat Q_L from the ground heat reservoir at temperature T_0 to a radiator in the house at temperature T_H. The total heat discharged into the radiator is the sum of the heat extracted from the ground and the work, $Q_H = Q_L + W$. (b) In the summer, the heat pump P reverses its function: it transfers heat from the rooms at temperature T_L into the ground heat reservoir at temperature T_0.

pump, it is possible to extract heat from within the ground at 15°C to a floor heating system in the house at $T_H = 50$°C. In the ideal case, if the machine is reversible, according to Eq. 6.19, the efficiency is

$$\eta_c = 1 - \frac{T_0}{T_H} = 1 - \frac{273 + 15}{273 + 50} \approx 0.108. \tag{6.61}$$

According to the definition of efficiency (Eq. 6.15), the mechanical work required to transfer heat from the ground at T_0 to the radiator at T_H to become Q_H is

$$W = \eta Q_H = 0.108 \times Q_H. \tag{6.62}$$

In other words, mechanical energy equal to a fraction of the heat is needed to transfer heat from the ground to the radiator.

In the summer, assuming that heat Q is extracted from an air-conditioning system at $T_L = -10$°C into the ground at $T_0 = 15$°C using a reversible Carnot engine. The machine is in fact a refrigerator. The Carnot efficiency is

$$\eta_c = 1 - \frac{T_L}{T_0} = 1 - \frac{273 - 10}{273 + 15} \approx 0.087. \tag{6.63}$$

Because the heat extracted from the air conditioning system is Q_L in Eq. 6.15, the mechanical work required is

$$W = \frac{\eta}{1 - \eta} Q_L = 0.095 \times Q_L. \tag{6.64}$$

Theoretically, if the refrigerator as a Carnot engine is reversable, the work needed to cool the space is less than one-tenth of the amount of heat taken. Practically, no machine is perfect. The actual work needed is more than that from the theoretical limit.

6.6.2 Coefficient of Performance

In the industry, a dimensionless number called the *coefficient of performance* (COP) is used to characterize the performance of the heat pump and refrigerator. For the heat pump, it is defined as

$$\text{COP} = \frac{Q_H}{W}, \tag{6.65}$$

and for the refrigerator, it is

$$\text{COP} = \frac{Q_L}{W}. \tag{6.66}$$

The theoretical limit looks extremely attractive: The conversion ratio from mechanical work to heat can be as high as 10. Practically, a conventional vapor-compression heat pump or refrigerator can achieve a COP of 3–4. Nevertheless, this is still highly desirable. Replacing an electrical heating system with a ground-source heat pump can achieve an energy saving up to 70%.

6.6.3 Vapor-Compression Heat Pump and Refrigerator

The commercial geothermal systems use the vapor compression cycle for both heat pumping and refrigeration. Most are reversible, which can be used for room heating and cooling by using a reversing valve. Figure 6.6 shows the cooling cycle. It is very similar to an ordinary central air-conditioning system based on the principle of the vapor compression cycle. A liquid with a low boiling point serves as the working medium, called the *refrigerant*. Since the early 20th century, ammonia (NH_3), also called R12, was the favorite refrigerant. Because of its toxicity, in air-conditioning systems, it was replaced by chlorodifluoromethane ($CHClF_2$), or R22. In spite of its flammability, propane (C_3H_8), or R290, is recently used in place of R22 because of its zero ozone depletion potential .

The central piece of a vapor compression refrigeration system is the *compressor*, (1 in Fig. 6.6). This is the point at which mechanical power is input. The compression process is approximately adiabatic. If the initial pressure is P_1 and the final pressure is P_2, according to Eq. 6.54,

$$P_1 V_1^\gamma = P_2 V_2^\gamma, \tag{6.67}$$

where γ is the ratio of specific heats (see Eq. 6.50), and the γ value of a typical refrigerant, R-22, is 1.26,

$$\gamma \equiv \frac{C_p}{C_v} = 1 + \frac{R}{C_v}. \tag{6.68}$$

Combining with the equation of state of an ideal gas, we find that the temperature is changed according to

$$\frac{T_2}{T_1} = \left(\frac{P_2}{P_1}\right)^{\frac{\gamma-1}{\gamma}} = \left(\frac{P_2}{P_1}\right)^{0.2857}. \tag{6.69}$$

If the gas is initially at room temperature (273 K), compressing it by a factor of 3, the temperature is raised to $3^{0.2857} \times 273 = 1.3687 \times 273 = 373$ K, which is about 100°C.

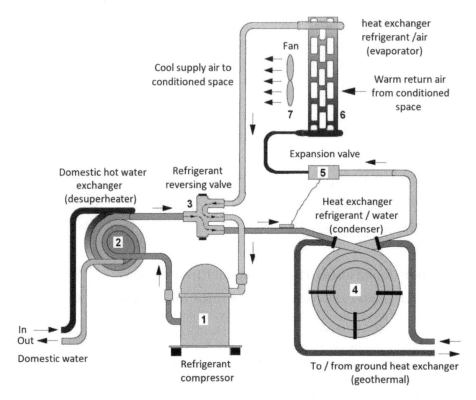

Figure 6.6 Ground-source heat pump: cooling mode. The refrigerant is compressed by a compressor (1) to become superheated. Through the heat exchanger (2), it first heats the domestic water. Then the still-hot refrigerant goes to the heat exchanger (4) to be cooled to underground temperature and condenses into liquid. Through an *expansion valve* (5) it becomes vapor. The evaporation process absorbs large quantity of heat from the space through the heat exchanger (6). The space is thus cooled. *Source*: Penwell Group.

Therefore, the refrigerant is a superheated gas. The heat thus generated is first utilized to heat the domestic water through the heat exchanger (2). Then, through reversing valve (3), the still hot refrigerant comes through the heat exchanger to contact with the water from the ground and is cooled to the underground temperature, for example, $T_0 = 15°C$, or 288 K. Because of the high pressure, the refrigerant condenses into liquid. Therefore, such a heat exchanger is called a *condenser*. The liquid refrigerant at T_0 and still under high pressure is letting through an *expansion valve* (5) to become vapor. The evaporation process absorbs large quantities of heat and the vapor temperature becomes very low, typically many degrees below 0°C. The cold vapor goes int the heat exchanger (6), a cold radiator. A fan blows warm air from the room onto the zigzag tubes and fins of the cold radiator (6). The cooled air then flows into the space to be conditioned. The warmed-up vapor of refrigerant is again sucked into the compressor (1) through the reversing valve (3). And the process goes on.

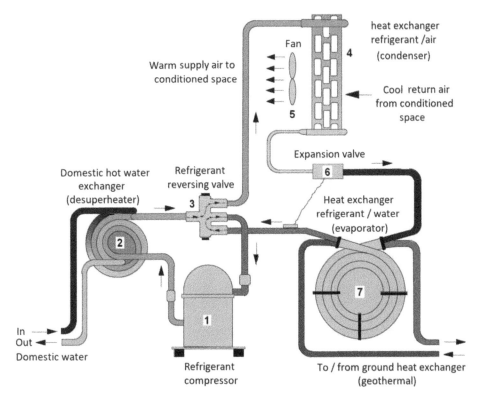

Warm supply air to
conditioned space

Fan

heat exchanger
refrigerant /air
(condenser)

Cool return air
from conditioned
space

Expansion valve

Domestic hot water
exchanger
(desuperheater)

Refrigerant
reversing valve

Heat exchanger
refrigerant / water
(evaporator)

In
Out
Domestic water

Refrigerant
compressor

To / from ground heat exchanger
(geothermal)

Figure 6.7   Ground source heat pump: heating mode. The refrigerant is compressed by a compressor (1) to become superheated. Through the heat exchanger (2), it first heats the domestic water. Then the still-hot pressurized gas-phase refrigerant goes to the radiator (4) to warm the space and become liquid. The liquid refrigerant goes through an expansion valve (5) to become supercooled vapor. The evaporation process absorbs large quantities of heat from the groundwater through the heat exchanger (6). *Source*: Penwell Group.

If domestic water heating is not required, the heat exchanger (2) can be eliminated. However, the fringe benefit of having this unit is obvious.

By turning the reversing valve (3) to another position, the system becomes a heat pump to be used in the winter; see Fig 6.7. Again, the compressor (1) squeezes the refrigerant into a superheated gas and transfers part of the heat to domestic water via the heat exchanger (2). Then, through the reversing valve (3), it goes into the heat exchanger (4), which is a heat radiator and refrigerant condenser. A fan (4) forces warm air into the space to be conditioned. At that time, the refrigerant is liquefied by the cool air. The liquid refrigerant goes through the expansion valve (6) and becomes supercooled vapor, often below the freezing point of water. By going through the heat exchanger (7) to contact the water at ground temperature, the refrigerant vapor picks up heat from the ground and the temperature recovers to T_0.

6.6.4 Ground Heat Exchanger

A ground heat exchanger can have many different configurations. Fig. 6.8 shows several of them. The requirement for a good ground heat exchanger is to access a large volume of soil, groundwater, or bedrock. Figs. 6.8(a), (b) show closed-loop heat exchangers. The vertical well does not require a large area, and so can be used in highly populated area. It is no surprise that most of the geothermal systems in New York City are vertical-well systems. However, drilling even a 100-m well is costly. For suburban or rural buildings, it is sufficient to bury a few hundred meters of copper coil in the ground a few meters deep. In places where groundwater is easily accessible, an open-loop system is even less expensive. By drilling two wells into the groundwater (Fig. 6.8(c)) or to access a pond (Fig. 6.8(d)), effective heat exchange could be established. In the United States, 46% are vertical closed-loop systems, 38% are horizontal closed-loop systems and 15% are open-loop systems.

Figure 6.9 shows the details of a vertical well at the Headquarters of the American Institute of Architects (AIA), Greenwich Village, New york City. It is a 1250-ft deep, 6-in-diameter well drilled into the sidewalk in front of the building. The underground water level is 40–50 ft down from the ground. A metal casing is placed to protect the well from collapsing due to the pressure of the soil. The lower part of the well is in the bedrock, which is strong enough to maintain the shape for probably hundreds of years. A 6-in-diameter polyvinyl chloride (PVC) tube, the so-called shroud straw, extends almost to the bottom of the well. Near the bottom, the shroud straw has 120

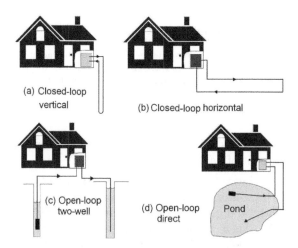

Figure 6.8 Heat-exchange configurations for ground-source heat pumps. The ground heat exchanger can have many different configurations. (a) and (b) are closed-loop heat exchangers. The vertical well in (a) does not require a large area, which means it can be used in a highly populated area. For suburban or rural buildings, the horizontal heat exchanger in (b) is much less expensive that the deep well. In places where groundwater is easily accessible, the open-loop system in (c) and (d) are even less expensive.

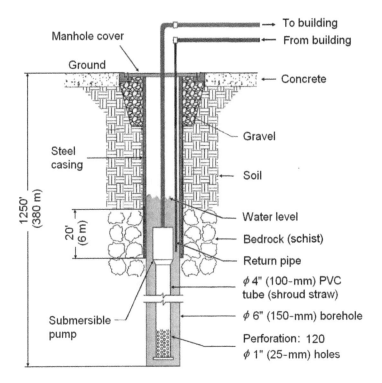

Figure 6.9 Vertical well in a heat pump system. Detail of the vertical well of a heat pump system at AIA. The depth is 1250 ft. About 20 ft below the underground water level, a submersible pump extracts water to the heat exchange unit. A 6-in-diameter PVC tube, the so-called shroud straw, extends almost to the bottom of the well. The water from the heat exchanger returns to the well through a return pipe and runs outside the shroud straw. *Source*: Courtesy of AIA.

1-in-diameter holes. The water from the deep well is extracted by a submersible pump to the heat exchange unit; see Fig. 6.7. The water from the heat exchanger returns to the well through a return pipe and runs outside the shroud straw. Therefore, the water runs all the way through the full depth of the vertical well and thus reaches thermal equilibrium with the ground bedrock, where the temperature is always 12°C.

Problems

6.1. The COP for heat pumps and refrigerators achieved in the industry is about 50% of the Carnot limit. Make estimates for the following cases:

1. For refrigerator to achieve $-10°C$ in the freezer area with an external temperature of 25°C

2. For the same refrigerator to achieve $-10°C$ in the freezer area but using the geothermal heat reservoir as the external source, at 10°C

6.2. A solar heat collector of 2 m^2 using the running water through underground pipes at 15°C to generate hot water at 55°C. If the efficiency of the solar collector is 80%, on average, how much hot water can this system generate per day?

1. In a region of 5 h of daily insolation

2. In a region of 3 h of daily insolation

Hint: The standard insolation is 1 kW/m^2.

Chapter 7

A Quantum Mechanics Primer

Quantum mechanics is the centerpiece of modern physics. It underlies much of modern science and technology, including condensed-matter physics, solid-state electronics, all of chemistry, and molecular biology. Quantum mechanics is the theoretical basis for solar cells, light-emitting diodes, photosynthesis, and rechargeable batteries.

Many quantum mechanics textbooks are formulated in a fictitious Hilbert space to describe the motions of fictitious point particles. The dynamic variables of the fictitious point particles, for example the position $\mathbf{r}(t)$ and the momentum $\mathbf{p}(t)$, are represented by Hermitian operators. Because of the complicated mathematics, it is often taught at a graduate level or senior level. Furthermore, the split personality of an electron as a particle localized at a geometrical point and as a continuous field spread out in space causes unresolvable conceptual conflicts. Richard Feynman famously said: "I think I can safely say that nobody understands quantum mechanics".

In the second half of the 20th century, quantum field theory, an advanced form of quantum theory, was established. Schrödinger's wavefunction as a non-relativistic approximation of the Dirac field is interpreted as a physical field.[1] In early 21st century, using the attosecond light pulses, which was awarded the Nobel Prize in Physics 2023;[2] and the scanning tunneling microscope, which was awarded the Nobel Prize in Physics 1986;[3] wavefunctions were observed and mapped. For details, see Appendix G and Section 7.2.4. By interpreting wavefunction as a physical field, quantum mechanics can be taught using elementary mathematics without paradoxes.

Historically, the foundation of non-relativistic quantum mechanics was established by Erwin Schrödinger in 1926 in six papers published on *Annalen der Physik* [95]. Regarding those papers, Paul Dirac commented: "The underlying physical laws necessary for the mathematical theory of a large part of physics and the whole of chemistry are thus completely known" [22]. Those papers, belonging to the defining publications of modern science, are worth reading. Here is a brief summary:

The first paper, *Quantization as an Eigenvalue Problem, Part I*, received by *Annalen der Physik* on January 27, 1926, defined a wavefunction $\psi(x, y, z)$ as a physical field in real space, that is "everywhere real, single-valued, finite, and continuously differentiable

---

[1] J. D. Bjorken and S. D. Drell, *Relativistic Quantum Fields*, McGraw-Hill, 1965.

[2] Press release: The Nobel Prize in Physics 2023, the Royal Swedish Academy of Sciences.

[3] C. J. Chen, *Introduction to Scanning Tunneling Microscopy*, Third Edition, Oxford University Press 2021, especially Chapter 8, *Imaging Wavefunctions*.

Figure 7.1 Austrian banknote with a portrait of Schrödinger. It is a rare honor for a scientist to have a portrait printed on a banknote. Note the large value. Born in 1887 in Erdberg, Vienna, Austria, Erwin Schrödinger studied at the University of Vienna between 1906 and 1910 for his Ph.D. degree, then spent his most productive years at University of Zurich from 1920 to 1927, and then succeeded Max Planck at the Friedrich Wilhelm University in Berlin. After working at many institutions in the world, in 1956, he returned to Vienna until he died in 1961. In addition to creating non-relativistic quantum mechanics, his 1944 monograph *What Is Life* motivated James Watson and Francis Crick in the research of the molecules carrying the genetic code.

up to the second order". A differential equation is introduced as a variation of the time-independent Hamilton–Jacobi equation. Applying that static Schrödinger equation to the hydrogen atom, the Rydberg formula was explained.

The second paper, *Quantization as an Eigenvalue Problem, Part II*, presents a simpler way to introduce the static Schrödinger equation based on the de Broglie relation and classical mechanics. Three further problems were treated: the harmonic oscillator, rigid rotor, and non-rigid rotor, all related to chemical physics.

The dynamic (time-dependent) Schrödinger equation was introduced in the sixth paper, received by *Annalen der Physik* on June 23, 1926. By treating radiation as an electromagnetic wave, its interaction with atomic systems was explained.

7.1 The Static Schrödinger Equation

Schrödinger's equation is one of the greatest discoveries in the history of science. In principle, it cannot be *derived* from classical physics. Nevertheless, Schrödinger's equation can be understood from a heuristic reasoning by applying the de Broglie relation, see Section 2.4.3, to the classical energy integral. If the force can be expressed as a gradience of a potential $\mathbf{F} = -\nabla \cdot V(\mathbf{r})$, Newton's second law is

$$m\frac{d^2\mathbf{r}}{dt^2} = -\nabla \cdot V(\mathbf{r}). \tag{7.1}$$

Multiplying both sides with $d\mathbf{r}$ and integrate from $\mathbf{r_1}$ to $\mathbf{r_2}$, the left-hand side becomes

$$
\begin{aligned}
\int_{\mathbf{r_1}}^{\mathbf{r_2}} m \frac{d^2\mathbf{r}}{dt^2} d\mathbf{r} &= \int_{\mathbf{r_1}}^{\mathbf{r_2}} m \frac{d\mathbf{r}}{dt} \frac{d}{dt}\left(\frac{d\mathbf{r}}{dt}\right) dt \\
&= \int_{\mathbf{r_1}}^{\mathbf{r_2}} \frac{d}{dt}\left[\frac{m}{2}\left(\frac{d\mathbf{r}}{dt}\right)^2\right] dt \\
&= \left. \frac{m}{2}\left(\frac{d\mathbf{r}}{dt}\right)^2 \right|_{\mathbf{r_1}}^{\mathbf{r_2}},
\end{aligned}
\tag{7.2}
$$

and the right-hand side becomes

$$
V(\mathbf{r_1}) - V(\mathbf{r_2}).
\tag{7.3}
$$

Therefore, the quantity in the left-hand side of the following equation is a constant, independent of position $\mathbf{r}$. We denote it as the *energy* E of the system,

$$
\frac{m}{2}\left(\frac{d\mathbf{r}}{dt}\right)^2 + V(\mathbf{r}) = E.
\tag{7.4}
$$

Introduce a *momentum* $\mathbf{p} = m\, d\mathbf{r}/dt$, Eq. 7.4 becomes

$$
\frac{1}{2m}\mathbf{p}^2 + V(\mathbf{r}) = E.
\tag{7.5}
$$

Although the starting point is Newton's second law for a material point, because the potential $V(\mathbf{r})$ is a *field*, Eq. 7.5 is a *field equation*. The first term in Eq. 7.5, the kinetic energy, is also a field quantity. By expressing the kinetic-energy term through the wavefunction, the static Schrödinger equation comes out naturally.

According to de Broglie, electron is a field. It was quantified by Schrödinger as a wavefunction. For a static electron, the wavefunction $\psi(\mathbf{r})$ is time independent. In a small region where the potential does not vary significantly, the spatial variation of the wavefunction follows the Helmholtz equation, Eq. 2.104,[4]

$$
\nabla^2\psi(\mathbf{r}) + k^2\psi(\mathbf{r}) = 0,
\tag{7.6}
$$

where k^2 is the square of the wave vector, related to the wavefunction through

$$
k^2 = -\frac{\nabla^2\psi(\mathbf{r})}{\psi(\mathbf{r})}.
\tag{7.7}
$$

According to the de Broglie's relation, Eq. 2.147,

$$
\mathbf{p} = \hbar\mathbf{k}.
\tag{7.8}
$$

---

[4]Eq. 2.104 is derived for electromagnetic waves. There is a time-dependent factor related to frequency. In quantum mechanics, that time-dependent factor alone has no physical meaning. Only the energy difference between the two quantum states is observable. See Section 7.5.3.

Combining Eqs. 7.7 and 7.8, one finds

$$\mathbf{p}^2 = -\frac{\hbar^2 \nabla^2 \psi(\mathbf{r})}{\psi(\mathbf{r})}. \tag{7.9}$$

Assuming that Eq. 7.9 also works with a varying potential, inserting it into Eq. 7.5, multiplying both sides by $\psi(\mathbf{r})$, a differential equation comes out naturally,

$$-\frac{\hbar^2}{2m_e} \nabla^2 \psi(\mathbf{r}) + V(\mathbf{r})\psi(\mathbf{r}) = E\,\psi(\mathbf{r}). \tag{7.10}$$

This is the *static Schrödinger equation of a single electron*. Its validity has been proved by comparing it with a large number of experiments.

7.1.1 Wavefunctions in a One-Dimensional Potential Well

To illustrate the meanings of the Schrödinger equation and wavefunctions, the problems of electrons in a one-dimensional potential well is analyzed.

Consider a one-dimensional potential well of length L. Within the well, the potential is zero. Schrödinger's equation is

$$-\frac{\hbar^2}{2m_e} \frac{d^2\psi(x)}{dx^2} = E\psi(x). \tag{7.11}$$

Assuming the potential walls at $x = 0$ and $x = L$ are impenetrable, the boundary conditions are: at the boundaries, the wavefunction must be zero.

Introduce a wave vector k defined as

$$k = \frac{\sqrt{2m_e E}}{\hbar}, \tag{7.12}$$

the Schrödinger's equation Eq. 7.11 becomes

$$\frac{d^2\psi(x)}{dx^2} = -k^2\psi(x). \tag{7.13}$$

The general solution of Eq. 7.13 is

$$\psi(x) = C\sin(kx + \phi), \tag{7.14}$$

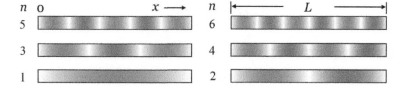

Figure 7.2 Wavefunctions in a one-dimensional potential well. The wavefunctions are labeled by quantum number n, the number of nodes in the wavefunction plus one.

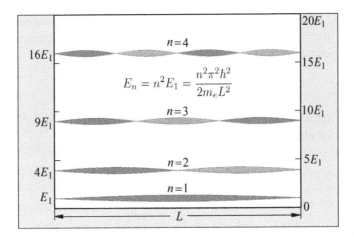

Figure 7.3 Energy levels in a one-dimensional potential well. The wavefunctions or eigenstates are labeled by quantum number n. The energy eigenvalue is proportional to n^2.

where C is a normalization constant, and ϕ is a phase angle. Because at $x = 0$, the wavefunction must be zero, the phase ϕ must be zero. We have

$$\psi(x) = C \sin(kx). \tag{7.15}$$

The boundary condition at the other end, $\psi(L) = 0$, requires that

$$k = \frac{n\pi}{L}, \tag{7.16}$$

where $n = 1, 2, 3, \ldots$ is an integer. The allowed wavefunctions are

$$\psi_n(x) = C \sin\left(\frac{n\pi x}{L}\right). \tag{7.17}$$

Figure 7.2 shows the wavefunctions.

In classical mechanics, as in Eq. 7.1, energy can take any value. In quantum mechanics, because of condition Eq. 7.16, energy is *quantized*: it can only take discrete values determined by Eqs. 7.12 and 7.16:

$$E_n = n^2 E_1 = \frac{n^2 \pi^2 \hbar^2}{2m_e L^2}, \tag{7.18}$$

Those allowed values of energy are called the *energy eigenvalues*. Similarly, the allowed wavefunctions, Eq. 7.17, are called the *eigenstates*.

Because the Schrödinger equation is linear, the constant C does not affect the energy eigenvalues. According to Schrödinger, the square of the wavefunction times $-e$ is the charge density distribution of an electron as a field:

$$\rho(x) = -e\psi^2(x). \tag{7.19}$$

Because the total charge of an electron over the space equals to one elementary charge $-e$, the integral of $\psi^2(x)$ over the entire space must equal to 1. In the current situation, the electron is confined in a well of width L,

$$\int_0^L \psi^2(x)dx = 1. \tag{7.20}$$

The average value of the square of sine function over any number of half periods is $1/2$. Therefore, for all quantum numbers, the constant is

$$C = \sqrt{\frac{2}{L}}. \tag{7.21}$$

The wavefunctions are

$$\psi_n(x) = \sqrt{\frac{2}{L}} \sin \frac{n\pi x}{L}. \tag{7.22}$$

It is straightforward to show that eigenstates with different quantum numbers n are *orthogonal*. In fact, using Eq. 7.22,

$$\int_0^L dx \, \psi_n(x)\,\psi_m(x) = \delta_{nm}, \tag{7.23}$$

where δ_{nm} is a mathematical symbol, the Kronecker delta, which is defined as zero when $n \neq m$ and as 1 when $n = m$.

Furthermore, the set of wavefunctions is *complete*. Any continuous function $f(x)$ in the interval $[0, L]$ can be expanded as a weighted sum of those wavefunctions,

$$f(x) = \sum_{n=1}^{\infty} b_n \psi_n(x), \tag{7.24}$$

with coefficients

$$b_n = \int_0^L f(x)\,\psi_n(x)\,dx. \tag{7.25}$$

This is a special case of the Fourier theorem. The proof can be found in any mathematics textbook containing a presentation of the Fourier series. Nonetheless, if Eq. 7.24 is true, it is easy to prove that the expression of the coefficients, Eq. 7.25, is correct. In fact, because of the orthonormal relation Eq. 7.23,

$$\int_0^L \psi_n(x)dx \sum_{m=1}^{\infty} b_m \psi_m(x) = \sum_{m=1}^{\infty} \delta_{nm} b_m = b_n. \tag{7.26}$$

7.1.2 The Bra-and-Ket Notations

The notations of wavefunctions, Eq. 7.22, and the integrals, Eqs. 7.20 and 7.23, occurs very often in quantum mechanics. In the third edition of *Principles of Quantum Mechanics*, Dirac introduced the bra-and-ket notations, that greatly simplifies mathematical notations in quantum mechanics. In the real formulation of quantum mechanics, bra and ket are equivalent. A wavefunction can be denoted either as a bra or as a ket. For the case of a one-dimensional potential well, the n-th wavefunction is

$$\langle n| = |n\rangle = \psi_n(x) = \sqrt{\frac{2}{L}} \sin\frac{n\pi x}{L} \quad 0 < x < L. \tag{7.27}$$

A complete bracket represents an integral of two wavefunctions,

$$\langle n|m\rangle = \int_{-\infty}^{\infty} dx\, \psi_n(x)\, \psi_m(x). \tag{7.28}$$

Obviously, the real Dirac notation is symmetric, or *commutative*,

$$\langle n|m\rangle = \langle m|n\rangle. \tag{7.29}$$

The orthogonal and normalizing condition in the bra-and-ket notation is

$$\langle n|m\rangle = \delta_{nm}, \tag{7.30}$$

and the completeness of the wavefunctions means that any continuous function $|f\rangle$ in the same interval $[0, L]$ can be expanded as a sum of the wavefunctions,

$$|f\rangle = \sum_{n=1}^{\infty} |n\rangle\langle n|f\rangle. \tag{7.31}$$

A linear operation on a wavefunction is represented by an operator. To distinguish it from a number, a hat is added, such as $\hat{o}$. Multiplying the wavefunction with the coordinate, for example x to a one-dimensional wavefunction, is an linear operation, denoted as $\hat{x}$. A differential operator is defined as

$$\hat{\delta}|f\rangle \equiv \frac{d}{dx}f(x). \tag{7.32}$$

If for all wavefunctions f and g, two operators α and β satisfying the following relation,

$$\langle \hat{\alpha}f|g\rangle = \langle f|\hat{\beta}g\rangle, \tag{7.33}$$

we say the operator α is an *adjoint operator* of β, denoted as

$$\hat{\alpha}^\dagger = \hat{\beta}, \quad \text{or} \quad \hat{\beta}^\dagger = \hat{\alpha}. \tag{7.34}$$

The operator $\hat{x}$ is obviously self-adjoint,

$$\hat{x} = \hat{x}^\dagger. \tag{7.35}$$

However, the differential operator is not. In fact, because

$$\int_{-\infty}^{\infty} \left[\frac{df(x)}{dx}g(x) + f(x)\frac{dg(x)}{dx}\right] dx = \int_{-\infty}^{\infty} \frac{d}{dx}\left[f(x)g(x)\right] dx = 0, \tag{7.36}$$

one has

$$\hat{\delta} = -\hat{\delta}^{\dagger}. \tag{7.37}$$

The Schrödinger equation, Eq. 7.10, can be written as

$$\hat{H}|\psi\rangle = E|\psi\rangle \tag{7.38}$$

by defining an *energy operator* or a Hamiltonian,

$$\hat{H} \equiv -\frac{\hbar^2}{2m_e}\nabla^2 + V(\mathbf{r}). \tag{7.39}$$

The energy operator is self-adjoint. The proof is left as an exercise.

7.1.3 The Harmonic Oscillator

In quantum mechanics, the harmonic oscillator is of fundamental importance. It describes the oscillation of molecules and solids near its equilibrium point. The electromagnetic wave can be decomposed into a number of simple harmonic oscillators. Using the quantization procedure presented here, the electromagnetic waves can be quantized. It is the basis of quantum electrodynamics.

In classical mechanics, the energy integral of a harmonic oscillator is

$$\frac{1}{2m}p^2 + \frac{1}{2}kx^2 = E, \tag{7.40}$$

where k is the constant of elasticity. The solution is

$$x = a\sin(\omega t + \phi), \tag{7.41}$$

where a is the amplitude and ϕ is a phase. The *circular frequency* ω is

$$\omega = \sqrt{\frac{k}{m}}. \tag{7.42}$$

By expressing the elasticity constant k through the circular frequency ω, following Eq. 7.10, the Schrödinger equation for a one-dimensional harmonic oscillator is

$$\left(-\frac{\hbar^2}{2m}\frac{d^2}{dx^2} + \frac{m}{2}\omega^2 x^2\right)\psi(x) = E\psi(x). \tag{7.43}$$

By introducing a dimensionless coordinate q defined as

$$q \equiv \sqrt{\frac{m\omega}{\hbar}}x, \tag{7.44}$$

the Schrödinger equation Eq. 7.43 is simplified to

$$\frac{1}{2}\left(-\frac{d^2}{dq^2} + q^2\right)\hbar\omega\psi(q) = E\psi(q). \tag{7.45}$$

Creation operator and annihilation operator

As we have presented in previous sections, the physical reality in quantum mechanics is the wavefunction, and it is governed by a partial differential equation, the Schrödinger equation. In some cases, the process of obtaining solutions of partial differential equations can be simplified by differential operators using algebraic methods. This is especially true for the harmonic oscillator.

In order to find an algebraic solution of the harmonic oscillator, a pair of operators are introduced: an *annihilation operator*,

$$\hat{a} = \frac{1}{\sqrt{2}} \left(q + \frac{d}{dq} \right),$$
(7.46)

and a *creation operator*
$$\hat{a}^\dagger = \frac{1}{\sqrt{2}} \left(q - \frac{d}{dq} \right).$$
(7.47)

From Eqs. 7.35 and 7.37, $\hat{a}$ and $\hat{a}^\dagger$ are adjoint of each other.

By acting on any function $f(q)$, a simple algebra shows that the two operators satisfy a *commutation relation*. On one hand, we have

$$\hat{a}\hat{a}^\dagger f(q) = \frac{1}{\sqrt{2}} \left(q + \frac{d}{dq} \right) \frac{1}{\sqrt{2}} \left(q - \frac{d}{dq} \right) f(q)$$

$$= \frac{1}{2} \left(q^2 - \frac{d^2}{dq^2} + \frac{d}{dq}q - q\frac{d}{dq} \right) f(q)$$
(7.48)

$$= \frac{1}{2} \left(q^2 - \frac{d^2}{dq^2} + 1 \right) f(q).$$

On the other hand, we have

$$\hat{a}^\dagger \hat{a} f(q) = \frac{1}{\sqrt{2}} \left(q - \frac{d}{dq} \right) \frac{1}{\sqrt{2}} \left(q + \frac{d}{dq} \right) f(q)$$

$$= \frac{1}{2} \left(q^2 - \frac{d^2}{dq^2} - \frac{d}{dq}q + q\frac{d}{dq} \right) f(q)$$
(7.49)

$$= \frac{1}{2} \left(q^2 - \frac{d^2}{dq^2} - 1 \right) f(q).$$

Here, the obvious identity is applied:

$$\frac{d}{dq}(q\,f(q)) = q\frac{d}{dq}f(q) + \frac{dq}{dq}f(q) = q\frac{d}{dq}f(q) + f(q).$$
(7.50)

Combining Eqs. 7.48 and 7.49, we find the commutation relation

$$[\hat{a}, \hat{a}^\dagger] \equiv \hat{a}\hat{a}^\dagger - \hat{a}^\dagger\hat{a} = 1.$$
(7.51)

Through a simple algebra, the Schrödinger equation Eq. 7.45 becomes

$$\left(\hat{a}^\dagger \hat{a} + \frac{1}{2} \right) \hbar\omega\psi(q) = E\psi(q).$$
(7.52)

Algebraic solution of the Schrödinger equation

In this Section, we show how to utilize Eq. 7.51 to solve the Schrödinger equation Eq. 7.52. Denoting the n-th eigenstate with Dirac notation $|n\rangle$ (see Section 7.1.2), it is sufficient to solve the following algebraic equation,

$$\hat{a}^\dagger \hat{a} |n\rangle = u_n |n\rangle. \tag{7.53}$$

Here the eigenstates are labeled by an integer index n with eigenvalue u_n. Note that the eigenvalues u_n must be non-negative, because

$$\langle n | \hat{a}^\dagger \hat{a} | n \rangle = \langle \hat{a} n | \hat{a} n \rangle \geq 0. \tag{7.54}$$

By comparing Eq. 7.52 with Eq. 7.53, the energy eigenvalues are

$$E_n = \left(u_n + \frac{1}{2} \right) \hbar\omega. \tag{7.55}$$

As a consequence of Eq. 7.51, if $|n\rangle$ is an eigenstate with eigenvalue u_n, then $\hat{a}|n\rangle$ is also an eigenstate,

$$\hat{a}^\dagger \hat{a}\, \hat{a} |n\rangle = (\hat{a}\hat{a}^\dagger - 1)\hat{a}|n\rangle = (u_n - 1)\, \hat{a}|n\rangle \tag{7.56}$$

with eigenvalue $u_n - 1$. Because $\langle n | \hat{a}^\dagger \hat{a} | n \rangle$ should not be negative, there must be an eigenstate $|0\rangle$ with eigenvalue 0,

$$\hat{a}^\dagger \hat{a} |0\rangle = 0. \tag{7.57}$$

On the other hand, also as a consequence of Eq. 7.51, if $|n\rangle$ is an eigenstate with eigenvalue u_n, then $\hat{a}^\dagger |n\rangle$ is also an eigenstate

$$\hat{a}^\dagger \hat{a}\hat{a}^\dagger |n\rangle = \hat{a}^\dagger (\hat{a}^\dagger \hat{a} + 1)|n\rangle = (u_n + 1)\hat{a}^\dagger |n\rangle \tag{7.58}$$

with eigenvalue $u_n + 1$. Starting with the lowest eigenstate $|0\rangle$, by applying $\hat{a}^\dagger$ many times, we have

$$\hat{a}^\dagger \hat{a} \left(\hat{a}^\dagger \right)^n |0\rangle = n \left(\hat{a}^\dagger \right)^n |0\rangle. \tag{7.59}$$

Because the eigenvalue of the zeroth eigenstate $|0\rangle$ of the operator $\hat{a}^\dagger \hat{a}$ is zero, and each time a creation operator $\hat{a}^\dagger$ is applied, the eigenvalue is added by 1, the eigenvalues of the operator $\hat{a}^\dagger \hat{a}$ equals to an integer, the number of times $\hat{a}^\dagger$ applied. Therefore, up to a constant, it is the n-th eigenstate of the operator $\hat{a}^\dagger \hat{a}$:

$$|n\rangle = C_n \left(\hat{a}^\dagger \right)^n |0\rangle. \tag{7.60}$$

The operator $\hat{a}^\dagger \hat{a}$ is deservedly called the *particle number operator*,

$$\hat{N} \equiv \hat{a}^\dagger \hat{a}, \tag{7.61}$$

because its eigenvalue is the number of energy quanta,

$$\hat{N}|n\rangle = n|n\rangle. \tag{7.62}$$

The normalization constant C_n can be determined as follows. The zeroth-order wavefunction is by definition normalized,

$$\langle 0|0 \rangle = 1. \tag{7.63}$$

By applying $\hat{a}^\dagger$ many times, we have

$$\langle 0| \left(\hat{a} \right)^n \left(\hat{a}^\dagger \right)^n |0 \rangle = n!. \tag{7.64}$$

Therefore, $C_n = (n!)^{-1/2}$, and

$$|n \rangle = \frac{1}{\sqrt{n!}} \left(\hat{a}^\dagger \right)^n |0 \rangle. \tag{7.65}$$

Following Eq. 7.55, the energy eigenvalues of the harmonic oscillator is

$$E_n = \left(n + \frac{1}{2} \right) \hbar\omega. \tag{7.66}$$

The energy level of the harmonic oscillator is thus quantized, with energy quanta $\hbar\omega$. The operator $\hat{a}^\dagger$ adds an energy quanta, thus named a *creation operator*; the operator $\hat{a}$ removes an energy quanta, thus named an *annihilation operator*.

Explicit expressions of wavefunctions

The above algebraic solution provides the simplest approach to find explicit expressions of the wavefunctions in terms of the coordinate q. First, from Eqs. 7.57 and 7.46, the

Table 7.1: Wavefunctions of the harmonic oscillator

State	Energy	wavefunction
$|0\rangle$	$\frac{1}{2}\hbar\omega$	$\frac{1}{\sqrt{\sqrt{\pi}}} e^{-q^2/2}$
$|1\rangle$	$\frac{3}{2}\hbar\omega$	$\frac{1}{\sqrt{2\sqrt{\pi}}} 2q e^{-q^2/2}$
$|2\rangle$	$\frac{5}{2}\hbar\omega$	$\frac{1}{\sqrt{2\sqrt{\pi}}} \left(2q^2 - 1 \right) e^{-q^2/2}$
$|3\rangle$	$\frac{7}{2}\hbar\omega$	$\frac{1}{\sqrt{3\sqrt{\pi}}} \left(2q^3 - 3q \right) e^{-q^2/2}$
$|4\rangle$	$\frac{9}{2}\hbar\omega$	$\frac{1}{\sqrt{24\sqrt{\pi}}} \left(4q^4 - 12q^2 + 3 \right) e^{-q^2/2}$
$|5\rangle$	$\frac{11}{2}\hbar\omega$	$\frac{1}{\sqrt{60\sqrt{\pi}}} \left(4q^5 - 20q^3 + 15q \right) e^{-q^2/2}$

zeroth-order wavefunction $|0\rangle = \psi_0(q)$ should satisfy

$$\left(q + \frac{d}{dq}\right)\psi_0(q) = 0. \tag{7.67}$$

The solution is

$$\psi_0(q) = C_0 \exp\left(-\frac{q^2}{2}\right). \tag{7.68}$$

The normalization constant C_0 can be determined directly by

$$\int_{-\infty}^{\infty} \psi_0^2(q)dq = C_0^2 \int_{-\infty}^{\infty} \exp(-q^2)dq = C_0^2\sqrt{\pi} = 1, \tag{7.69}$$

which gives

$$C_0 = \pi^{-1/4}. \tag{7.70}$$

All wavefunctions of the harmonic oscillator $|n\rangle = \psi_n(q)$ can be obtained from Eqs. 7.47 and 7.65 using Eqs. 7.68 and 7.70:

$$\psi_n(q) = \frac{1}{\sqrt{2^n n! \sqrt{\pi}}}\left(q - \frac{d}{dq}\right)^n \exp\left(-\frac{q^2}{2}\right). \tag{7.71}$$

The first few wavefunctions are listed in Table 7.1, and graphically displayed in Fig. 7.4. The solid curve, a parabola, represents the potential energy as a function of

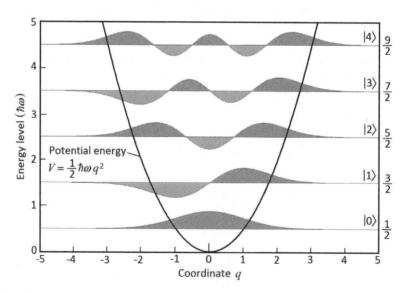

Figure 7.4 Energy levels and wavefunctions of a harmonic oscillator. The solid curve is the potential energy. The wavefunctions of the first five eigenstates are shown. The red shade indicates positive phase, and blue shade indicates negative phase. The y-position of the energy eigenvalue is the baseline for the wavefunction.

the dimensionless coordinate q. The horizontal lines represent the energy levels of the eigenstates. For the region inside the potential curve, energy level E is greater than the potential energy. The wavefunction resembles a sinusoidal wave. Outside the potential curve, the energy level is lower than the potential curve. The wavefunction resembles an exponential function decaying into the barrier. The lowest eigenstate has an energy value of $\frac{1}{2}\hbar\omega$. The wavefunction has no node. The energy eigenvalue increases by $\hbar\omega$ each step, while a new node is added. Those nodal structures make the wavefunctions orthogonal to each other.

7.1.4 The Hydrogen Atom

Hydrogen atom is a central subject in quantum mechanics. The hydrogen wavefunctions are also the foundation for the understanding of complex atoms and atomic systems. There are only two real-world systems that the Schrödinger equation has analytic solutions: the hydrogen atom, and the hydrogen molecular ion, H_2^+, which is the basis to understand the chemical bond and condensed matter.

In the SI unit system, the potential energy function of an electron in a hydrogen atom $V(r)$ is the attractive force from the positively charged proton,

$$V(\mathbf{r}) = -\frac{e^2}{4\pi\epsilon_0 r}. \tag{7.72}$$

To simplify notations, a *potential constant* K is introduced,

$$K = \frac{e^2}{4\pi\epsilon_0} = 2.306 \times 10^{-28} \text{ J} \cdot \text{m}. \tag{7.73}$$

Because the electron mass m_e is much smaller than the proton mass, to a good approximation, the Schrödinger equation Eq. 7.10 is

$$-\frac{\hbar^2}{2m_e}\nabla^2\psi - \frac{K}{r}\psi = E\psi. \tag{7.74}$$

Figure 7.5 Hydrogen atom in spherical polar coordinates. The center of the coordinate system is the positively charged proton. The force and potential energy V only depends on radius r. Therefore, to resolve the Schrödinger equation of the hydrogen atom, it is natural to use spherical polar coordinates, with radius r, polar angle θ, and azimuth ϕ. For mathematical details, see Appendix D.

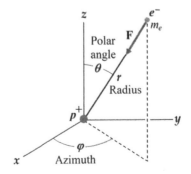

The potential only depends on r. It is natural to use polar coordinates, in terms of radius r, polar angle θ, and azimuth ϕ, see Fig. 7.5. Equation 7.74 becomes

$$-\frac{\hbar^2}{2m_e}\left[\frac{1}{r^2}\frac{\partial}{\partial r}\left(r^2\frac{\partial\psi}{\partial r}\right) - \frac{1}{r^2}\mathbf{L}^2\psi\right] - \frac{K}{r}\psi = E\psi. \tag{7.75}$$

where the *angular momentum operator* $\mathbf{L}^2$ is defined as

$$\mathbf{L}^2\psi \equiv -\frac{1}{\sin^2\theta}\left[\frac{\partial^2\psi}{\partial\phi^2} + \sin\theta\frac{\partial}{\partial\theta}\left(\sin\theta\frac{\partial\psi}{\partial\theta}\right)\right]. \tag{7.76}$$

The solutions of Eq. 7.76 are spherical harmonics, see Appendix D.

The ground state

First, we study the ground state, where the wavefunction ψ only depends on r, and $\mathbf{L}^2$ is zero. The Schrödinger equation Eq. 7.75 becomes

$$-\frac{\hbar^2}{2m_e}\frac{1}{r^2}\frac{d}{dr}\left(r^2\frac{d\psi}{dr}\right) - \frac{K}{r}\psi = E\psi. \tag{7.77}$$

Intuitively, since the electron is being attracted by the positively charged proton, the wavefunction reaches maximum near the proton, and decays with distance r. To resolve Schrödinger's equation Eq. 7.77, we use the following trial function

$$\psi = Ce^{-r/a}, \tag{7.78}$$

where a is a decay length to be determined, and C is a normalization constant. Insert Eq. 7.78 into Eq. 7.77, eliminate the common factor ψ, we obtain

$$\frac{\hbar^2}{m_e ar} - \frac{\hbar^2}{2m_e a^2} - \frac{K}{r} = E. \tag{7.79}$$

The solution should be valid for all values of r. The two terms with common factor $1/r$ must cancel each other. It implies

$$a = \frac{\hbar^2}{m_e K}. \tag{7.80}$$

The decay length of the exponential function is a, given by Eq. 7.80, which is the *Bohr radius*,

$$a = a_0 \equiv \frac{\hbar^2}{m_e K} \approx 52.9 \text{ pm}. \tag{7.81}$$

The rest of Eq. 7.79 provides an expression of the energy eigenvalue

$$E = -\frac{\hbar^2}{2m_e a_0^2} = -\frac{K}{2a_0}. \tag{7.82}$$

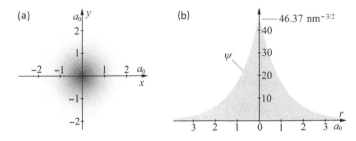

Figure 7.6 Wavefunction of ground-state hydrogen atom. (a) The density plot of the wavefunction. (b) The amplitude profile of the wavefunction.

The absolute value of the energy eigenvalue in Eq. 7.82 is the *Rydberg constant*, which agrees well with experimental findings,

$$\text{Ry} = \frac{K}{2a_0} \approx 2.178 \times 10^{-18} \text{ J} \approx 13.6 \text{ eV}. \tag{7.83}$$

In order to determine the normalization constant C, we note that according to Schrödinger, the square of the wavefunction is proportional to the charge density distribution of an electron as a field spread out in space,

$$\rho = -e\psi^2. \tag{7.84}$$

Because the total change is $-e$, the space integration of ψ^2 must be 1,

$$\int_0^\infty 4\pi\psi^2 r^2 dr = \int_0^\infty 4\pi C^2 e^{-2r/a_0} r^2 dr = 1, \tag{7.85}$$

which yields

$$C = \frac{1}{\sqrt{\pi a_0^3}}. \tag{7.86}$$

The ground-state wavefunction of hydrogen atom is then

$$|1s\rangle \equiv \psi_{1s} = \frac{1}{\sqrt{\pi a_0^3}} e^{-r/a_0}. \tag{7.87}$$

The subscript $1s$ indicates that it is the state of lowest energy and spherically symmetric. A density plot and an amplitude contour of the ground-state wavefunction of hydrogen atom are shown in Fig. 7.6. Wavefunction is the form of existence of the electron. The ground state represents an electron at rest. There is no moving point charge. The electrical charge density of an electron is spread out in space around the proton, with the highest charge density in the immediate vicinity of the proton.

Energy eigenvalues of excited States

By writing the wavefunction as a product of a radial function $R(r)$ and a function of angle variables $Y(\theta, \phi)$

$$\psi(\mathbf{r}) = R(r)Y(\theta, \phi), \tag{7.88}$$

the differential equation for the radial function $R(r)$ is

$$-\frac{\hbar^2}{2m_e}\left[\frac{1}{r^2}\frac{d}{dr}\left(r^2\frac{dR(r)}{dr}\right) - \frac{l(l+1)R(r)}{r^2}\right] + \frac{K}{r}R(r) = ER(r). \tag{7.89}$$

The solution of the angle-dependent factor $Y(\theta, \phi)$ is presented in Appendix D, which are *spherical harmonics*. The parameter l is the *orbital quantum number*, an integer parameter coming out from the solution of $Y(\theta, \phi)$, see Appendix D. To resolve Eq. 7.89, we use the following trial function

$$R(r) = C\,r^b e^{-r/a}, \tag{7.90}$$

where the parameters a and b are to be determined by Eq. 7.89, and C is a normalization constant. Insert Eq. 7.90 into Eq. 7.89, eliminate the common factor $R(r)$, we obtain an algebraic equation

$$-\frac{\hbar^2}{2m_e}\left[\frac{b(b+1)}{r^2} - \frac{l(l+1)}{r^2} + \frac{2(b+1)}{ar} - \frac{1}{a^2}\right] + \frac{K}{r} = E. \tag{7.91}$$

To cancel the two terms with $1/r^2$, by assuming $b > 0$, one obtains

$$b = l. \tag{7.92}$$

To cancel the two terms with $1/r$, one must have

$$a = (b+1)\frac{\hbar^2}{m_e K} = (b+1)a_0. \tag{7.93}$$

The remaining terms in Eq. 7.91 determine the energy eigenvalue

$$E = -\frac{\hbar^2}{2m_e a^2} = -\frac{\text{Ry}}{(b+1)^2}. \tag{7.94}$$

Because l is an integer, we can define an integer parameter n for $(b+1)$,

$$n = b+1, \tag{7.95}$$

which is called the *principal quantum number*. The wavefunction is

$$\psi = Cr^{n-1}e^{-r/na_0}\,Y_{lm}(\theta, \phi). \tag{7.96}$$

The energy eigenvalue depends on the principal quantum number n, perfectly explains the experimentally discovered Rydberg formula,

$$E_n = -\frac{\text{Ry}}{n^2}. \tag{7.97}$$

The length scale of the wavefunction, $a = na_0$, is a useful parameter for further study of the hydrogen wavefunction. It is

$$a = na_0 = \frac{\hbar}{\sqrt{-2m_e E}}. \tag{7.98}$$

Similar to Eq. 7.85, the normalization constant is found to be

$$C = \sqrt{\frac{2^{2n+1}}{\pi(2n+2)!\,(na_0)^{2n+3}}}. \tag{7.99}$$

In Fig. 7.7, we show some interesting cases. Following Appendix D, for $m = l$, the spherical harmonics up to a normalization constant is

$$Y_{ll}^g(\theta, \phi) \propto \sin^l \theta \cos l\phi, \tag{7.100}$$

and

$$Y_{ll}^u(\theta, \phi) \propto \sin^l \theta \sin l\phi. \tag{7.101}$$

The wavefunctions shown in Fig. 7.7 are on the plane $z = 0$, where $\sin\theta = 1$. Using Eqs. 7.92 and 7.95, $l = n - 1$. The explicit expressions are

$$\psi_g \propto r^{n-1} e^{-r/na_0} \cos(n-1)\phi, \tag{7.102}$$

and

$$\psi_u \propto r^{n-1} e^{-r/na_0} \sin(n-1)\phi. \tag{7.103}$$

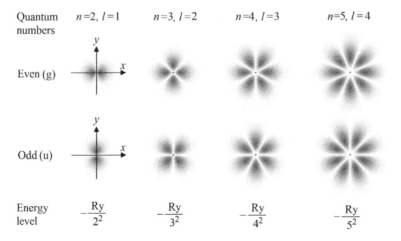

Figure 7.7 Wavefunctions of excited-states of hydrogen atom. Wavefunctions in Eqs. 7.102 and 7.103 on the plane $z = 0$. For each principal quantum number, there is an even wavefunction symmetric to the x-axis, and an odd wavefunction antisymmetric to the x-axis. The energy level, Eq. 7.97, depends on principal quantum number n, explains the Rydberg formula.

In the above equations, the subscript g and u indicates even and odd with respect to x-axis, respectively. Those wavefunctions represent standing waves on the $z = 0$ plane. Following de Broglie and Schrödinger, the energy-level quantization in hydrogen atom resembles the nodes of a vibrating string. Different patterns of standing waves with different number of nodes result in different energy levels.

Other useful wavefunctions

In the previous Subsection, following an elementary mathematical procedure, several special solutions of Eq. 7.89 are derived. Because the energy eigenvalue only depends on the principal quantum number n, all energy eigenvalues are obtained. Especially, for the cases of $l = n - 1$ and $m = l$, wavefunctions are shown in Figs 7.6 and 7.7. Nevertheless, Eq. 7.89 does have other useful solutions as follows.

Inspired by the special solutions, a dimensionless variable ρ is introduced by scaling the radius r using the length $a = na_0$ in Eq. 7.98,

$$\rho = \frac{r}{a}. \tag{7.104}$$

Equation 7.89 becomes

$$\frac{1}{\rho^2}\frac{d}{d\rho}\left(\rho^2\frac{dR(\rho)}{d\rho}\right) + \left[-1 - \frac{l(l+1)}{\rho^2} + \frac{2n}{\rho}\right]R(\rho) = 0. \tag{7.105}$$

In the process of the algebra, Eq. 7.81 is applied to obtain

$$\frac{2m_e a K}{\hbar^2} = \frac{a}{a_0} = n. \tag{7.106}$$

Equation 7.89 becomes

$$\frac{1}{\rho^2}\frac{d}{d\rho}\left(\rho^2\frac{dR(\rho)}{d\rho}\right) + \left[-1 - \frac{l(l+1)}{\rho^2} + \frac{2n}{\rho}\right]R(\rho) = 0. \tag{7.107}$$

Analogous to Eqs. 7.90 and 7.92, we make a substitution

$$R(\rho) = \rho^l\, e^{-\rho}\, F(\rho). \tag{7.108}$$

Insert Eq. 7.108 to Eq. 7.107, the differential equation for $F(\rho)$ is obtained:

$$\rho\frac{d^2 F(\rho)}{d\rho^2} + 2(l + 1 - \rho)\frac{dF(\rho)}{d\rho} + 2(n - l - 1)F(\rho) = 0. \tag{7.109}$$

It is the differential equation for associate Laguerre polynomials, well known for two centuries. The general formula is cumbersome. In condensed-matter physics, chemistry and molecular biology, only three more are needed. The expressions can be verified by inserting those polynomials $F(r)$ into Eq. 7.109. Here is a complete list.

$$F(\rho) = 1 - \frac{1}{2}\rho, \qquad\qquad n = 2, \quad l = 0;$$

$$F(\rho) = 1 - \frac{2}{3}\rho + \frac{2}{27}\rho^2, \qquad n = 3, \quad l = 0; \qquad (7.110)$$

$$F(\rho) = 1 - \frac{1}{6}\rho, \qquad\qquad n = 3, \quad l = 1.$$

By combining the radial functions in Eq. 7.108, using Eqs. 7.110 and the spherical harmonics in Table D.1, the first nine wavefunctions of the hydrogen atom are listed in Table 7.2, and displayed in Fig. 7.8. In Table 7.2, the first column is the chemist's name of the wavefunction. The second column is in Dirac notation. The expression of the wavefunction, column 3, is in cartesian coordinates with Bohr radius a_0 as the unit. The last column is the wavefunction in angular variables.

What is the meaning of wavefunctions? According to Schrödinger and quantum field theory, wavefunction is a physical field. At a microscopic scale, Born's statistical

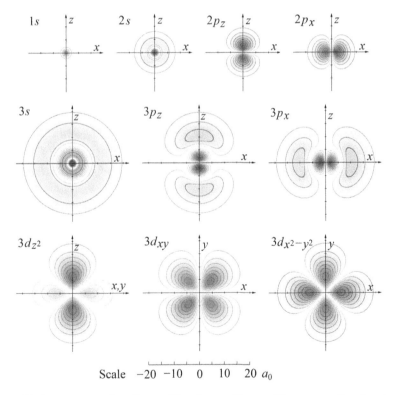

Figure 7.8 Hydrogen wavefunctions. The phase, either positive or negative, is colored as red or blue. The amplitude of the wavefunction is shown by the intensity of color, with several equal-amplitude contours to guide the eyes. The scale is shown at the bottom of the Figure.

Table 7.2: Wavefunctions of the hydrogen atom in Bohr radius a_0

Name	State	In cartesian coordinates	In angular parameters
$1s$	$\|1s0\rangle$	$\dfrac{1}{\sqrt{\pi}}\,e^{-r}$	$\dfrac{1}{\sqrt{\pi}}\,e^{-r}$
$2s$	$\|2s0\rangle$	$\dfrac{1}{2\sqrt{2\pi}}\left(1-\dfrac{1}{2}r\right)e^{-r/2}$	$\dfrac{1}{2\sqrt{2\pi}}\left(1-\dfrac{1}{2}r\right)e^{-r/2}$
$2p_x$	$\|2p1u\rangle$	$\dfrac{1}{4\sqrt{2\pi}}\,x\,e^{-r/2}$	$\dfrac{1}{4\sqrt{2\pi}}\,r\sin\theta\,\cos\phi\,e^{-r/2}$
$2p_y$	$\|2p1g\rangle$	$\dfrac{1}{4\sqrt{2\pi}}\,y\,e^{-r/2}$	$\dfrac{1}{4\sqrt{2\pi}}\,r\sin\theta\,\sin\phi\,e^{-r/2}$
$2p_z$	$\|2p0\rangle$	$\dfrac{1}{4\sqrt{2\pi}}\,z\,e^{-r/2}$	$\dfrac{1}{4\sqrt{2\pi}}\,r\cos\theta\,e^{-r/2}$
$3s$	$\|3s0\rangle$	$\dfrac{1}{3\sqrt{3\pi}}\left(1-\dfrac{2}{3}r+\dfrac{2}{27}r^2\right)e^{-r/3}$	$\dfrac{1}{3\sqrt{3\pi}}\left(1-\dfrac{2}{3}r+\dfrac{2}{27}r^2\right)e^{-r/3}$
$3p_x$	$\|3p1u\rangle$	$\dfrac{4}{27\sqrt{2\pi}}\left(1-\dfrac{1}{6}r\right)x\,e^{-r/3}$	$\dfrac{4}{27\sqrt{2\pi}}\left(r-\dfrac{1}{6}r^2\right)\sin\theta\,\phi\,e^{-r/3}$
$3p_y$	$\|3p1g\rangle$	$\dfrac{4}{27\sqrt{2\pi}}\left(1-\dfrac{1}{6}r\right)y\,e^{-r/3}$	$\dfrac{4}{27\sqrt{2\pi}}\left(r-\dfrac{1}{6}r^2\right)\sin\theta\,\sin\phi\,e^{-r/3}$
$3p_z$	$\|3p0\rangle$	$\dfrac{4}{27\sqrt{2\pi}}\left(1-\dfrac{1}{6}r\right)z\,e^{-r/3}$	$\dfrac{4}{27\sqrt{2\pi}}\left(r-\dfrac{1}{6}r^2\right)\cos\theta\,e^{-r/3}$
$3d_{z^2}$	$\|3d0\rangle$	$\dfrac{1}{27\sqrt{6\pi}}\left(z^2-\dfrac{1}{3}r^2\right)e^{-r/3}$	$\dfrac{1}{27\sqrt{6\pi}}\,r^2\left(\cos^2\theta-\dfrac{1}{3}\right)e^{-r/3}$
$3d_{xz}$	$\|3d1u\rangle$	$\dfrac{2}{81\sqrt{2\pi}}\,xz\,e^{-r/3}$	$\dfrac{1}{81\sqrt{2\pi}}\,r^2\sin2\theta\,\cos\phi\,e^{-r/3}$
$3d_{yz}$	$\|3d1g\rangle$	$\dfrac{2}{81\sqrt{2\pi}}\,yz\,e^{-r/3}$	$\dfrac{1}{81\sqrt{2\pi}}\,r^2\sin2\theta\,\sin\phi\,e^{-r/3}$
$3d_{xy}$	$\|3d2u\rangle$	$\dfrac{2}{81\sqrt{2\pi}}\,xy\,e^{-r/3}$	$\dfrac{1}{81\sqrt{2\pi}}\,r^2\sin^2\theta\,\sin2\phi\,e^{-r/3}$
$3d_{x^2-y^2}$	$\|3d2g\rangle$	$\dfrac{1}{81\sqrt{2\pi}}\left(x^2-y^2\right)e^{-r/3}$	$\dfrac{1}{81\sqrt{2\pi}}\,r^2\sin^2\theta\,\cos2\phi\,e^{-r/3}$

interpretation, that $|\psi(\mathbf{r})|^2 d^3\mathbf{r}$ is the probability of finding an electron in an elementary volume $d^3\mathbf{r}$, makes no sense. For the ground state of the hydrogen atom, according to Born's rule, the probability of finding an electron at the Bohr radius, 53.9 pm from the proton, should be $e^{-2} = 0.135$ times smaller than at the vicinity of the proton. That statement cannot been tested experimentally. The position of an electron was never measured at a subatomic scale. All measurements of the location of an electron, including a cell in a light-sensitive chip, a Wilson chamber tract, or a Geiger counter, are at a macroscopic scale, where Born's statistical interpretation works.

A measuring instrument with subatomic resolution is the scanning tunneling microscope (STM). STM experiments showed that at a subatomic scale, the wavefunction can be observed and mapped, verifying the view of quantum field theory. No material points were observed. We will come back to STM experiments in Section 7.2.4.

7.1.5 The Stern–Gerlach Experiment

As shown in Sections 2.1.4, electromagnetic waves have two polarizations, responding to crystals with anisotropic refractive indices. In 1922, Stern and Gerlach showed experimentally that electron waves are similarly polarized. An electron wave has two polarizations, similar to electromagnetic waves. Instead of anisotropic optical crystals, it responds to an inhomogeneous magnetic field, as shown in Fig. 7.9.

Using an oven, silver atoms are evaporated and streamed through a slot, then letting through an inhomogeneous magnetic field between two magnetic poles of different shapes. A single beam split into two, similar to the case of radiation.

Although the experiment was performed with silver atoms, the subject is the electron. A silver atom is a closed core with a single outer electron. And the electrons are described by a wavefunction. Later, the same experiment was performed with hydrogen atoms, and the same results are produced.

To describe the Stern–Gerlach experiment, the Pauli equation is applied. Instead of a single-component Schrödinger wavefunction, a two-component *Pauli spinor* is required. For electromagnetic waves propagating in the z-direction, the two polarizations are often identified as x-polarized and y-polarized, respectively. For electrons, the two polarizations are often called spin up and spin down, respectively.

If the spatial function of the two spin states is the same, the two polarization states

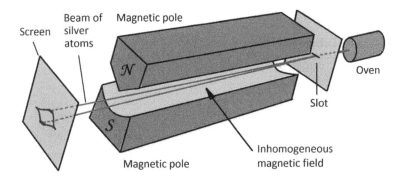

Figure 7.9 Schematics of Stern-Gerlach experiment. A beam of silver atoms is generated by an oven then goes through an inhomogeneous magnetic field. the beam of silver atoms split into two, representing different polarizations of the electron wave.

can be written as a pair of Pauli spinors, marked as spin up and spin down

$$u_\uparrow = \begin{pmatrix} \psi(\mathbf{r}) \\ 0 \end{pmatrix} = \psi(\mathbf{r}) \begin{pmatrix} 1 \\ 0 \end{pmatrix} = \psi(\mathbf{r})\,\alpha(\xi), \tag{7.111}$$

$$u_\downarrow = \begin{pmatrix} 0 \\ \psi(\mathbf{r}) \end{pmatrix} = \psi(\mathbf{r}) \begin{pmatrix} 0 \\ 1 \end{pmatrix} = \psi(\mathbf{r})\,\beta(\xi). \tag{7.112}$$

The symbols $\alpha(\xi) = \binom{1}{0}$ and $\beta(\xi) = \binom{0}{1}$ are the basic one-electron spin functions [81]. The variable ξ takes two discrete values, labeled $+\frac{1}{2}$ and $-\frac{1}{2}$.

7.1.6 Nomenclature of Atomic States

The wavefunctions of the hydrogen atom are the foundation of the nomenclature of electron states in many-electron atoms, see Table 7.3.

The principal quantum number n identifies the shells, labeled by K, L, M, etc. The orbital quantum number l identifies the subshells. The labels, s, p, d, etc., have its origin in the atomic spectrum. The spectral lines starting from $l = 0$ states are often sharp, thus named s, and it has only one azimuthal quantum number, $m = 0$. The spectral lines starting from $l = 1$ states are often intensive, thus named p, means principal, having 3 azimuthal quantum numbers. The spectral lines starting from $l = 2$ are often diffuse, thus named d, having 5 azimuthal quantum numbers.

As shown in Section 7.1.5, for each set of n, l, m, there are two spin states, similar to the two polarizations in radiation. See Table 7.3.

7.1.7 Degeneracy and Wavefunction Hybridization

The energy eigenvalues of the hydrogen atom only depend on the principal quantum number n. For each principal quantum number, there are n^2 different wavefunctions: it is n^2-fold degenerate. For $n = 2$, there are four states, $|2s\rangle$, $|2px\rangle$, $|2py\rangle$, and $|2pz\rangle$. Because the Schrödinger equation is linear, any linear superposition of wavefunctions with the same energy eigenvalue is also a good wavefunction with the same

Table 7.3: Nomenclature of atomic states

n	Shell	Maximum electrons	l	Subshell	Maximum electrons
1	K	2	0	1s	2
2	L	8	0	2s	2
			1	2p	6
3	M	18	0	3s	2
			1	3p	6
			2	3d	10

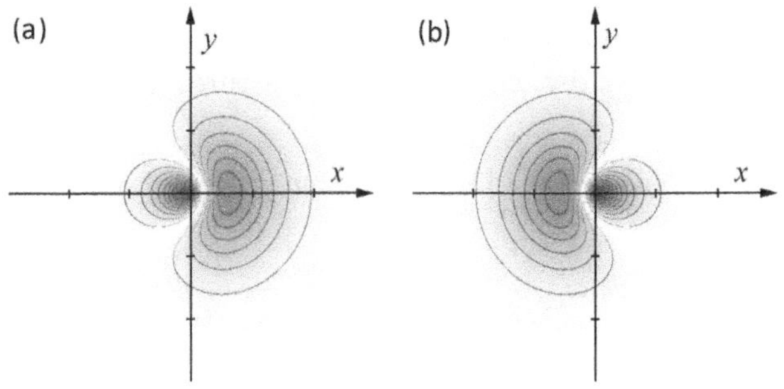

(a) y (b) y

x x

Figure 7.10 Hybrid $sp1$ wavefunctions. The phase of the wavefunction is shown by color, positive in red, negative in blue. The equal-value contours are also shown. The main lobe spans a wide angle. It is difficult to show both on the same figure.

energy eigenvalue. Especially, an s-wavefunction can make linear superposition with p-wavefunctions, to form *hybrid* wavefunctions. This concept is fundamental in chemistry, such as for carbon and silicon, especially in organic chemistry. Here we show the concept of hybridization using hydrogen wavefunctions.

Figure 7.10 shows two $sp1$ hybrid wavefunctions. (a) is an s-wavefunction superposed with a positive $2px$ wavefunction,

$$|2sp1+\rangle = \frac{1}{\sqrt{2}}|2s\rangle + \frac{1}{\sqrt{2}}|2px\rangle, \qquad (7.113)$$

resulting in a wavefunction preferentially concentrated in the $+x$ direction; and (b) is with a negative $2px$-wavefunction,

$$|2sp1-\rangle = \frac{1}{\sqrt{2}}|2s\rangle - \frac{1}{\sqrt{2}}|2px\rangle, \qquad (7.114)$$

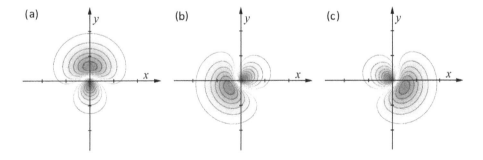

Figure 7.11 Hybrid $sp2$ wavefunctions. The three $sp2$ hybrid wavefunctions are pointing to the three vertices of a regular triangle, $120°$ apart in the same plane. The phase, positive or negative, is shown by color. The equal-value contours are also shown.

resulting in a wavefunction preferentially concentrated in the $-x$ direction. This happens for example in acetylene C_2H_2, for both the σ-bond between the two carbon atoms, and the two C–H bonds. Figure 7.11 shows three $sp2$ hybrid wavefunctions. The formulas are

$$|2sp2u\rangle = \frac{1}{\sqrt{3}}|2s\rangle + \sqrt{\frac{2}{3}}|2py\rangle, \tag{7.115}$$

$$|2sp2l\rangle = \frac{1}{\sqrt{3}}|2s\rangle - \frac{1}{\sqrt{2}}|2px\rangle - \frac{1}{\sqrt{6}}|2py\rangle, \tag{7.116}$$

and

$$|2sp2r\rangle = \frac{1}{\sqrt{3}}|2s\rangle + \frac{1}{\sqrt{2}}|2px\rangle - \frac{1}{\sqrt{6}}|2py\rangle, \tag{7.117}$$

respectively. It is the basic structure of graphene and carbon nanotubes.

A prevailing hybridization is the $sp3$ mode, where one s-wavefunction is superposed with three p-wavefunctions to form four hybrid wavefunctions pointing to the four vertices of a tetrahedron, see Fig. 7.12. Here are the mathematical formula:

$$\begin{aligned}
|t111\rangle &= \frac{1}{2}\left(|2s\rangle + |2px\rangle + |2py\rangle + |2pz\rangle\right), \\
|t\bar{1}1\bar{1}\rangle &= \frac{1}{2}\left(|2s\rangle - |2px\rangle + |2py\rangle - |2pz\rangle\right), \\
|t\bar{1}\bar{1}1\rangle &= \frac{1}{2}\left(|2s\rangle - |2px\rangle - |2py\rangle + |2pz\rangle\right), \\
|t1\bar{1}\bar{1}\rangle &= \frac{1}{2}\left(|2s\rangle + |2px\rangle - |2py\rangle - |2pz\rangle\right).
\end{aligned} \tag{7.118}$$

It is the backbone of all alkanes, including methane, ethane, propane, butane, etc., and the crystalline structures of diamond and silicon.

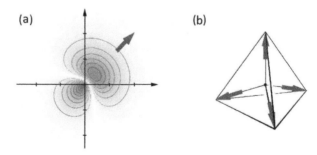

(a) (b)

Figure 7.12 Hybrid $sp3$ wavefunctions. (a) one of the four hybrid wavefunctions. The other three have the same shape but different orientation, pointing to the four vertices of a regular tetrahedron. (b) a regular tetrahedron.

7.2 Many-Electron Systems

Following the procedure leading to the single-electron Schrödinger's equation, a many-electron wave equation can also be heuristically derived from the classical energy integral and the de Broglie relation. Because the mass of a nucleus is thousands of times greater than the mass of an electron, in quantum-chemistry computations, the coordinates of the nuclei are treated as fixed input parameters. The accuracy of such approximation was analyzed mathematically by Born and Oppenheimer in 1927. For a system with M electrons with coordinates $\mathbf{r}_j$ in a potential field formed by N nuclei, the classical energy integral, similar to that for a single electron, Eq. 7.5, is

$$\sum_{j=1}^{j=M} \left[\frac{\mathbf{p}_j^2}{2m_e} + v(\mathbf{r}_j) \right] + \sum_{i>j>0}^{i=M} \frac{K}{|\mathbf{r}_i - \mathbf{r}_j|} = E. \tag{7.119}$$

the external potential $v(\mathbf{r}_j)$ is the total attractive potential of all nuclei on the j-th electron at position $\mathbf{r}_j$,

$$v(\mathbf{r}_j) = -\sum_{l=1}^{l=N} \frac{Z_l\,K}{|\mathbf{r}_j - \mathbf{R}_l|}, \tag{7.120}$$

where $\mathbf{R}_l$ is the position of the l-th nucleus with atomic number Z_l. The second sum in Eq. 7.119 is the repulsive potential from the negative change density of the rest of electrons, each one is $-e|\psi(\mathbf{r})|^2$. The condition $i > j$ is to avoid double counting.

The wavefunction as a function of the positions of the electrons is

$$\psi = \psi(\mathbf{r}_1, \mathbf{r}_2, \ldots, \mathbf{r}_M). \tag{7.121}$$

Similar to Eq. 7.7, the wave vector associated to the j-th electron is

$$k_j^2 = -\frac{1}{\psi} \frac{\hbar^2 \nabla_j^2 \psi}{2m_e}, \tag{7.122}$$

where the Laplacian for the j-th electron is

$$\nabla_j^2 = \frac{\partial^2}{\partial x_j^2} + \frac{\partial^2}{\partial y_j^2} + \frac{\partial^2}{\partial z_j^2}. \tag{7.123}$$

Using the de Broglie relation,

$$p_j^2 = \hbar^2 k_j^2, \tag{7.124}$$

we obtain the Schrödinger equation for the wavefunction ψ,

$$\sum_{j=1}^{j=M} \left[-\frac{\hbar^2 \nabla_j^2}{2m_e} + v(\mathbf{r}_j) \right] \psi + \sum_{i>j>0}^{i=M} \frac{K}{|\mathbf{r}_i - \mathbf{r}_j|} \psi = E\psi. \tag{7.125}$$

7.2.1   The Self-Consistent Field (SCF) Method

The many-electron wavefunction Eq. 7.121 is complicated. In 1928, Douglas Hartree invented a method to resolve Eq. 7.125 by writing the wavefunction of M electrons as a product of M single-electron wavefunctions,

$$\psi(\mathbf{r}_1, \mathbf{r}_2, \ldots, \mathbf{r}_M) = \phi_1(\mathbf{r}_1)\,\phi_2(\mathbf{r}_2)\ldots\phi_M(\mathbf{r}_M). \tag{7.126}$$

Equation 7.125 is then decomposed into M differential equations. The wavefunction of each individual electron is determined by a single-electron Schrödinger equation in a potential field formed by the positive changes of the nuclei and the negative change distribution of all electrons except the one under consideration.

Take an example of a two-electron system, either a helium atom or a hydrogen molecule. The wavefunction is

$$\psi(\mathbf{r}_1, \mathbf{r}_2) = \phi_1(\mathbf{r}_1)\,\phi_2(\mathbf{r}_2). \tag{7.127}$$

In an elementary volume of $d^3\mathbf{r}_2$, according to Schrödinger, the electrical charge is

$$d\rho = -e|\phi_2(\mathbf{r}_2)|^2 d^3\mathbf{r}_2, \tag{7.128}$$

The potential energy of the second electron in that elementary volume acting on the first electron is

$$V_{12}(\mathbf{r}_1) = \frac{e^2}{4\pi\epsilon_0} \int \frac{|\phi_2(\mathbf{r}_2)|^2}{|\mathbf{r}_1 - \mathbf{r}_2|} d^3\mathbf{r}_2. \tag{7.129}$$

The Schrödinger equation for the wavefunction $\phi_1(\mathbf{r}_1)$ is

$$\left[-\frac{\hbar^2 \nabla_1^2}{2m_e} + v(\mathbf{r}_1) + V_{12}(\mathbf{r}_1) \right] \phi_1(\mathbf{r}_1) = E_1 \phi_1(\mathbf{r}_1). \tag{7.130}$$

It is a differential equation for the first electron in a potential field of the nuclei and the charge distribution of the second electron. It can be extended to M electrons.

The process of computation is as follows. First, assuming a starting set of single-electron wavefunctions for all electrons. Solve the single-electron Schrödinger equation for each electron to obtain a new set of single-electron wavefunctions. Then repeat the solution of the single-electron Schrödinger equations using the new set of single-electron wavefunctions. After a number of iterations, the single-electron wavefunctions converge. Therefore, the method is termed self-consistent field (SCF) method. A single-electron wavefunction is called an *orbital*. Applying the SCF method to molecules, those single-electron molecular wavefunctions are called *molecular orbitals* (MO).

A good example of the application of the SCF method is the alkali metal atoms, for example Li and Na. Both have core electrons that form a spherical charge density and thus a spherical potential. The wavefunction of the valence electron is the solution of the Schrödinger equation in a spherically symmetric potential field.

The SCF method follows Schrödinger's definition that the wavefunctions are physical fields and the charge distribution of an electron is $\rho(\mathbf{r}) = -e|\psi(\mathbf{r})|^2$. The wavefunction of each electron is a continuous field. There are no material points.

7.2.2 Slater Determinates and the Hartree-Fock Method

According to the *Pauli exclusion principle*, no two electrons can occupy a single state. That principle has consequences on the form of wavefunctions. The standard representation of the exclusion principle, as proposed by John Slater in 1930, is to write the wavefunction as a determinant. As we have discussed in Section 7.1.5, the electron has two polarizations, or spin states. Using the following notation,

$$
\begin{aligned}
u_1(1) &= \psi_1(1)\,\alpha(1) \\
u_2(1) &= \psi_1(1)\,\beta(1) \\
u_3(1) &= \psi_2(1)\,\alpha(1) \\
u_4(1) &= \psi_2(1)\,\beta(1) \\
\ldots &= \ldots
\end{aligned}
\tag{7.131}
$$

a Slater determinate can be written as [57]

$$
\Psi(1,2,...,N) = \frac{1}{\sqrt{N!}}
\begin{vmatrix}
u_1(1) & u_2(1) & \ldots & u_N(1) \\
u_1(2) & u_2(2) & \ldots & u_N(2) \\
\ldots & \ldots & \ldots & \ldots \\
u_1(N) & u_2(N) & \ldots & u_N(N)
\end{vmatrix}.
\tag{7.132}
$$

By exchanging any two rows or any two columns, the wavefunction changes sign. Therefore, if any two rows are identical, the Slater determinant becomes zero. Because there are altogether $N!$ terms, the sum is divided by a square root of $N!$ for normalization. For general definition and properties of determinates, see Appendix C.

For the single-electron Schrödinger equation similar to Eq. 7.130, besides the mutual repulsion term $V_{12}(\mathbf{r})$, an *exchange term* is added. The computation becomes more complicated. Nevertheless, the result becomes more accurate.

7.2.3 Density-Functional Theory (DFT)

Although the Hartree–Fock method produced good agreement with experimental observations, the computational load is enormous. In the 1980s, Walter Kohn and his collaborators discovered that for the ground state of a many-electron system, the *electron density function over the space*

$$
n(\mathbf{r}) = \sum |\psi_k(\mathbf{r})|^2
\tag{7.133}
$$

contains full information on the quantum-mechanical system. In Eq. 7.133, the sum is over all electrons in the system. It is based on Schrödinger's definition that for each single-electron wavefunction, the electron density distribution is

$$
n_k(\mathbf{r}) = |\psi_k(\mathbf{r})|^2.
\tag{7.134}
$$

The computation consists of the solution of the Kohn–Sham wavefunctions $\psi_k(\mathbf{r})$ that satisfies an eigenvalue equation similar to the single-electron Schrödinger equation following the self-consistent method. However, because the electron density function $n(\mathbf{r})$

in Eq. 7.133 contains full information, the potential energy and the exchange potential can be derived from the density function. The computation load is substantially less that the Hartree–Fock method. Nevertheless, in most cases, the accuracy of the DFT results are comparable to or even better than the Hartree–Fock method.

The Kohn–Sham wavefunctions, or Kohn–Sham MOs in chemistry, are similar to the wavefunctions, or MOs following the Hartree–Fock method. In both cases, each electron is a continuous fields spread out in space, not material points.

7.2.4   HOMO and LUMO

The solution of the SCF method or DFT yields a series of single-electron wavefunctions, also called as molecular orbitals (MO). The total number of MOs is infinite. From a base skeleton of nuclei, the states can be filled up from the lowers energy and on. When the number of single-electron states equals to the number of protons in the nuclei, the molecule becomes neutral. The single-electron wavefunction or the molecular orbital with the highest energy level is called the highest occupied molecular orbital (HOMO). The next molecular orbital above the HOMO is the lowest unoccupied molecular orbital (LUMO). Both HOMO and LUMO are active in chemical reactions, and are called collectively as *frontal orbitals* (FO).

Using a scanning tunneling microscope (STM), the frontal molecular orbitals were observed and mapped in real space, confirming that MOs are physical reality by direct experiments [18]. A schematics of STM is shown in Fig. 7.13. The heart of the STM is a piezoelectric drive. By applying programmed voltages on the electrodes of the piezoelectric drive, the tip can scan the sample wavefunction over many nanometers at an accuracy better than 1 picometer. The apex of the STM tip is made of an atom or a molecule. In (a), the tip apex is a copper atom, which has a $3s$ wavefunction protruding out to the sample. In (b), the apex of the tip is a CO molecule. The oxygen atom has two $2p$ wavefunctions reaching out to the sample.

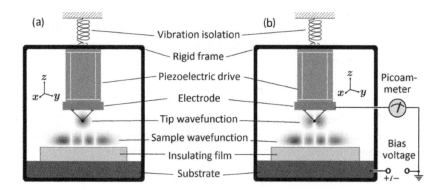

Figure 7.13 Scanning tunneling microscope. With a piezoelectric drive, the tip can scan over the sample with a subpicometer resolution. (a) with an s-type tip, the square of the wavefunction is probed. (b) with a p-type tip, the lateral derivatives of the wavefunction are probed.

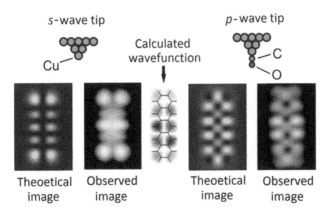

Figure 7.14 HOMO of pentacene imaged by STM. The HOMO of pentacene is computed using quantum mechanics. With a *s*-type tip, the STM image is the square of the wavefunction. With a *p*-type tip, the STM image is the square of the lateral derivatives of the wavefunction.

The tunneling process was studied by the Schrödinger's equation under general operational conditions. For an *s*-type tip, such as a Cu atom, the tunneling current is proportional to the square of the sample wavefunction at the center of the tip atom. For a *p*-type tip, such as a CO molecule, the tunneling current is proportional to the square of the *x*- and *y*-lateral derivatives of the sample wavefunction at the center of the oxygen atom. The energy level of the sample wavefunction can be selected by using different bias voltages of the power supply, shown at the lower right corner of Fig. 7.13. By using a positive bias, for example +1.25 V, the electron is tunneling from the tip to the unoccupied wavefunction of the sample, that is the LUMO. By using a negative bias, for example −2.15 V, the electron is tunneling from the occupied wavefunction of the sample, that is the HOMO, to the tip. By scanning the tip over the samples, both the HOMO and LUMO are observed and mapped.

In the first quarter of the 21st century, many molecules have been probed. One of the first molecules under observation is the pentacene, an organic molecule made of five linearly-fuzed benzene rings. The electronic structures of the molecules have been computed in details using computational quantum mechanics.

Figure 7.14 shows the STM images of the HOMO of pentacene, observed using an *s*-type tip wavefunction from a Cu tip, and using a *p*-type tip wavefunction from a CO-functionalized tip. Both the theoretical images from numerical quantum-mechanical computations and the observed images from STM experiments are shown. In the middle of Fig. 7.2.4, the skeleton of the molecular structure (in lines) and the calculated wavefunction (in shades) are shown. The STM image with an *s*-type tip wavefunction should be proportional to the square of the sample wavefunction. The predicted image should show a pair of rows, each has five dots. The observed image confirms the theoretical prediction. For a *p*-type tip wavefunction, the image, or the tunneling current intensity distribution, should be proportional to the square of the lateral derivative of

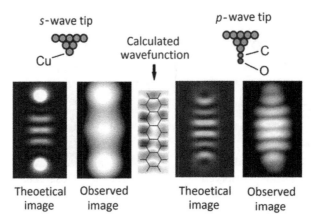

Figure 7.15 LUMO of pentacene imaged by STM. The LUMO of pentacene ia computed using quantum mechanics. With a s-type tip, the STM image is the square of the wavefunction. With a p-type tip, the STM image is the square of the lateral derivatives of the wavefunction.

he sample wavefunction,

$$I_p(x,y) \propto \left| \frac{\partial \psi(x,y,z_0)}{\partial x} \right|^2 + \left| \frac{\partial \psi(x,y,z_0)}{\partial y} \right|^2. \tag{7.135}$$

Because the oxygen atom has p-type wavefunctions in both x- and y-directions, the squares of the lateral derivatives in both x- and y-directions are added. Again, the observed STM image reproduces the theoretical prediction nicely.

Figure 7.15 shows the STM images of the LUMO of pentacene, observed using an s-type tip wavefunction from a Cu tip, and using a p-type tip wavefunction from a CO-functionalized tip. In the middle of Fig. 7.15, the skeleton of the molecular structure (in lines) and the calculated wavefunction (in shades) are shown. Again, the STM image with an s-type tip wavefunction is proportional to the square of the sample wavefunction, and with a p-type tip wavefunction, the image is proportional to the sum of the square of the lateral derivative of the sample wavefunction.

The peak values of the lateral derivatives, Eq. 8.2, usually occur at places where the wavefunction changes sign. The totality of the points of sign change constitutes the *nodal structure* of the wavefunction. Although the nodal structures of the wavefunctions, or the orbitals, have been described and discussed in textbooks of quantum mechanics and especially quantum chemistry for many decades, those experimental observations constitute a direct verification of those nodal structures in real space.

The experiments were repeated with many other molecules with similar results. As shown by many experiments with STM, at a subatomic scale, Schrödinger's wavefunctions are observable physical fields [18]. No individual electrons as point particles were observed. According to quantum field theory, the universe consists of continuous fields. There are no point particles. That statement was verified by STM experiments.

7.3 The Chemical Bond

There are four types of forces between atoms and molecules. The van der Waals force is a ubiquitous, long ranged but weak force. Some atoms can be ionized, such as Na to become Na^+ and Cl to become Cl^-, then form ionic molecules and crystals by electrostatic forces, such as rock salt, NaCl. At very short distances, the core–core repulsion due to Pauli exclusion dominates, which sets a limit of how close two atoms can approach each other. Electrostatic attraction is much stronger than the van der Waals force. However, it cannot explain a basic fact in chemistry: two identical atoms can form a bond often much stronger than the ionic bond, such as the nitrogen molecule and the oxygen molecule. The strongest bonds are formed between identical atoms, for example in diamond. Such type of bond is called a *covalent bond*, or *chemical bond*, which is a central concept of modern chemistry.

7.3.1 Bonding Energy and Antibonding Energy

The origin of chemical bond can be understood in terms of linear superposition of atomic orbitals to form molecular orbitals. Figure 7.16 (a) and (b) show two atoms, each has a wavefunction, or atomic orbital (AO), at the same energy E_0. When the two atoms come together to become a diatomic molecule, molecular wavefunctions, or molecular orbitals (MO), are formed. Because the Schrödinger equation is linear, a molecular orbital can be a linear combination of atomic orbitals (LCAO), see Fig. 7.16(c) and (d). The term *orbital* refers to a single-electron wavefunction. Because the overall sign of a wavefunction is irrelevant, for intuitive understanding, the sign of the two atomic wavefunctions near the interface is chosen to be positive.

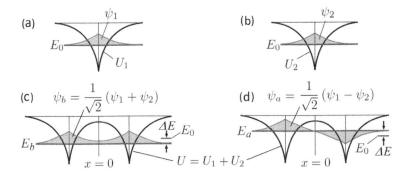

Figure 7.16 Concept of chemical bond. When two atoms (a) and (b) approach each other to become a diatomic molecule, the MOs are linear superpositions of AOs. Two types of superpositions are possible. The symmetric superposition, (c), makes a *bonding MO*. A bonding energy ΔE is gained. The antisymmetric superposition, (d), makes an *antibonding MO*. The bonding MO does not have additional node, and its energy is lower than the energy of individual AOs. The antibonding MO creates an additional node, and its energy is higher.

There are two types of MOs. The MO in Fig. 7.16(c) is the sum of the AOs, or a bonding MO ψ_b. In the interface region, the amplitude of the MO is the sum of the AOs. The productive linear superposition of AOs lowers the energy level from E_0 to the energy level of the MO, E_b. It gains a bonding energy,

$$\Delta E = E_0 - E_b. \qquad (7.136)$$

In Fig. 7.16(d), the MO is the difference of the two AOs, an antibonding MO ψ_a. The MO changes sign near the interface. The surface where the amplitude of wavefunction vanishes is called a nodal plane or simply a *node*. Due to the destructive linear superposition, the energy level of the MO E_a is higher than that of the atoms. The difference, an antibonding energy, approximately equals the bonding energy,

$$\Delta E = E_a - E_0. \qquad (7.137)$$

Both the bonding energy and the antibonding energy can be expressed as a surface integral on a separation surface:

$$\Delta E = E_0 - E_b = E_a - E_0 = \frac{\hbar^2}{2m} \int_{x=0} [\psi_1 \nabla \psi_2 - \psi_2 \nabla \psi_1] \cdot d\mathbf{S}. \qquad (7.138)$$

By applying those formulas to the hydrogen molecular ion H_2^+, very simple expressions are obtained. The approximate expression is shown to be the first two terms of the asymptotic expansions of the accurate analytic solution [18].

7.3.2 The Hydrogen Molecular Ion

The hydrogen molecular ion is the simplest molecule. It is also one of two real problems in nature for which exact analytic solutions of Schrödinger equation exist. It is no surprise that John Slater considered it the cornerstone for the understanding of interatomic forces and even all of condensed-matter physics.

The bonding energy is the sum of the van der Waals energy and the covalent bond energy. The van der Waals energy is universal, varies with the minus fourth power of the internuclear distance. The chemical bonding energy depends on the type of MO, which can be calculated from Eq. 7.138. Here is the results of the perturbation method. The interatomic distance ρ is in the unit of Bohr radius. For the bonding MO,

$$\Delta E(1\sigma_g) = -\frac{9\,\mathrm{Ry}}{2\,\rho^4} - 4\,\mathrm{Ry}\rho\, e^{-\rho - 1}; \qquad (7.139)$$

and for the antibonding MO,

$$\Delta E(1\sigma_u) = -\frac{9\,\mathrm{Ry}}{2\,\rho^4} + 4\,\mathrm{Ry}\rho\, e^{-\rho - 1}. \qquad (7.140)$$

The hydrogen molecular ion problem has an analytic solution. The asymptotic expansion of the interaction energy was obtained [20]. Up to the sixth term, it is

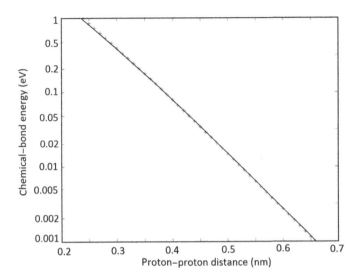

Figure 7.17 Accuracy of the perturbation treatment of hydrogen molecular ion. The exact chemical bond energy is shown as the solid curve. The approximate values represented by Eq. 7.140 are shown as crosses. For the proton–proton distance over the range of 0.25–0.65 nm, the error is less than 2%.

$$\Delta E_{\pm} = \text{Ry}\left[-\frac{9}{2\rho^4} - \frac{15}{\rho^6} - \frac{213}{2\rho^7} + \ldots\right]$$
$$\pm 2\,\text{Ry}\rho\,e^{-\rho-1}\left[1 + \frac{1}{2\rho} - \frac{25}{8\rho^2} - \ldots\right]. \tag{7.141}$$

For hydrogen molecular ion, the perturbation method reproduces the first two terms of the exact analytic solution, verifying the accuracy of the perturbation method. The perturbation method can be applied to more complicated cases.

7.3.3 Types of Chemical Bonds

Using the perturbation theory of the chemical bond based on the surface integral, the chemical bonds in dimers of the atoms in the first row of the periodic table (including H, He, Li, Be, B, C, N, O, F, and Ne) can be explained. To calculate the surface integrals, only the wavefunction outside the cores of the atoms are required, see Fig. 7.18. (a) is the s-type wavefunction, (b) is the p_z-type wavefunction, (c) is the p_x-type wavefunction. The p_y-type wavefunction is similar to the p_x-type.

Based on the published data of atomic orbitals from numerical computations, the surface integral in Eq. 7.138 can be evaluated, and compare with the experimental data for dissociation energies and vibrational frequencies. Good agreements are found. Here, rather than presenting the quantitative details, the conceptual understanding of the chemical bonds is presented.

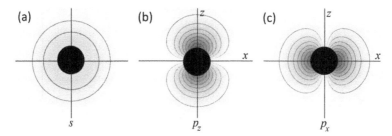

Figure 7.18 Wavefunctions outside the atomic core. The bonding and antibonding energies depend on the wavefunctions or atomic orbitals outside the cores of the participating atoms. Three types of atomic orbitals are shown: (a), s-type. (b), p_z-type. (c), p_x-type.

Chemical bonds from s-type atomic orbitals

In this subsection, we discuss the chemical bonds formed from two s-type AOs. The atoms include hydrogen with configuration $1s^1$, helium with configuration $1s^2$, lithium with configuration $1s^2\, 2s^1$, and beryllium with configuration $1s^2\, 2s^2$.

The electron configuration of hydrogen is $1s^1$. By bringing two hydrogen atoms together, $s\sigma$ MOs are formed, see Fig. 7.19. Because electron has two spin versions, both $s\sigma$ MOs are occupied. A hydrogen molecule with a strong bond is formed.

The electron configuration of helium is $1s^2$. Because two helium atoms have four electrons, two $s\sigma$ MOs and two $s\sigma^*$ MOs are occupied. Because the binding energy of the antibonding orbital is equal but in opposite sign, the net chemical bond energy is zero. Therefore, there is no He_2 molecule. The electron configuration of Li is $1s^2\, 2s^1$, or $[He]\, 2s^1$. Therefore, it is similar to hydrogen. When evaporated at high temperature in a good vacuum, a stable molecule Li_2 appears in the gas phase. The electron configuration of beryllium is $[He]\, 2s^2$. Similar to helium, the four $2s$ electrons form two $s\sigma$ MOs and two $s\sigma^*$ MOs. The net bonding energy is zero.

Chemical bonds from p-type atomic orbitals

The elements B, C, N, O, and F have something in common: the chemical bonds are formed by $2p$ AOs. By assigning the z-axis as the line connecting the nuclei, there are

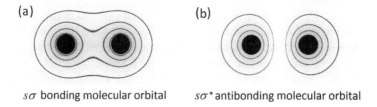

$s\sigma$ bonding molecular orbital $s\sigma^*$ antibonding molecular orbital

Figure 7.19 Molecular orbitals built from two s-type AOs. The bonding MO $s\sigma$ lowers the energy to form a stable molecule. The antibonding MO $s\sigma^*$ has a node, which raises the energy.

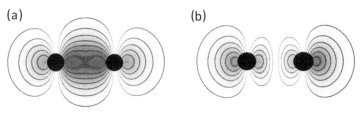

$p\sigma$ bonding molecular orbital $p\sigma^*$ antibonding molecular orbital

Figure 7.20 The $p\sigma$ and $p\sigma^*$ MOs. The bonding MO1 $p\sigma$ lowers the energy to form a stable molecule. The antibonding MO1 $p\sigma^*$ creates a new node, which raises the energy.

two types of chemical bonds, either $p\sigma$ type or $p\sigma^*$ type.

For the $p\sigma$ MO, a constructive superposition lowers the energy level of the system. For the $p\sigma^*$ MO, a node makes the energy level higher. See Fig. 7.20.

For p_x-type or p_y-type AOs, the MOs are either of $p\pi$ type or $p\pi^*$ type. Wavefunctions of $p\pi$ bonding MO and $p\pi^*$ antibonding MO are shown in Fig. 7.21. Intuitively, the bonding energy of the $p\sigma$ MO is greater than the bonding energy of the $p\pi$ MO.

The experimental data for the dimers of first-row elements are shown in Fig. 7.22. Theoretical values from quantum mechanical computation are also shown. The general features can be understood through intuitive arguments. As we have presented, the bonds in Be$_2$ is negligible. The electron configuration of boron is $1s^2\,2s^2\,2p^1$. The single $2p$ AOs forms a $p\sigma$ bonding MO. The electron configuration of carbon is $1s^2\,2s^2\,2p^2$. In addition to the two $p\sigma$ bonding MOs, the p_x- and p_y-type AOs form two $p\pi$ MOs. The total binding energy is higher than boron. The electron configuration of nitrogen is $1s^2\,2s^2\,2p^3$. It has the greatest number of binding MOs, thus have the highest binding energy. The electron configuration of oxygen is $1s^2\,2s^2\,2p^4$. In addition to the four $p\pi$ MOs, there are two $p\pi^*$ antibonding MOs, which neutralize two $p\pi$ bonds. The net binding energy is less than that of N$_2$. The electron configuration of fluorine is $1s^2\,2s^2\,2p^5$. Now four $p\pi^*$ antibonding MOs neutralize four $p\pi$ bonds. The net binding

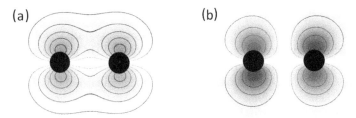

$p\pi$ bonding molecular orbital $p\pi^*$ antibonding molecular orbital

Figure 7.21 The $p\pi$ and $p\pi^*$ MOs. The bonding MO $p\pi$ lowers the energy to form a stable molecule. The antibonding MO $p\pi^*$ creates a new node, which raises the energy.

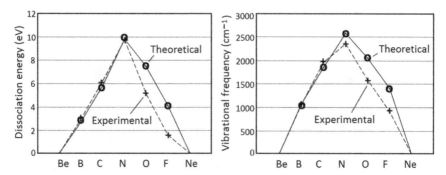

Figure 7.22 Chemical bonds of seven first-row elements. Comparing theoretical results with experimental data. (a) Dissociation energy. (b) Vibrational frequency. The elements with no dimers, Be and Ne, are also marked for convenience. See Chapter 4 of [18]/ Oxford University Press.

energy is even less than O_2. The electron configuration of neon is $1s^2\,2s^2\,2p^6$. The $p\sigma^*$ antibonds cancel the $p\sigma$ bonds completely. There is no neon dimer Ne_2.

7.4 The Solid State

A substantial part of modern technology relies on the quantum mechanics of a special type of solid-state material: the semiconductor. The list includes transistors, integrated circuits, light-emitting diodes, solid-state lasers, infrared light sensors, and so on. Especially, most of the solar cells are based on semiconductors.

7.4.1 Bloch Waves and Energy Bands

The central concept of quantum theory of solids is the energy band. For atoms and molecules, the energy levels are *discrete*. The interaction of atoms and molecules with radiation is through the transition between different discrete quantum states. In solid state, the energy levels are in the form of continuous *energy bands*. Here is an intuitive presentation explaining how energy bands are formed based on the interatomic forces. For more details, see the standard references.[5]

The general solution of the Schrödinger equation in a crystalline solid is the Bloch wavefunction (also called Bloch wave or Bloch function). It is an extension of the molecular wavefunctions. When two identical atoms come together to form a molecule, a linear combination of the AOs makes an MOs. A chemical bond is formed. There are two types of chemical bonds. If the signs of the two AOs near the border are the same, an addition of the two AOs forms a *bonding MO*, the energy of the system

---

[5]Neil W. Ashcroft and N. David Mermin, *Solid State Physics*, Saunders College, Chapter 10; Charles Kittle, *Introduction to Solid State Physics*, John Wiley and Sons, 8th ed., 2005, Chapter 9; and John M. Ziman, *Principles of the Theory of Solids*, Cambridge University Press, 2nd ed, 1979.

is lowered. A subtractive combination makes an *antibonding MO*, the energy of the system is increased. See Section 7.3.3.

Figure 7.23 shows a one-dimensional solid formed by N identical atoms. The distance between adjacent atomic nuclei, the *lattice constant*, is a. The total length is $L = Na$. The wavefunction is formed by linear combinations of AOs, shown here as of s-type. For a diatomic molecule, the coefficients is either $(1, 1)$ or $(1, -1)$. For a solid with N atoms, the coefficients are *sinusoidal functions* of the coordinate $0 < x \leq L$. In 1928, Felix Bloch proved that the wavefunctions in a crystalline solid is a product of a sinusoidal function and a periodic function. It is called the *Bloch theorem*. Those *Bloch wavefunctions* are eigenfunctions of the solid, and the energy level depends on the wave vector of the sinusoidal function. Here, we make an intuitive introduction to the Bloch wavefunctions.

As shown in Fig. 7.23, at the two ends of the solid, $x = 0$ and $x = L$, there is no atoms. The m-th envelope sine function should be

$$f_m(x) = \sin(k_m x) = \sin\left(\frac{m\pi x}{L}\right). \tag{7.142}$$

The wave vector is

$$k_m = \frac{m\pi}{L}, \tag{7.143}$$

The x-coordinate of the nucleus of the n-th atom is

$$X_n = \left(n - \frac{1}{2}\right)a. \tag{7.144}$$

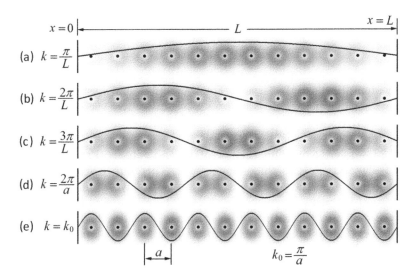

Figure 7.23 Bloch wavefunctions. The wavefunctions in a solid is formed by linear combinations of AOs. The wavefunctions with well-defined energy levels formed by linear combinations of AOs using coefficients of a sinusoidal profile function with a wavevector **k** are called *Bloch wavefunctions*.

which is part of the three-dimensional coordinate $\mathbf{R}_n$.

$$\mathbf{R}_n = (X_n, 0, 0). \tag{7.145}$$

The n-th AO is $u(\mathbf{r} - \mathbf{R}_n)$. The Bloch wave function is

$$\psi_m(\mathbf{r}) = \sum_{n=1}^{N} \sin(k_m X_n)\, u(\mathbf{r} - \mathbf{R}_n). \tag{7.146}$$

The first Bloch wave in Fig. 7.23(A) is

$$\psi_1(\mathbf{r}) = \sum_{n=1}^{N} \sin\left(\frac{\pi X_n}{L}\right) u(\mathbf{r} - \mathbf{R}_n). \tag{7.147}$$

As shown in Fig. 7.23, at the middle plane of two adjacent atoms, the wavefunctions of the two atoms are of the same sign. A chemical bond is formed to lower the energy. Between each pair of adjacent atoms, there is a chemical bond. The number of bonds is the maximum. The energy of the system is the lowest. In the case of Fig. 7.23(B), $k = 2\pi/L$, an antibond is formed. The energy is higher than the case (A). The case of Fig. 7.23(C) has two antibonds, the energy level is even higher. In general, the greater the value of the wave vector, the higher the energy.

The wave vector has an upper limit. In the case of Fig. 7.23(E), when

$$k = k_0 = \frac{\pi}{a}, \tag{7.148}$$

the Bloch wave is

$$\psi(\mathbf{r}) = \sum_{n=1}^{N} (-1)^n\, u(\mathbf{r} - \mathbf{R}_n). \tag{7.149}$$

The number of antibonds is the largest. The energy is the highest.

On the other hand, it is obvious that for a negative wave vector, the energy is the same as the positive wave vector with the same absolute value, because the numbers of bonds and antibonds are identical.

In Fig. 7.23, we show a one-dimensional solid with 12 atoms. Practically, the number is huge. And all real crystals are three-dimensional. Therefore, it should be a continuous three-dimensional wave vector $\mathbf{k}$. These wave vectors have a dimension of L^{-1} and form a *reciprocal space*, as shown in Fig. 7.24. The effective range of the reciprocal space is $-\pi/a < k < \pi/a$. The interval $(-\pi/a, \pi/a)$ is called *the first Brillouin zone*. Accordingly, $-\pi/a$ and π/a are the *edges of the first Brillouin zone*.

In Fig. 7.23, the case of s-type AOs are shown. The energy band curve in Fig. 7.24 is U-shaped, where the lowest energy is at $k = 0$. If the AOs are of the p_x-type, at $k = 0$, the number of antibonds is maximized, and the energy is the highest. On the other hand, at the $k = \pi/a$ point, the number of bonds is maximized. The energy is the lowest. The band curve should have an inverse U-shape.

7.4.2 Effective Mass

The sinusoidal waves in a solid is similar to free electrons in space. According to the de Broglie relation, the momentum of the electron is

$$p = \hbar k. \tag{7.150}$$

Recall that in classical mechanics, the kinetic energy is proportional to the square of the momentum,

$$E = \frac{p^2}{2m} \tag{7.151}$$

where m is the mass. The parabolic dependence of band energy with the wave vector, see Fig. 7.24, can be interpreted as having an effective mass m^*,

$$E = E_0 + \frac{\hbar^2 k^2}{2m^*}. \tag{7.152}$$

The effective mass of electron in a solid can be expressed as

$$m^* = \frac{\hbar^2}{d^2 E / dk^2}. \tag{7.153}$$

7.4.3 Conductor, Semiconductor, and Insulator

In an atom, the number of discrete quantum states is infinite. Starting from the bare nuclei, the electrons fill up the states from the lowest and up. When the number of electrons equal to the number of protons, the atom becomes neutral.

When atoms are combined to a solid, each discrete energy level of an atom becomes a continuous energy band. An energy band formed by one atomic state may overlap the energy band formed by another atomic state. If there is no overlap between two adjacent bands, an *energy gap* takes place, see Fig. 7.25. Similar to atoms, the electrons fill up the states from the bottom up, until the number of electrons equals the number of protons in the system. There are three cases:

Figure 7.24 Reciprocal space and the first Brillouin zone. For real crystals, the vector k in Eq. 7.143 is continuous and form a *reciprocal space*. The interval $-\pi/a < k < \pi/a$ is *the first Brillouin zone*. The dependence of energy with k makes an energy-band diagram. At $k = 0$, the energy is at a minimum, denoted as the bottom of the energy band E_0. The curve near the bottom is approximately parabolic. An effective mass can be defined, see Eqs. 7.152 and 7.153.

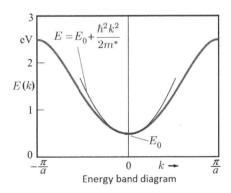

Energy band diagram

1. In Fig. 7.25(a), the highest occupied state is in the middle of an energy band. The solid is called as a *conductor*. The electrons can move freely from the filled states to the unfilled states. The highest energy level of the filled states is the *Fermi energy level*, E_F.

2. In Fig. 7.25(c), the highest occupied energy level matches the top of an energy band, the *valence band*, marked as E_v. If the distance to the next energy band is large, the electrons can hardly excited to a higher band, the *conduction band*, with a bottom marked as E_c. This type of solid is an *insulator*.

3. In Fig. 7.25(b) shows an important case between those two, the *semiconductor*, where the gap between the top of the valance band E_v and the bottom of the next energy band E_c is small such that when the temperature is not too low, electrons can be excited to a higher energy band, the *conduction band*. Typically the energy gap is less than a few electron volts. Once the electrons are excited to the conduction band, some conduction can take place. And in the valence band, a number of unoccupied Bloch states are left out. Those unoccupied states near the top of the valence band can also make electrical current, which are called *holes*, see Section 8.1.1.

As an analogy to chemistry, the top of the valence band is similar to the HOMO, and the bottom of the conduction band is similar to the LOMO.

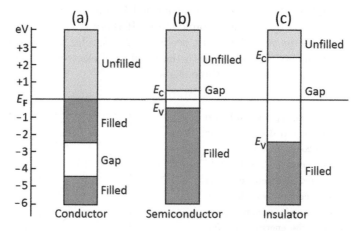

Figure 7.25 Conductor, semiconductor, and insulator. (a) For conductors, the highest occupied energy level is in the middle of an energy band. (b) For semiconductors, the highest occupied energy level matches the top of the valance band but the energy gap to the conduction band is small. (c) If the energy gap is big, the solid is an insulator.

7.4.4 Semiconductors

As shown in Fig. 7.24, the energy of a Bloch wavefunction depends on its wave vector $\mathbf{k}$. For some semiconductors, the top of the valence band and the bottom of the conduction band share the same wave vector $\mathbf{k}$. For other semiconductors, the wave vector of the top of the valence band is different from the wave vector of the bottom of the conduction band. It has a dramatic effect on the optical property of the semiconductor.

Figure 7.26 shows the band structures of the two types of semiconductors. In Fig. 7.26(a), the top of the valence band and the bottom of the conduction band are at the point $\mathbf{k} = 0$. It has a *direct band-gap*. In Fig. 7.26(b), the top of the valence band is at point $\mathbf{k} = 0$, but the bottom of the conduction band are at $\mathbf{k} = \mathbf{k}_0$. It has an *indirect band-gap*. The two types of band-gaps behave differently during the process of interacting with radiation. When a photon with energy greater than the energy gap reaches a piece of semiconductor with a direct gap, an electron near the top of the valence band can be easily lifted to the conduction band, leaves a *hole* in the valence band. However, for semiconductors with an indirect gap, because the wave vectors are different, it takes a process involving in the lattice vibration to mediate. The process is much weaker. On the other hand, an electron near the bottom of the conduction band can recombine with a hole near the top of the valence band to emit a photon. Such a process is much easier for semiconductors with a direct band-gap, and much weaker for semiconductors with an indirect band-gap.

Figure 7.27 shows the band-gaps and the types of a number of important semiconductors. A red bar indicates a direct band-gap. A blue bar indicates an indirect gap. The bottom scale shows the wavelengths of light corresponding to the band-gap energy. The most important semiconductors, silicon and germanium, are indirect semiconduc-

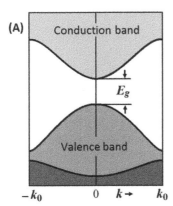
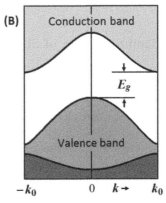

Figure 7.26 Direct semiconductors and indirect semiconductors. (a) For some semiconductors, the top of the valence band and the bottom of the conduction band have the same wave vector $\mathbf{k}$. Such semiconductor is called *direct*. (b) For some other semiconductors, the top of the valence band and the bottom of the conduction band have different wavevectors. To make a transition with a photon, it takes another step through the lattice vibration. The combined process makes the transition weak. Such semiconductor is called *indirect*.

Figure 7.27 Band gaps of a number of semiconductors. In the Figure, the energy quantum of radiation corresponding to the energy gap is shown in color (top) and wavelength (bottom). Most of the semiconductors used in solar cells have an energy gap corresponding to the photons of near-infrared light.

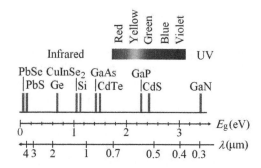

tors. From Figs. 7.23 and 7.24, we can understand why they have indirect band gaps. The atomic wavefunctions of silicon and germanium are basically p-type. Therefore, at the $\mathbf{k} = 0$ point of the reciprocal space, for both the valence band and the conduction band, the energy is the highest, see Fig. 7.24. Silicon is the most important semiconductor for digital electronics and solar cells. Nevertheless, silicon has an indirect gap, resulting in a smaller absorption coefficient. A thicker wafer is required. Nevertheless, the perfect fit of the bandgap with the spectrum of solar radiation and the abundance of resources makes silicon the best choice.

Many compound semiconductors have a direct band-gap, make them suitable for light-related applications. In early years, the most important materials were GaAs and GaP, where the band-gaps are in visible-light range. The early light-emitting diodes generate red light. Currently, the focus is on GaN, which enables the emission of blue light. By mixing blue light with green light and red light, white light can be produced. The LED-based white light generators have replaced incandescent light and fluorescent light because of much higher efficiency and much longer life.

7.4.5   The Band Structure of Silicon

Crystalline silicon is the most important semiconductor. Almost all transistors, integrated circuits, and solar cells are made from silicon. As a semiconductor of a single element, it is also one of the simplest. The electronic configuration of the Si atom is $1s^2\, 2s^2\, 2p^6\, 3s^2\, 3p^2$. The core electrons, $1s$, $2s$, and $2p$, are at deep energy levels and well confined in a sphere of one Bohr radius. They are inactive. The two $3s$ orbitals and the two $3p$ orbitals have similar radius and similar energy level. The linear combinations of those orbitals could make four sp^3 hybridized orbitals, see Fig. 7.12.

The process of energy band formation in crystalline silicon is shown in Fig. 7.28. When the distances of the silicon atoms are large, the outer shell of each free silicon atom has two $3s$ electrons and two $3p$ electrons. When the distance between silicon atoms becomes smaller, the four outer electrons become four tetrahedral hybrid orbitals. Each silicon atom is bonded with four nearest-neighbor silicon atoms to form four strong covalent bonds. Those bonded pairs of electrons make up the valence band. The four antibonding orbitals are unoccupied, and form the conduction band.

Figure 7.28 Formation of energy bands in crystalline silicon. The origin of the energy bands in crystalline silicon is the $3s^2$ and $3p^2$ atomic orbitals. After an sp^3 hybridization, eight orbitals are formed. The four lower tetrahedral orbitals make the valence band. For a crystal of N silicon atoms, there are $4N$ states filled by $4N$ electrons. The four higher tetrahedral orbitals make the conduction band with no electronic states filled.

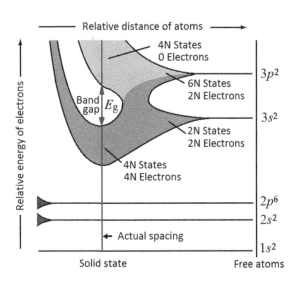

7.5 The Dynamic Schrödinger Equation

In the sixth 1926 paper, *Quantization as an Eigenvalue Problem, Part IV*, Schrödinger introduced his dynamic wave equation. The external electromagnetic wave is presented as a classical field. A good understanding of the interaction of electromagnetic wave and electron system can be achieved.

7.5.1 A Heuristic Derivation

Similar to the static Schrödinger equation, the starting point is the classical energy integral, the de Broglie's postulate, and the Planck–Einstein relation. For a one-electron system in three-dimensional space, the classical energy integral is

$$\frac{\mathbf{p}^2}{2m_e} + V = E, \tag{7.154}$$

where V is the potential energy function. According to de Broglie, electron is a wave. In a small region of space, a typical wavefunction is

$$\Psi = C \sin(\mathbf{k} \cdot \mathbf{r} - \omega t + \phi). \tag{7.155}$$

The wave vector $\mathbf{k}$ can be obtained from Eq. 7.155,

$$\mathbf{k}^2 = -\frac{\nabla^2 \Psi}{\Psi}. \tag{7.156}$$

According to de Broglie, the momentum $\mathbf{p}$ of an electron is associated with the wave vector $\mathbf{k}$ as

$$\mathbf{p} = \hbar\mathbf{k}. \tag{7.157}$$

Equation 7.156 now becomes

$$\mathbf{p}^2 = -\hbar^2 \frac{\nabla^2 \Psi}{\Psi}. \tag{7.158}$$

Insert Eq. 7.158 into Eq. 7.154, multiply both sides by Ψ, one obtains the static wave equation

$$E\Psi = \left[-\frac{\hbar^2}{2m_e} \nabla^2 + V \right] \Psi. \tag{7.159}$$

Applying Eq. 7.159 on itself, it becomes

$$E^2\Psi = \left[-\frac{\hbar^2}{2m_e} \nabla^2 + V \right]^2 \Psi. \tag{7.160}$$

From Eq. 7.155, the circular frequency is

$$\omega^2 = -\frac{1}{\Psi} \frac{\partial^2 \Psi}{\partial t^2}. \tag{7.161}$$

The energy E is associated with ω through the Planck–Einstein relation

$$E = \hbar\omega. \tag{7.162}$$

Equation 7.161 becomes

$$E^2\Psi = -\hbar^2 \frac{\partial^2 \Psi}{\partial t^2}. \tag{7.163}$$

Combining Eqs. 7.160 and 7.163, one obtains

$$\hbar^2 \frac{\partial^2 \Psi}{\partial t^2} = -\left[-\frac{\hbar^2}{2m_e} \nabla^2 + V \right]^2 \Psi. \tag{7.164}$$

According to Schrödinger, Eq. 7.164 is the prototypical dynamic wave equation. Schrödinger made a highlighted comment in his sixth 1926 paper that it is *the uniform and general wave equation for the field scalar ψ*. However, it includes fourth-order differentiations, which is inconvenient. To reduce the order of differentiation, Schrödinger's proposed two options. First, to split up Eq. 7.164 into a pair of real differential equations. Second, to make the wavefunction complex, with a real part and an imaginary part. The two approaches are mathematically equivalent.

By introducing a pair of real wavefunctions Ψ_1 and Ψ_2, Eq. 7.164 is then split up into a pair of second-order real differential equations,

$$\hbar \frac{\partial \Psi_1}{\partial t} = \left[-\frac{\hbar^2}{2m_e} \nabla^2 + V \right] \Psi_2, \tag{7.165}$$

and

$$\hbar \frac{\partial \Psi_2}{\partial t} = -\left[-\frac{\hbar^2}{2m_e} \nabla^2 + V \right] \Psi_1. \tag{7.166}$$

Equations 7.165 and 7.166 are the dynamic Schrödinger equations for a pair of real time-dependent wavefunctions, Ψ_1 and Ψ_2.

Regarding the validity of such arguments, Schrödinger made the following explanations. Strictly speaking, the energy integral Eq. 7.154 is valid only for a time-independent potential function V. Now, using Eq. 7.163, the energy constant E is eliminated. Although Eqs. 7.165 and 7.166 are initially derived for a time-independent potential function V, those equations now do not contain the constant energy E thus could be valid even if the potential function V is time dependent. Eventually, the general validity of Eqs. 7.165 and 7.166 can only be verified by comparing its consequences with experiments. Actually this has always been true up to today.

The above differential equations are invariant under a linear transformation with a phase θ, thus the two wavefunctions are identical twins,

$$\begin{pmatrix} \Psi_1 \\ \Psi_2 \end{pmatrix} \implies \begin{pmatrix} \cos\theta & -\sin\theta \\ \sin\theta & \cos\theta \end{pmatrix} \begin{pmatrix} \Psi_1 \\ \Psi_2 \end{pmatrix}. \tag{7.167}$$

In the sixth 1926 Schrödinger paper, another solution to the Eq. 7.164 was proposed. By taking a square root of Eq. 7.164, a pair of outcomes appear,

$$\pm i\hbar\frac{\partial\Psi}{\partial t} = \left[-\frac{\hbar^2}{2m_e}\nabla^2 + V\right]\Psi. \tag{7.168}$$

The $\pm$ sign before the imaginary unit i means that there is no difference between $+i$ and $-i$. From a mathematical point of view, positive $+i$ and negative $-i$ are equivalent. As a historical fact, the positive sign becomes the convention of physicists. In terms of a Hamiltonian operator, Eq. 7.39, the dynamic Schrödinger equation is

$$i\hbar\dot{\Psi} = \hat{H}\Psi. \tag{7.169}$$

Because a complex number has a real component and an imaginary component, from a physics point of view, there is no difference with the real wavefunctions and the real Schrödinger equations presented in Eqs. 7.165 and 7.166. By letting

$$\Psi = \Psi_1 + i\Psi_2, \tag{7.170}$$

Equation 7.169 splits into a pair of real equations, Eqs. 7.165 and 7.166. Using the complex conjugate of Ψ,

$$\Psi^* = \Psi_1 - i\Psi_2, \tag{7.171}$$

it is easy to show that the charge density is

$$\rho(\mathbf{r}) = -e\Psi^*\Psi, \tag{7.172}$$

the normalization condition is

$$\int \Psi^*\Psi = 1; \tag{7.173}$$

and the current density is

$$\mathbf{j} = \frac{e\hbar}{2m_e i} \left[\Psi^* \nabla \Psi - \Psi \nabla \Psi^* \right]. \tag{7.174}$$

By multiplying the complex wavefunction with an arbitrary complex factor $e^{i\theta}$, all observable quantities remain unchanged. In terms of group theory, the complex Ψn is invariant under group SU(1), which is equivalent to SO(2) for the real version.

7.5.2 Reduction to Static Schrödinger's Equation

If the potential V is time independent, the dynamic Schrödinger's equation can be reduced to a static Schrödinger's equation.

By expressing the time-dependent wavefunction as a product of a spatial wavefunction and a time-dependent factor $f(t)$,

$$\Psi(\mathbf{r}, t) = \psi(\mathbf{r}) \, f(t), \tag{7.175}$$

and insert to Eq. 7.169,

$$i\hbar \frac{\partial(\psi(\mathbf{r}) \, f(t))}{\partial t} = \hat{H} \psi(\mathbf{r}) \, f(t), \tag{7.176}$$

we obtain

$$\frac{i\hbar}{f(t)} \frac{\partial f(t)}{\partial t} = \frac{1}{\psi(\mathbf{r})} \hat{H} \psi(\mathbf{r}). \tag{7.177}$$

The left-hand side only depends on t, and the right-hand side only depends on $\mathbf{r}$. It must be a constant. By denoting it as E, we have

$$i\hbar \frac{df(t)}{dt} = E \, f(t) \tag{7.178}$$

and the wavefunction of the coordinates only, $\psi(\mathbf{r})$, satisfies the time-independent Schrödinger's equation,

$$\hat{H} \psi(\mathbf{r}) = E \, \psi(\mathbf{r}). \tag{7.179}$$

7.5.3 Meaning of the Time-Dependent Phase Factor

The solution of the time-dependent equation, Eq. 7.178, is

$$f(t) = \exp\left(-\frac{iEt}{\hbar} \right). \tag{7.180}$$

Because the wavefunction is normalized, the integration constant is 1.

The exponential factor does not mean the electron is in constant motion. First of all, as also in classical mechanics, energy is a result of integration of the Newtonian equation of motion. The absolute value of energy contains the integration constant, which is arbitrary. Only the difference of energies of two different states has physical significance, see Section 7.5.4. Second, the observable physical quantities, such as the charge distribution of the electron in a static state, is expressed by Eq. 7.173, where the complex phase factors cancels each other to become

$$\rho(\mathbf{r}) = -e\Psi^*\Psi = -e\psi^*(\mathbf{r})\psi(\mathbf{r}). \tag{7.181}$$

In quantum field theory, the situation becomes even more explicit. The interaction picture is applied exclusively in quantum field theory, that all individual states of the atomic system are treated as time independent. Only during the interaction with other fields, the energy difference between two atomic states becomes explicit.

7.5.4 Interaction with Radiation

In this Section, the problem of the interaction of an atomic system with radiation is treated using the dynamic Schrödinger's equations, Eqs. 7.165 and 7.166. According to the Wigner theorem, for a system with time-reversal symmetry, all static wavefunctions can be real. For notational brevity, the Dirac bra and ket scheme is applied. For real wavefunctions in coordinate space, such as in Eq. 7.71, a ket and a bra are identical. A matrix element can be represented succinctly as a Dirac bracket.

Denoting the energy operator of an atomic system as

$$\hat{H}_0 = -\frac{\hbar^2}{2m_e}\nabla^2 + V. \tag{7.182}$$

At $t < 0$, there is no external disturbance. The spatial wavefunctions of the atomic system are eigenstates of the energy operator $\hat{H}_0$,

$$\hat{H}_0 \left| n \right\rangle = E_n \left| n \right\rangle. \tag{7.183}$$

The time-dependent unperturbed wavefunctions are

$$\Psi_n = \left| n \right\rangle e^{-iE_n t/\hbar}. \tag{7.184}$$

After $t > 0$, the radiation is turned on. Typically, the wavelength is much larger than the atomic systm, radiation can be represented as an sinusoidal electric field

$$\mathbf{E} = \mathbf{E}_0 \cos \omega t. \tag{7.185}$$

The interaction can be represented in complex number using the Euler formula,

$$v = v_0 \left(e^{i\omega t} + e^{-i\omega t} \right). \tag{7.186}$$

The dynamic Schrödinger's equation for the system is

$$i\hbar \frac{\partial \Psi}{\partial t} = \left[\hat{H}_0 + v \right] \Psi. \tag{7.187}$$

Assuming that at $t < 0$, the system is in an initial state $|i\rangle$. After the radiation perturbation v is turned on, the wavefunction starts to transmit into other states with coefficients $c_n(t)$,

$$\Psi = |i\rangle e^{-iE_i t/\hbar} + \sum_{n \neq i} c_n(t) |n\rangle e^{-iE_n t/\hbar}. \tag{7.188}$$

Insert Eq. 7.188 into Eqs. 7.187, eliminating equal terms in both sides, writing the complex exponent term in Eq. 7.186 separately, the significant terms are

$$i\hbar \sum_{n \neq i} \frac{dc_n(t)}{dt} |n\rangle e^{-iE_n t/\hbar} = v_0 |i\rangle e^{-iE_i t/\hbar \pm i\omega t}. \tag{7.189}$$

To look for the coefficients for a final state $|f\rangle$, multiplying both sides of Eq. 7.189 by $\langle f|$, we find

$$i\hbar \frac{dc_f(t)}{dt} e^{-iE_f t/\hbar} = \langle f|v_0|i\rangle e^{-iE_i t/\hbar \pm i\omega t}. \tag{7.190}$$

Multiplying Eq. 7.190 by $e^{-iE_f t/\hbar}$ we obtain

$$i\hbar \frac{dc_f(t)}{dt} = \langle f|v_0|i\rangle e^{-i(E_f - E_i)t/\hbar \pm i\omega t}. \tag{7.191}$$

The coefficient $c_f(t)$ can be obtained by integration with initial condition $c_f(0) = 0$,

$$c_f(t) = \langle f|v_0|i\rangle \frac{e^{(E_f - E_i \pm \hbar\omega)t/\hbar} - 1}{E_f - E_i \pm \hbar\omega}. \tag{7.192}$$

The probability of finding a final state $|f\rangle$ at time t is

$$p_{fi} \equiv |c_f(t)|^2 = \frac{2\pi}{\hbar} t |\langle f|v_0|i\rangle|^2 \left[\frac{\sin^2 \left[(E_f - E_i \pm \hbar\omega)(t/2\hbar) \right]}{\pi \left[E_f - E_i \pm \hbar\omega \right]^2 (t/2\hbar)} \right]. \tag{7.193}$$

Denote $E_f - E_i \pm \hbar\omega = u$ and $t/2\hbar = a$; the function in square brackets has a sharp peak near $u = 0$, as shown in Fig. 7.7. The area under it is 1:

$$\int_{-\infty}^{\infty} \frac{\sin^2 au}{\pi a u^2} du = 1. \tag{7.194}$$

As $a \to \infty$, it approaches a delta function,

$$\lim_{a \to \infty} \frac{\sin^2 au}{\pi a u^2} = \delta(u). \tag{7.195}$$

Therefore, when $t \to \infty$, the function in square brackets in Eq. 7.193 approaches a delta-function $\delta(E_f - E_i \pm \hbar\omega)$. Also, from Eq. 7.193, the probability of $|f\rangle$ is proportional to t, and therefore, the transition rate is

$$R_{fi} \equiv \frac{p_{fi}}{t} = \frac{2\pi}{\hbar} |\langle f|v_0|i\rangle|^2 \delta(E_f - E_i \pm \hbar\omega). \tag{7.196}$$

Figure 7.29 Condition of energy conservation. The integrand in Eq. 7.195 has a sharp peak near $u = 0$. The area below the curve is 1. If time t is long, the value of the function near $u = 0$ approaches infinity, and the curve approaches a delta function. In this case, the *condition of energy conservation* is valid, represented by the delta function in Eq. 7.196.

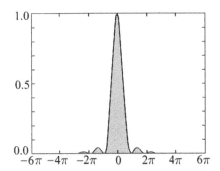

Equation 7.196 is the Golden Rule for atomic systems with discrete energy levels. The Bohr frequency condition for both absorption and stimulated emission,

$$\hbar\omega = \pm(E_f - E_i), \tag{7.197}$$

comes naturally. The radiation field can only transfer an energy quantum of $\hbar\omega$ to the atomic system, which is the essence of Einstein's theory of photons and provides an explanation of the line spectra, especially the Ritz combination principle.

The transition probabilities thus obtained are equivalent to Einstein's absorption coefficnent B_{12} and the stimulated emission coefficnent B_{21}. The sponteneous emission coefficnet is not available from the semiclassiocal treatment of the interaction of radiation with atomic systems. Using Einstein's elementary theory of radiation, presented in Section 2.5, the spontaneous emission coefficient A can be inferred from the absorption coeffieicnt using Eq. (2.159),

$$A = \frac{8\pi h\nu^3}{c^3}B = \frac{2\hbar\omega^3}{\pi c^3}B. \tag{7.198}$$

The interaction energy of radiation and the atomic system is usually through a dipole interaction. The interaction matrix element is

$$\langle f|v|i \rangle = -e\mathbf{E} \cdot \langle f|\mathbf{r}|i \rangle. \tag{7.199}$$

The measurement of polarization of radiation is mostly based on such a mechanism. Note that polarization is not a property of the non-existing point-like particle as a misconception of the term photon.

Problems

7.1. A one-dimensional harmonic oscillator in Newtonian mechanics is described by

$$m\frac{d^2x}{dt^2} = -kx, \tag{7.200}$$

where m is the mass, and k is the spring constant. Using the energy integral, find the solution as a sinusoidal function, and find the frequency.

7.2. In a two-dimensional circular potential well of radius R, solve the Schrödinger's equaiton and give the wavefunctions and energy levels.

7.3. Prove the following identities:

$$\begin{aligned} (\hat{\alpha}^\dagger)^\dagger &= \hat{\alpha}, \\ (a\hat{\alpha} + b\hat{\beta})^\dagger &= a\hat{\alpha}^\dagger + b\hat{\beta}^\dagger, \\ (\hat{\alpha}\hat{\beta})^\dagger &= \hat{\beta}^\dagger\hat{\alpha}^\dagger. \end{aligned} \tag{7.201}$$

7.4. Prove that the Hamiltonian, or energy operator, Eq. 7.39, is self-adjoint to each other.

7.5. Prove that the creation operator and the annihilation operator, Eqs. 7.46 and 7.47, are adjoint to each other.

7.6. Prove the form of Schrödinger's equation Eq. 7.52 in term of creation operator and the annihilation operator, Eqs. 7.46 and 7.47.

7.7. Prove that the three hydrogen-atom radial wavefunctions in Eq. 7.110 satisfy the differential equation Eq. 7.109.

7.8. Resolve the Schrödinger's equation outside a spherical potential well.

7.9. By defining a complex field intensity

$$\mathbf{F} = \mathbf{E} + ic\mathbf{B}, \tag{7.202}$$

in vacuum, we have

$$\epsilon_0\mu_0 = \frac{1}{c^2}. \tag{7.203}$$

Prove that the Maxwell's equaitons in vacuum can be written as a single complex equation similar to the complex Schrödinger's equation,

$$\nabla\times\mathbf{F} = \frac{i}{c}\frac{\partial\mathbf{F}}{\partial t}. \tag{7.204}$$

7.10. In terms of the complex field $\mathbf{F}$, prove that the energy density is

$$W = \frac{\epsilon_0}{2}\left|\mathbf{F}\right|^2. \tag{7.205}$$

7.11. Using the two-component time-dependent wavefunction, prove the interaction of radiation with atomic systems.

Chapter 8

pn-Junctions

To date, most solar cells are made of semiconductors. A semiconductor is characterized by a relatively narrow energy gap, typically a fraction of an electron volt to a few electron volts. In this chapter, we present the basic physics of semiconductors especially the pn-junction, primarily for the understanding of semiconductor solar cells. The theory of pn-junction is also the basis of another important application of semiconductor, the light-emitting diodes (LED).

8.1 Semiconductors

In Section 7.4.4, we presented an introduction on semiconductors. For pure semiconductors, at a low temperature, the conductivity is low. The conductivity increases with temperature. By doping semiconductor with impurities, free electrons can be generated in the conduction band to make n-type semiconductors, and holes can be left over in the valence band to make p-type semiconductors. In both cases, the conductivity is increased. Furthermore, by illuminating the semiconductor with radiation or passing a electric current to a semiconductor device, an electron can be excited from the *valence band* to the *conduction band* and form an *electron–hole pair*. The electron–hole pair stores a substantial portion of the applied energy. The electron–hole pairs can either recombine to emit a photon as a *radiative recombination*, which is the principle of LEDs; or dissipate the energy to the lattice vibration as a *non-radiative recombination*. By separating the electrons and holes through a *pn-junction*, electric current and power can be generated. This is the principle of solar cells.

8.1.1 Electrons and Holes

At low temperature, pure semiconductors have almost no mobile electrons, and the conductivity is low. By raising the temperature, electrons in the valence band can be excited to the conduction band, and the holes leftover in the valence band can also conduct electrical current; see Fig. 8.1. Therefore, pure semiconductor has an important property: the higher the temperature, the higher the conductivity.

According to Fermi–Dirac statistics (see Appendix F), at temperature T, the con-

189

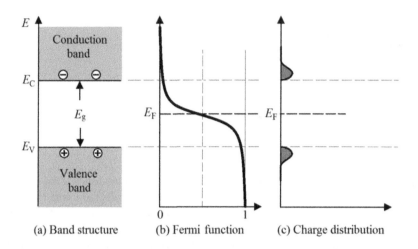

(a) Band structure   (b) Fermi function   (c) Charge distribution

Figure 8.1 Intrinsic semiconductors: Free electrons and holes. Thermal excitation could raise electrons from the valence band to the conduction band, to make electron–hole pairs. Charge neutrality requires that the concentration of electrons equals the concentration of holes.

centration of electrons n_0 at the bottom of the conduction band is

$$n_0 = N_c\, f(E_c), \tag{8.1}$$

where N_c is the effective density of states of the conduction band, a quantity determined by the property of the semiconductor, and $f(E_c)$ is the Fermi function (Eq. F.19 in Appendix F), the distribution function of electrons at absolute temperature T. At room temperature, $k_\mathrm{B}T \approx 0.026$ eV, the value of $E_c - E_F$ is of the order of 1 eV, the factor 1 in the Fermi function can be neglected. We have

$$f(E_c) = \frac{1}{e^{(E_c - E_F)/k_\mathrm{B}T} + 1} \approx e^{-(E_c - E_F)/k_\mathrm{B}T}. \tag{8.2}$$

Therefore, the concentration of electrons in the conduction band is

$$n_0 = N_c\, e^{-(E_c - E_F)/k_\mathrm{B}T}. \tag{8.3}$$

In the valance band, there is a deficiency of electrons from the full occupation. The deficiency of electrons in the valence band forms mobile carriers of positive elementary charge, the *holes*. Similarly, the concentration of holes, p_0, is given as

$$p_0 = N_v[1 - f(E_v)] \approx N_v\, e^{-(E_F - E_v)/k_\mathrm{B}T}, \tag{8.4}$$

where N_v is the effective density of states in the valence band and E_v is the energy level of the top of the valence band.

It is an interesting and important fact that the product $n_0 p_0$ is *independent of the Fermi level.* Actually, by combining Eqs. 8.3 and 8.4, we obtain

$$
\begin{aligned}
n_0 p_0 &= \left[N_c\, e^{-(E_c - E_F)/k_{\mathrm{B}}T} \right] \left[N_v\, e^{-(E_F - E_v)/k_{\mathrm{B}}T} \right] \\
&= N_c\, N_v\, e^{-(E_c - E_v)/k_{\mathrm{B}}T} = N_c\, N_v\, e^{-E_g/k_{\mathrm{B}}T}.
\end{aligned}
\tag{8.5}
$$

For intrinsic semiconductors, or semiconductors without impurities, charge neutrality requires that $n_0 = p_0$. An *intrinsic carrier concentration* n_i can be defined,

$$
n_i = \sqrt{N_c\, N_v}\, e^{-E_g/2k_{\mathrm{B}}T},
\tag{8.6}
$$

with the general property

$$
n_0 p_0 = n_i^2.
\tag{8.7}
$$

Obviously, the equation is valid regardless of the position of the Fermi level.

8.1.2 p-Type and n-Type Semiconductors

Semiconductors have another important property: the conductivity depends on the concentration of *impurities*. According to the position of the energy level of the atoms in the band gap of a semiconductor, there are two types of impurities.

The energy level of *donor* atoms is a few meV below the bottom of the conduction band. An impurity atom can be ionized to contribute an electron to the conduction band. For silicon and germanium, atoms from group V of the periodic table (N, P,

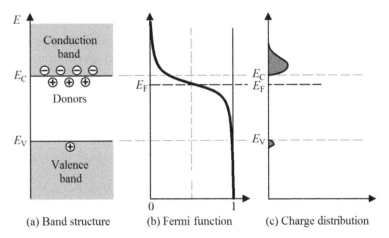

(a) Band structure (b) Fermi function (c) Charge distribution

Figure 8.2 The n-type semiconductor. The donor atoms release electrons into the conduction band. The Fermi level moves to the bottom of the conduction band. The concentration of free electrons approximately equals the concentration of donor atoms.

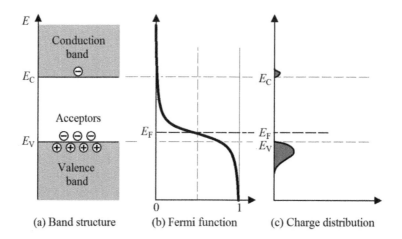

(a) Band structure   (b) Fermi function   (c) Charge distribution

Figure 8.3 The *p*-type semiconductor. The acceptor atoms grab electrons from the valence band to create holes. The Fermi level shifts toward the valence band. The concentration of holes approximately equals the concentration of acceptor atoms.

As and Sb) are effective donors. The Fermi level moves to near the bottom of the conduction band, as shown in Fig. 8.2. If the temperature is not too low, all donor atoms are ionized. The concentration of free electrons in an *n*-type semiconductor, n_n, approximately equals the concentration of donor atoms N_D,

$$n_n \cong N_D. \tag{8.8}$$

On the other hand, the energy level of *acceptor* atoms is a few meV above the top of the valence band. An electron in the valence band can be trapped by the acceptor atoms and leave a mobile hole. For silicon and germanium, atoms from Group IIIA (B, Al, Ga, and In) are effective acceptors. The Fermi level moves to near the top of the valence band, as shown in Fig. 8.3. If the temperature is not too low, all acceptor atoms become negative ions. The concentration of holes in a *p*-type semiconductor, p_p, approximately equals the acceptor concentration N_A,

$$p_p \cong N_A. \tag{8.9}$$

In both cases, the product of the concentrations of free electrons and holes equals the square of the intrinsic carrier concentration,

$$n_n \, p_n = p_p \, n_p = n_i^2, \tag{8.10}$$

where p_n is the hole concentration in an *n*-type semiconductor and n_p is the free-electron concentration in a *p*-type semiconductor. Each is a *minority-carrier concentration*.

Silicon is the most important semiconductor for both electronic devices and solar cells. Here by some more specific information about silicon is provided. Figure 8.4

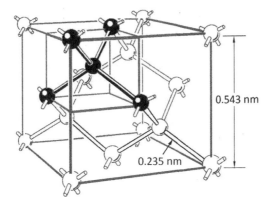

Figure 8.4 Unit cell in crystalline silicon. The crystallographic structure of silicon is the diamond cubic structure with a lattice parameter of 0.543 nm. The nearest neighbor distance is 0.235 nm. Each silicon atom has four nearest neighbor silicon atoms, forming four sp^3 hybridized orbitals. Counting spin, there are eight states. The four lower states are occupied, forming the valence band. The four higher states are unoccupied, forming the conduction band.

shows how silicon atoms come together to form crystalline silicon. Each silicon atom has four nearest neighbors, each neighbor silicon atom is bonded with an hybrid $sp3$ wavefunction to form a tetrahedral structure, see Section 7.1.7. Those tetrahedral bond electrons become the valence band. The antibonding wavefunctions from the hybrid $sp3$ wavefunctions form the conduction band, which is unoccupied.

Schematically, the role of the donors and acceptors is shown in Fig. 8.5. Each silicon atom forms four covalent bonds with four neighboring silicon atoms. For an impurity of a group V atom of the periodic table (N, P, As and Sb), there is an extra electron on the outer shell. It forms a free electron in the crystal. For a group IIIA atom (B, Al, Ga, and In), there are only three electron in the outer shall. To form a covalent bond, one electron is taken from the crystal to create a hole.

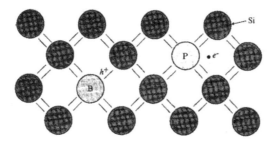

Figure 8.5 Roles of doners and acceptors in silicon. Different types of impurity atoms play different roles. A group V atom (N, P, As, and Sb), has an extra electron on the outer shell to become a free electron. A group IIIA atom (B, Al, Ga, and In) has only three electrons in the outer shall. To form a covalent bond, one electron is taken from the crystal to create a hole.

8.2 Formation of a *pn*-Junction

When a *p*-type semiconductor and an *n*-type semiconductor are brought together, a *built-in potential* is established. Because the Fermi level of a *p*-type semiconductor is close to the top of the valence band and the Fermi-level of an *n*-type semiconductor is close to the bottom of the conduction band, there is a difference between the Fermi levels of the two sides. When the two pieces are combined to form a single system, the Fermi levels must be aligned. As a result, the energy levels of the two sides undergo a shift with a *diffusion potential* V_0. Letting E_{cp} be the energy level of the bottom of the conduction band of the *p*-type semiconductor, and E_{cn} be the bottom of the conduction band of the *n*-type semiconductor, the built-in potential is

$$eV_0 = E_{cp} - E_{cn}; \tag{8.11}$$

see Fig. 8.6. The establishment of a built-in potential in a *pn*-junction can be understood from another point of view, the flow of carriers. Because in the *n*-region the hole concentration is almost zero, the holes diffuse from the *p*-region to the *n*-region. After a number of holes move to the *n*-region, an electrical field is formed to drive the holes back to the *p*-region. At equilibrium, the net current density $J_p(x)$ must be zero,

$$J_p(x) = e \left[\mu_p p(x) E_x(x) - D_p \frac{dp(x)}{dx} \right] = 0, \tag{8.12}$$

where μ_p is the mobility of the holes, $p(x)$ is the concentration of holes as a function of x, $E_x(x)$ is the x-component of electric field intensity as a function of x, and D_p is the diffusion coefficient of the holes. Using Einstein's relation,

$$\frac{D_p}{\mu_p} = \frac{k_B T}{e}, \tag{8.13}$$

and the relation between the potential $V(x)$ and electric field intensity, $E_x(x) = -dV(x)/dx$, Eq. 8.12 becomes

$$-\frac{e}{k_B T} \frac{dV(x)}{dx} = \frac{1}{p(x)} \frac{dp(x)}{dx}. \tag{8.14}$$

Integrating Eq. 8.14 over the entire transition region yields

$$-\frac{e}{k_B T} (V_n - V_p) = \ln \frac{p_n}{p_p}. \tag{8.15}$$

Because $V_n - V_p = V_0$, Eq. 8.15 can be rewritten as

$$p_n = p_p \exp \left(\frac{-eV_0}{k_B T} \right). \tag{8.16}$$

Similarly, because in the *p*-region the free-electron concentration is almost zero, the free electrons diffuse from the *n*-region to the *p*-region. As a result, an electrical field is formed to drive the free electrons back to the *n*-region. At equilibrium, the net current of free electrons must be zero. A similar equation is found,

$$n_p = n_n \exp\left(\frac{-eV_0}{k_\mathrm{B}T}\right). \tag{8.17}$$

By applying Eqs. 8.8–8.10 to Eqs. 8.16 and 8.17, we find an approximate expression of the diffusion potential V_0,

$$V_0 = \frac{k_\mathrm{B}T}{e}\ln\frac{N_D\,N_A}{n_i^2}. \tag{8.18}$$

Equations 8.16–8.18 are essential to the derivation of the Shockley diode equation and to understand the current-voltage behavior of the *pn*-junction.

Here is an order-of-magnitude estimate of the two equations. A typical value of the built-in potential is $V_0 \approx 0.75$ eV. At room temperature, $k_\mathrm{B}T \approx 0.026$ eV. The factor $\exp(-0.75/0.026) \approx 10^{-12.5}$. Therefore, the absolute values are very small. For obvious reasons, both p_n and n_p are called *minority carriers*.

To better understand the *pn*-junction, we need to establish a mathematical model for the space charge and the potential curve. A very effective and fairly accurate model is based on the *depletion approximation*; see Fig. 8.7. Under such an approximation, in the *p*-region near the junction boundary there is a layer of thickness x_p where all

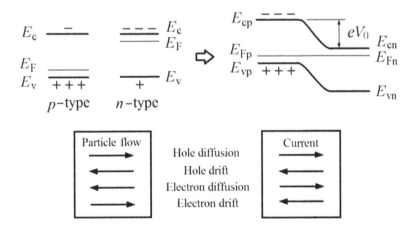

Figure 8.6 Formation of a *pn*-junction. By bringing a piece of *p*-type semiconductor and a piece of *n*-type semiconductor together to form a junction, the Fermi levels must be aligned. This would happen naturally as follows: The holes in the *p*-side diffuse to the *n*-side, and the free electrons in the *n*-side diffuse to the *p*-side, to form a double charged layer, until an equilibrium is established. The dynamic equilibrium is established by a balance of *drift current* driven by the electrical field, and a *diffusion current* driven by the gradient of carrier density.

the holes are removed and the charge density ρ_p is determined by the density of the acceptors N_A, which are negatively charged,

$$\rho_p = -eN_A. \tag{8.19}$$

The electric field intensity in this region is given by the Gauss law,

$$\frac{dE_x}{dx} = -\frac{e}{\epsilon}N_A. \tag{8.20}$$

where ϵ is the dielectric constant of the semiconductor and is the product of the dielectric constant of a vacuum, ϵ_0, and the relative dielectric constant ϵ_r of the semiconductor. The dielectric constant of the vacuum is $\epsilon_0 = 8.85 \times 10^{-14}$ F/cm. For example, for silicon, $\epsilon_r = 11.8$, $\epsilon \approx 1.04 \times 10^{-12}$ F/cm.

Similarly, there is a slab of thickness x_n where all the free electrons are removed, and the charge density ρ_n is determined by the density of the donors, N_D, which are positively charged,

$$\rho_n = eN_D. \tag{8.21}$$

the electric field intensity is

$$\frac{dE_x}{dx} = \frac{e}{\epsilon}N_D. \tag{8.22}$$

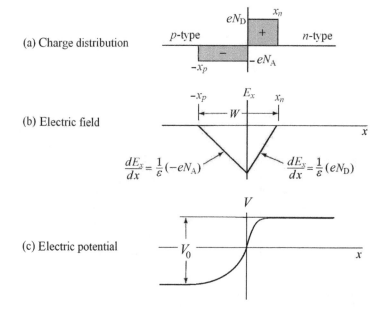

(a) Charge distribution

(b) Electric field

(c) Electric potential

Figure 8.7　The depletion model of pn-junction. Charge, field, and potential distributions in the depletion model of pn-junction. In the p-region near the junction boundary, all holes are depleted. The negatively charged acceptor ions form the space charge. In the n-region near the junction boundary, all electrons are depleted. The positively charged donor ions form the space charge.

The boundary conditions for Eqs 8.20 and 8.22 are as follows. First, the charge neutrality of the entire transition region requires that

$$N_A x_p = N_D x_n. \tag{8.23}$$

Second, outside the transition region, the electric field should be zero:

$$E_x = 0 \quad \text{for} \quad x \leq -x_p \quad \text{and} \quad x \geq x_n. \tag{8.24}$$

The solution s of Eqs 8.20 and 8.22 with boundary condition Eq. 8.24 are

$$E_x = -\frac{eN_A}{\epsilon}(x + x_p), \quad \text{for} \quad -x_p \leq x < 0;$$

$$E_x = \frac{eN_D}{\epsilon}(x - x_n), \quad \text{for} \quad 0 < x \leq x_n. \tag{8.25}$$

The potemtial is determined by the electrical field intensity as

$$\frac{d\phi}{dx} = -E_x. \tag{8.26}$$

The boundary conditions are

$$\phi = 0, \quad \text{at } x = -x_p,$$

$$\phi = V_0, \quad \text{at } x = x_n. \tag{8.27}$$

The solution of Eq. 8.26 with boundary conditions 8.27 is

$$V_0 = \frac{eN_A}{2\epsilon}x_p^2 + \frac{eN_D}{2\epsilon}x_n^2. \tag{8.28}$$

Using relation Eq. 8.23 and the definition of the width of the transition region W,

$$W = x_p + x_n, \tag{8.29}$$

the following relations are obtained:

$$x_p = \frac{N_D W}{N_A + N_D} \quad \text{and} \quad x_n = \frac{N_A W}{N_A + N_D}. \tag{8.30}$$

Equation 8.28 finally becomes

$$V_0 = \frac{e}{2\epsilon} \frac{N_A N_D}{N_A + N_D} W^2. \tag{8.31}$$

The width of the transition region as a function of V_0 is

$$W = \sqrt{\frac{2\epsilon V_0}{e}\left(\frac{1}{N_A} + \frac{1}{N_D}\right)}. \tag{8.32}$$

8.3 Analysis of *pn*-Junctions

As shown in Section 8.2, in the absence of external applied voltage, there is no current running through a *pn*-junction, because the diffusion current and the drift current cancel each other for both holes and free electrons. By applying an external voltage on a *pn*-junction, the equilibrium is broken and a net current is generated.

Qualitatively, the mechanism can be explained as follows; see Fig. 8.8. At equilibrium, as shown in Fig. 8.8(a), for both electrons and holes, there is a concentration gradient which gives rise to diffusion and an electrical field pointing to $-x$-direction

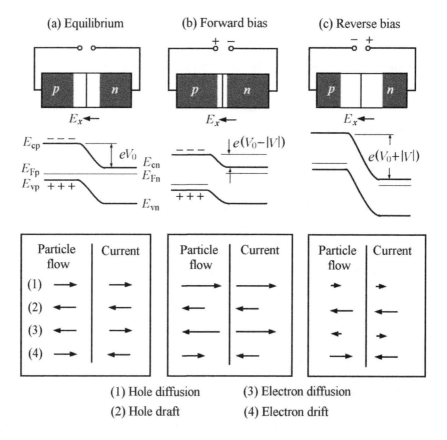

(1) Hole diffusion   (3) Electron diffusion

(2) Hole draft   (4) Electron drift

Figure 8.8 Effect of bias in a *pn*-junction. (a) At equilibrium, without external bias, the diffusion current and the drift current cancel each other. (b) A positive bias voltage pushes the holes to the *n*-side and the free electrons to the *p*-side. The potential barrier is reduced. Both diffusion currents of holes and free electrons are increased. The drift currents, depending on the available carriers, are unchanged. The net current is nonzero. (c) By applying a reversed bias, the holes are pushed further back into the *p*-region and the free electrons are pushed further back into the *n*-region. The diffusion current is reduced. Only the drift current persists.

which drives the carriers in the opposite direction. The net current is zero. By applying a positive bias voltage, namely, connecting the positive terminal of a battery to the *p*-side and the negative terminal to the *n*-side, as shown in Fig. 8.8(b), the external potential pushes the holes to the *n*-side and the free electrons to the *p*-side. The potential barrier is reduced. Diffusion currents of both holes and free electrons are increased. The drift current, depending on the available carriers, is unchanged. The net current is nonzero. On the other hand, by applying a reversed bias, as shown in Fig. 8.8(c), the holes are pushed further back into the *p*-region and the free electrons are pushed further back into the *n*-region. The diffusion current is further reduced. The drift currents are unchanged and become the dominant factor. Eventually, the current reaches a saturated value determined by the drift currents.

8.3.1 Effect of Bias Voltage

To account for the current quantitatively, we should further develop the concepts in Section 8.2. By applying a forward bias, as shown in Fig. 8.8(b), the potential difference across a *pn*-junction becomes $V_0 - V$. The electron concentration in the *p*-region, n_p, changes:

$$n_p \longrightarrow n_n \exp\left(\frac{-e(V_0 - V)}{k_B T}\right). \tag{8.33}$$

Comparing with Eq. 8.16, one finds an *excess free-electron concentration at the border of the neutral p-region,*

$$\delta n_p(x = 0) = n_p \left[\exp\left(\frac{eV}{k_B T}\right) - 1\right]. \tag{8.34}$$

Similarly, the external forward bias voltage V generates an *excess hole concentration at the border of the neutral n-region,*

$$\delta p_n(x = 0) = p_n \left[\exp\left(\frac{eV}{k_B T}\right) - 1\right]. \tag{8.35}$$

The excess carrier concentrations generate an excess diffusion current, which is the main part of the forward-bias current of a diode.

To obtain an explicit expression of the current as a function of the bias voltage, we notice first that even with a substantial forward bias, for example, 0.5 V, the excess minority carriers δp_n and δn_p are still much smaller than the majority carrier concentrations p_p and n_n. For example, using $V = 0.5$ V, from Eq. 8.33, $\delta p_n \approx \exp(-250/26) \times p_p \approx 10^{-5.8} \times p_p$. Therefore, the concentration of majority carriers can be treated as a constant even with a substantial bias voltage.

8.3.2 Lifetime of Excess Minority Carriers

Diffusion of excess minority carriers is the origin of junction current. However, there is a competing process which limits the junction current. The excess minority carriers are surrounded by a sea of majority carriers which are constantly courting for recombination. Because the concentration of majority carriers, p_p or n_n, is several orders of magnitude greater than the concentration of excess minority carriers, even with recombination, p_p or n_n is virtually a constant. The rate of decay of excess minority carriers is thus proportional to its concentration, which can be characterized by a *lifetime*. The combined effect of diffusion and lifetime of the excess minority carriers can be summarized in the following equations. For free electrons,

$$\frac{\partial \delta n_p(x,t)}{\partial t} = -\frac{\delta n_p(x,t)}{\tau_n} + D_n \frac{\partial^2 \delta n_p(x,t)}{\partial x^2}, \tag{8.36}$$

where D_n is the diffusion coefficient, and τ_n is the lifetime of free electrons. For holes,

$$\frac{\partial \delta p_n(x,t)}{\partial t} = -\frac{\delta p_n(x,t)}{\tau_p} + D_p \frac{\partial^2 \delta p_n(x,t)}{\partial x^2}, \tag{8.37}$$

where D_p is the diffusion coefficient and τ_p is the lifetime of holes.

8.3.3 Junction Current

At equilibrium, the concentration of carriers is independent of time. For example, Eq. 8.36 becomes

$$D_n \frac{d^2 \delta n_p(x)}{dx^2} = \frac{\delta n_p(x)}{\tau_n}. \tag{8.38}$$

It is equivalent to two first-order differential equations in the same sense that $x^2 = 4$ is equivalent to $x = 2$ and $x = -2$,

$$\frac{d \delta n_p(x)}{dx} = \frac{1}{\sqrt{\tau_n D_n}} \delta n_p(x) \tag{8.39}$$

$$\frac{d \delta n_p(x)}{dx} = -\frac{1}{\sqrt{\tau_n D_n}} \delta n_p(x), \tag{8.40}$$

which represent decays to the $+x$ and $-x$ directions, respectively. To account for junction current, for each side of the junction, only one of the two is needed. The diffusion current of electrons is

$$I_n = e D_n \frac{d \delta n_p(x)}{dx} = e \sqrt{\frac{D_n}{\tau_n}} \delta n_p(x). \tag{8.41}$$

At $x = 0$, using Eq. 8.34, the junction current of electrons is

$$I_n(x_p = 0) = -e \sqrt{\frac{D_n}{\tau_n}} n_p \left[\exp\left(\frac{eV}{k_B T}\right) - 1 \right]. \tag{8.42}$$

Similarly, for holes,

$$I_p(x_n = 0) = e \sqrt{\frac{D_p}{\tau_p}} p_n \left[\exp\left(\frac{eV}{k_B T}\right) - 1 \right]. \tag{8.43}$$

The total junction current is

$$I = e \left(\sqrt{\frac{D_n}{\tau_n}} \, n_p + \sqrt{\frac{D_p}{\tau_p}} \, p_n \right) \left[\exp\left(\frac{eV}{k_B T}\right) - 1 \right]. \tag{8.44}$$

Furthermore, using the approximate relations

$$p_n = \frac{n_i^2}{N_D}, \quad n_p = \frac{n_i^2}{N_A}, \tag{8.45}$$

Eq. 8.44 can be reduced to

$$I = e n_i^2 \left(\frac{1}{N_A} \sqrt{\frac{D_n}{\tau_n}} + \frac{1}{N_D} \sqrt{\frac{D_p}{\tau_p}} \right) \left[\exp\left(\frac{eV}{k_B T}\right) - 1 \right]. \tag{8.46}$$

8.3.4 Shockley Equation

Denoting a constant

$$I_0 \equiv e n_i^2 \left(\frac{1}{N_A} \sqrt{\frac{D_n}{\tau_n}} + \frac{1}{N_D} \sqrt{\frac{D_p}{\tau_p}} \right), \tag{8.47}$$

Eq. 8.46 is simplified to the well-known form of the *diode equation*, also known as the *Shockley equation*,

$$I = I_0 \left[\exp\left(\frac{eV}{k_B T}\right) - 1 \right]. \tag{8.48}$$

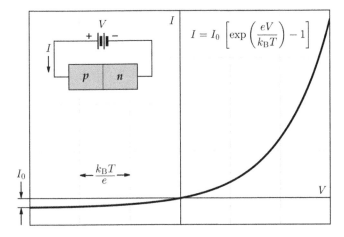

Figure 8.9 Current-voltage behavior of a *pn*-junction. With a forward bias, at room temperature, the current is growing exponentially about one order of magnitude per 60 mV. In the third quadrant, as the applied reverse voltage exceeds 100 mV, the current reaches a saturate value, which is dominated by the drift current.

Figure 8.9 shows the current–voltage behavior of a *pn*-junction. According to the diode equation, Eq. 8.48, in the first quadrant with a forward bias at room temperature, the current is growing exponentially, about one order of magnitude per 60 mV. In the third quadrant, as the applied voltage exceeds 100 mV the current reaches a saturated value, which is dominated by the drift current. By applying a large reversed bias voltage on the diode, the exponential term vanishes. The current I approaches $-I_0$. Therefore, the constant I_0 is the *reverse saturation current density*. The two sides are highly asymmetric, which makes the *pn*-junction the most widely used rectifier.

The quality of a rectifier critically depends on the magnitude of the reverse saturation current density I_0. As shown in Eq. 8.47, the decisive factor is the minority-carrier lifetimes τ_n and τ_p. The longer the lifetime, the smaller the reverse saturation current density, the better the quality of the rectifier. This fact is equally important for solar cells. As we will show in Chapter 9, the most significant limiting factor of solar cell efficiency is minority-carrier lifetime.

8.4 Light-Emitting Diodes for Illumination

Since mankind discovered fire in prehistory time, oil lamp was invented to conquer the darkness at night. Candle, a more stable light source, was invented in the Bronze age in many cultures in the world independently. In the 19th century, kerosene and paraffin wax were discovered as the energy source to replace the animal or vegetable sources. Nevertheless, the efficiency of generating light was less than 1%.

After the invention of light bulbs in late 19th century, incandescent light has been

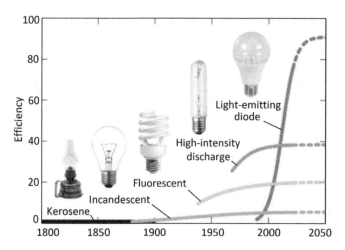

Figure 8.10 Evolution of the efficiency of light sources. The efficiency of kerosene lamp was less than 1%. For incandescent light, it is less than 5%. LED brought about the most dramatic improvement in efficiency of converting electrical energy to visible radiation. Currently, the efficiency reaches about 80%, and still have space for improvement. After a 2021 NASA Report [53].

the main means of illumination until the late 20th century. The efficiency is less than 5%. More than 95% of the electricity is wasted, see Fig. 8.10. In the 20th century, about a quarter of electricity was used for illumination. Because most electricity was generated by burning fossil fuel, illumination was a significant origin of pollution and global warming. The introduction of fluorescent light and high intensity discharge lamp only slightly improves efficiency. Because the inclusion of toxic mercury, the bulkiness and fragility of the devices, they have not been used extensively.

The invention and mass-production of light-emitting diodes (LED) as light source brought to a dramatic improvement, see Fig. 8.10. Currently, the efficiency reaches about 80%, more than one order of magnitude greater than the incandescent light. Furthermore, the device is rugged, small, and having a very long life.

8.4.1 Invention of the Blue LED

On October 7, 2014, the Royal Swedish Academy of Sciences announced that year's Nobel Prize in Physics was awarded to Isamu Akasaki, Hiroshi Amano, and Shuji Nakamura "for the invention of efficient blue light-emitting diodes which has enabled bright, long-lasting, and energy-saving white light sources". See Fig. 8.11. Although red and green light-emitting diodes have been available since the 1950s, without blue light, white light could not be generated. Despite considerable efforts in both scientific community and industry, the blue LED had remained a challenge for three decades.

In early 1990s, the three scientists succeeded in producing high-efficiency blue LEDs where everyone else had failed. Akasaki worked together with Amano at the University of Nagoya. Nakamura was employed at Nichia Corporation, a small company in Tokushima, then moved to University of California Santa Barbara (UCSB) in 1999 and became a United States citizen. Their inventions were revolutionary. Incandescent light bulbs lit the 20th century; the 21st century will be lit by LED lamps.

Figure 8.11   Nobelists for the invention of blue-light LED. Isamu Akasaki (left), Hiroshi Amano (middle), and Shuji Nakamura (right) were awarded the 2014 Nobel Prize in Physics for the invention of efficient blue light-emitting diodes which have enabled bright white light sources.

8.4.2 The Working Principle

An light-emitting diode (LED) is a *pn*-junction made from direct-gap semiconductors working in a forward-biased mode. An external power supply injects minority carriers into the region near the *pn*-junction. The minority carriers recombine with the majority carriers and emit light through *spontaneous emission of radiation.*

A formal theory of spontaneous emission of radiation requires quantum electrodynamics, where the electromagnetic waves are quantized. Within elementary quantum mechanics, the rate of absorption is treated through time-dependent perturbation theory, see Section 7.5.4. The rate of spontaneous emission of radiation can be inferred through Einstein's elementary theory of radiation in Section 2.5.

According to Einstein, the coefficient of spontaneous emission of radiation A is proportional to the absorption coefficient B,

$$A = \frac{8\pi h\nu^3}{c^3}B. \tag{8.49}$$

According to Eq. 7.88

$$B \propto |\langle f|\mathbf{r}|i\rangle|^2. \tag{8.50}$$

Therefore, A is also proportional to the square of $\langle f|\mathbf{r}|i\rangle$. One consequence of that statement is that direct semiconductors works much better than indirect semiconductors, as shown in Section 7.4.4 especially Fig. 7.26. In a semiconductor, both the initial state and the final state are Block states with a wave vector. The integral in Eq. 8.50 is a product of the initial Bloch wave and the final Bloch wave. If the initial state and the final state have different wave vectors, the integral is zero. Any transition must be mediated by lattice vibration. The intensity is much weaker.

Electrical excitation

An LED is a *pn*-junction operated under a forward-biased condition. In Eq. 8.46, the exponential term is much greater than 1. It can be simplified to

$$I = en_i^2\left(\frac{1}{N_A}\sqrt{\frac{D_n}{\tau_n}} + \frac{1}{N_D}\sqrt{\frac{D_p}{\tau_p}}\right)\exp\left(\frac{eV}{k_\mathrm{B}T}\right). \tag{8.51}$$

The term n_i^2 is the most significant factor for the current. Using Eq. 8.18,

$$n_i^2 = N_D N_A \exp\left(-\frac{eV_0}{k_\mathrm{B}T}\right) \tag{8.52}$$

Equation 8.51 can be further simplified to

$$I = e\left(N_D\sqrt{\frac{D_n}{\tau_n}} + N_A\sqrt{\frac{D_p}{\tau_p}}\right)\exp\left(\frac{e}{k_\mathrm{B}T}(V - V_0)\right). \tag{8.53}$$

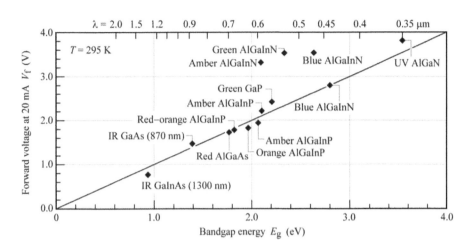

Figure 8.12 The relation between driving voltage and bandgap. Experimental results of the relation between driving voltage and bandgap. Except for some cases of extra voltage dur to parasitic serial resistance, the driving voltage is basically the value of the bandgap. After Schubert [97].

As shown in Fig. 8.6, the difference in diffusion voltage V_0 and the bandgap voltage E_g/e is only a few $k_B T$, but the difference of bandgaps of different semiconductors is much greater than $k_B T$, to a good approximation, we have

$$V_0 \cong \frac{E_g}{e}. \tag{8.54}$$

Equation 8.53 can be simplified to

$$I = I_0^* \exp\left(\frac{eV - E_g}{k_B T}\right), \tag{8.55}$$

where I_0^* is the factor before the exponential term that does not vary significantly for different semiconductors. Therefore, the voltage needed to generate a significant current is roughly the value of the semiconductor bandgap. That conclusion is well supported by experiments. Figure 8.12 shows the forward diode voltage at a diode current of 20 mA for LEDs in ultraviolet, visible, and infrared wavelengths. Most semiconductor diodes follow the rule in Eq. 8.54. For several semiconductors, the driving voltage is higher, mainly due to parasitic resistance in the device [97].

Internal quantum efficiency and external quantum efficiency

The observed relation between driving voltage and bandgap exhibits the law of energy conservation. The light quantum emitted by the LED is from the electron-hole pair with the energy of the bandgap. The electron-hole pair requires external energy to create. Each electron injected could emit one photon. Practically, there are also losses. To measure the efficiency quantitatively, two indices are defined.

Ideally, the active region of an LED emits one photon for every electron injected. the *internal quantum efficiency* (IQE) should be unity. For actual LEDs, that efficiency is always less than unity, and defined as

$$\eta_{\text{int}} = \frac{\text{\# of photons emitted from active region per second}}{\text{\# of electrons injected into LED per second}} = \frac{P_{\text{int}}/(h\nu)}{I/e}, \quad (8.56)$$

where η_{int} is IQE, P_{int} is the optical power emitted from the active region, and I is the injected current. Furthermore, not every photon emitted from the active region can reach the free space due to internal reflection and absorption from the packaging, which gives rise to a *light extraction efficiency* which is always smaller than unity. For evaluating the useful output, an *external quantum efficiency* (EQE) is defined as

$$\eta_{\text{ext}} = \frac{\text{\# of photons emitted into free space per second}}{\text{\# of electrons injected into LED per second}} = \frac{P/(h\nu)}{I/e}. \quad (8.57)$$

Obviously, EQE is the product of the IQE and the light extraction efficiency.

After decades of hard work, the value of EQE has advanced from less than 1% to above 50%. Further progress is continuously being achieved.

8.4.3 Wavelength Engineering

White light is a mixture of several pure-color lights, each has a well-defined wavelength. Typically, a mixture of three pure-color lights are needed to make white light: red, green, and blue. Since the 1950s, red-light LEDs and green-light LEDs already exist. Nevertheless, without blue-light LEDs, white light cannot be generated.

After the availability of blue-light LED, there are two ways to generate white light, see Fig. 8.13 [73]. The first method uses a blue-light LED with a *wavelength converter* to generate radiations of longer wavelengths by a *phosphor*. A typical phosphor is $Y_3Al_5O_{12}$:Ce, see Fig 8.14. It generates a spectrum of light including red and green, the combination of those pure lights looks yellow. The structure of such a white LED is shown in Fig 8.14(a). The figure shows a blue-light emitting LED and the YAG phosphor surrounding the blue LED. The YAG phosphor is made as a powder, suspended in epoxy resin surrounding the blue LED. The emission spectrum of the phosphor-based

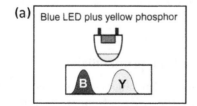
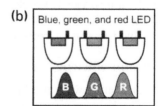

Figure 8.13 Two main methods to produce white light from LEDs. (A) Creating white light using blue luminescence and yellow phosphorescence. (B) Creating white light using three monochromic LEDs. After Schubert [97].

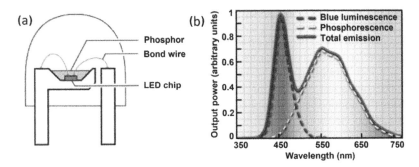

Figure 8.14 Generating white light by blue luminescence and yellow phosphorescence. (a) Structure of the device. (b) Spectrum emitted by the device. Schubert [97].

white LED thus consists of the blue light originating from the semiconductor LED and longer-wavelength phosphorescence, as shown in Fig 8.14(b).

A fundamental disadvantage of the phosphor-based white LEDs is the energy loss when converting blue light to longer-wavelength light. The energy efficiency is the ratio of the luminescence wavelength and the phosphorescence wavelength,

$$\eta = \frac{\epsilon_{\text{phosphorescence}}}{\epsilon_{\text{luminescence}}} = \frac{\lambda_{\text{luminescence}}}{\lambda_{\text{phosphorescence}}}. \tag{8.58}$$

For a higher energy efficiency, a combination of three LEDs is preferred.

As shown in Fig. 8.15, by changing the composition index x of $\text{In}_x\text{Ga}_{1-x}\text{N}$, the wavelength of the light emitted can be adjusted from near UV to infrared. An empirical formula for the dependence of bandgap on the index x is [97]

$$E_g = [3.42 - 2.65x - 2.4x(1 - x)] \, (\text{eV}). \tag{8.59}$$

Also shown in Fig. 8.15, within the range of visible light, the difference in lattice constant is minimal. By using freestanding crystalline GaN as the substrate, as shown in the next section, blue LED and green LED can be grown side by side. With an easily available red LED, white light can be generated with high efficiency.

Figure 8.15 Wavelength and lattice constant for $\text{In}_x\text{Ga}_{1-x}\text{N}$. By changing the composition index x of $\text{In}_x\text{Ga}_{1-x}\text{N}$, the wavelength of light emitted can be adjusted from near UV to infrared. Within the wavelength range of visible light, the lattice constant does not vary too much. Structures of such materials for different wavelengths can be grown on the same GaN substrate.

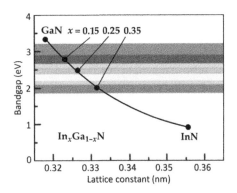

8.4.4   The Freestanding GaN Substrate

The early blue GaN LEDs were grown on sapphire substrates. The severe lattice mismatch causes a lot of dislocations. Since 2003, freestanding GaN substrates were produced [73]. It has all the ideal properties for making high-intensity blue LEDs. First, its lattice constant closely matches the useful materials which are basically GaN. Second, even if undoped, it appears to be *n*-type with a high electrical conductivity, ready to act as an electrode. Third, it has a high thermal conductivity to enable heat dissipation. Fourth, it is completely transparent to visible light. Finally, GaN has a high hardness and chemically inert, almost as good as sapphire. The active device can be fabricated on the back side of the GaN substrate with a silver conductor and reflector on the bottom. Figure 8.16 is a diagram of a modern blue-light LED.

8.4.5   A Brief Sketch of History

All important scientific discoveries and technological inventions have prior arts. The blue-light LED is no exception [97, 62]. The phenomenon of generating blue light by electricity, electroluminescence, was first discovered by Henry J. Round in 1907 using carborundum, SiC, a large-bandgap semiconductor. The device has rectifying current-voltage characteristics, which is a *pn*-junction, and emits blue light. After studied in detail by Soviet physicist Oleg V. Lossev during the 1920s, it was referred to as the Lossev radiation. Nevertheless, the efficiency was less than 0.005%.

In the early 1960s, RCA was the world's largest producer of CRT-based television monitors. At that time, red-light and green-light LEDs based on III-IV semiconductors were already available. The headquarters of RCA envisioned that by adding blue-light LEDs, the bulky CRT can be replaced by flat-panel displays with three colored LEDs. James J. Tietjen, an engineer and manager at the RCA Research Laboratory in Princeton, New Jersey, thought that GaN, with its wide but direct bandgap, would be a better material than SiC to fabricate blue LEDs. He further reasoned that GaN-based blue LEDs would have great commercial potential [62]. With the support of Jacques Pankove, a senior physicist at RCA and a visiting professor at Stanford University, he hired Herbert P. Maruska, then a PhD student of Pankove, to work on GaN material and blue-light LEDs in the Princeton RCA laboratory [62].

In 1968, single-crystal films of GaN were fabricated on sapphire substrates. Because the GaN film is completely transparent and shiny, it looked no difference from the bare

Figure 8.16   Blue-light LED fabricated on a GaN substrate. Modern blue-light LEDs are fabricated on the back of highly conducting freestanding GaN substrates as the negative electrode. On the bottom of the device it is a silver-based film as the positive electrode and a reflector. After Nakamura and Kramers [73].

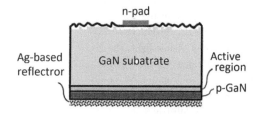

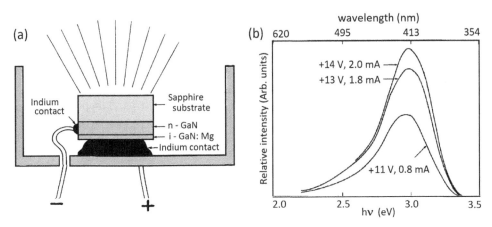

Figure 8.17 **The first blue-light LED built in 1972.** After a 1973 *Applied Physics Letter* by Maruska, Stevenson, and Pankove [63]. The device, a semiconductor-metal junction, is fabricated on the back side of a sapphire substrate. The structure is deceivingly similar to the modern blue-light LED, shown in Fig. 8.16, except the sapphire substrate is replaced by a highly conducting GaN substrate, and a *p*-type GaN layer is included.

sapphire substrate by naked eyes. Nevertheless, X-ray diffraction and light absorption experiments showed a high-quality single-crystal film with a sharp absorption edge, indicating a direct bandgap of 3.39 eV. Maruska and Tietjen reported that discovery on *Applied Physics Letters* in November 1969 [64]. Three years later, the first blue-light LED based on GaN was fabricated. A report by Maruska, Stevenson, and Pankove was published in *Applied Physics Letters* in March 1973 [63]. Figure 8.17(a) is adapted from that paper, showing a GaN-based blue-light LED with the sapphire side up. The peak wavelength was 2.9 eV, as shown in Fig. 8.17(b). Nevertheless, the efficiency was low, apparently caused by two seemingly unresolvable problems. First, there is a severe lattice mismatch between sapphire and GaN. Second, the GaN films appeared to be always *n*-type. Many attempts to make *p*-type GaN failed.

In late 1960s, RCA decided to defocus on consumer electronics and to become a "conglomerate", diversifying its business to various fields, such as carpet making, car rental, and book publishing. Meantime, RCA spent $100 million to develop its mainframe computer 601, trying to outcompete IBM. Only four units were sold. To absorb the $250 million loss, in 1974, RCA eliminated a number of expensive and risky research projects [41]. The GaN-based blue LED project was among them.

The remaining problems for creating high-efficiency blue-light LEDs were difficult. It requires a combined resolution and persistence of government, industry, and personals to overcome [97, 62]. After two decades of sweating and sacrifice, the blue LED enters into billions of families as illumination and flat-panel displays. On the other hand, a series of disastrous business decisions caused the demise of RCA in 1986, just before the fulfillment of one of the greatest inventions of mankind [41].

Problems

8.1. A typical silicon solar cell has the following parameters: The *p*-type material has acceptor concentration $N_A = 1 \times 10^{16} \mathrm{cm}^{-3}$, hole diffusion coefficient $D_p = 40 \ \mathrm{cm}^2/\mathrm{s}$, and lifetime $\tau_p = 5 \ \mu\mathrm{s}$. The *n*-type material has donor concentration $N_D = 10^{19} \mathrm{cm}^{-3}$, electron diffusion coefficient $D_n = 40 \ \mathrm{cm}^2/\mathrm{s}$, and lifetime $\tau_n = 1 \ \mu\mathrm{s}$. The permittivity of free space is $\epsilon_0 = 8.85 \times 10^{-14} \ \mathrm{F/cm}$, and the relative permittivity of silicon is $\epsilon_r = 11.8$. The intrinsic carrier concentration in silicon is $n_i = 1.5 \times 10^{10} \ \mathrm{cm}^{-3}$. Assuming the built-in potential is $qV_0 = 0.75 \ \mathrm{V}$, calculate the following:

1. Width of transition layer, W

Hint: The donor concentration is too high, thus the *n*-layer is very thin. The width of transition layer is effectively that of the *p*-layer. The dielectric constant (permittivity) of silicon is the product of the permittivity of free space and the relative permittivity of silicon.

2. Diffusion lengths for holes and electrons
3. Saturation current density I_0
4. The pn-junction capacitance of the 10-cm × 10-cm solar cell.

The *diffusion lengths* are defined as

$$L_p = \sqrt{D_p \, \tau_p} \tag{8.60}$$

$$L_n = \sqrt{D_n \, \tau_n}. \tag{8.61}$$

8.2. Show that excess minority carriers decay with x by

$$\delta p_n(x) = \delta p_n(0) e^{-x/L_p} \tag{8.62}$$

and

$$\delta n_p(x) = \delta n_p(0) e^{-x/L_n}. \tag{8.63}$$

8.3. Using diffusion length as a parameter, show that the junction current is

$$I = e \left(\frac{D_n \, n_p}{L_n} + \frac{D_p \, p_n}{L_p} \right) \left(e^{eV/k_B T} - 1 \right). \tag{8.64}$$

8.4. The mobility of holes in silicon is $\mu_p = 480 \ \mathrm{cm}^2/\mathrm{V \cdot s}$ at room temperature. What is its diffusion coefficient D_p?

Hint: use Einstein's relation.

Chapter 9

Semiconductor Solar Cells

The modern solar cell was invented in the 1950s by Gerald Pearson, Darryl Chapin, and Calvin Fuller at Bell Labs. Using crystalline silicon, they demonstrated a solar cell of efficiency of 5.7%, ten times greater than that of the selenium solar cell; see Section 1.3. In the 1950s through 1970s, solar cells were very expensive. Most of the applications were in space, where a high cost is tolerable. Since 1970s, solar cells has entered the civilian market. The efficiency has also been improving gradually. Currently, the efficiency of silicon cells is about 24%, very close to the theoretical limit.

To date, semiconductor solar cells account for roughly 95% of the market share. Especially, crystalline silicon solar cells account for more than 90% of the solar cell market. Thin-film solar cells, especially those based on CIGS (copper indium gallium selenide) and CdTe-CdS, are second to silicon solar cells in market share.

In addition to those well-established solar cells, since early 21st century, extensive research work has been conducted into two new fields: the organic solar cells, and perovskite-based solar cells. Those novel types of solar cells have the potential advantages of inexpensive manufacturing and flexible forms. Nevertheless, as the traditional solar cells based in silicon, CIGS, and CdTe-CdS have proven to have excellent stability under severe environmental conditions and can work for at least 20 years, the long-term stability of those noval solar cells is still an open question.

9.1 Basic Concepts

The solar cell is a solid-state device which converts sunlight, as a stream of quantized electromagnetic waves, into electrical power. Figure 1.11(b) shows the structure of a silicon solar cell. The base is a piece of p-type silicon, lightly doped with boron, a fraction of a millimeter thick. A highly doped n-type silicon, with a thickness of a fraction of one micrometer, was generated by doping with phosphorus of much higher concentration. After external radiation generates electron–hole pairs, electrons and holes are separated to generate electric power for the external circuit.

According to the theory of quantum transitions presented in Chapter 7, radiation, as a stream of quantized electromagnetic waves, or photons, interacts with a semiconductor in two ways; see Fig. 9.1. A photon with energy greater than the gap energy of the semiconductor material can be absorbed and create an electron–hole pair, with

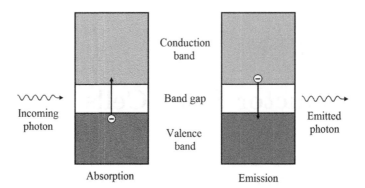

Figure 9.1 Interaction of radiation with semiconductors. According to the theory of quantum transitions, a photon with energy greater than the gap energy of the semiconductor material can be absorbed and create an electron–hole pair. An electron–hole pair can recombine and emit a photon of energy roughly equal to the energy gap of the semiconductor.

an energy equals to the band gap. An electron–hole pair can be separated and move to the external circuit, with a large part of the band gap energy converts to electrical energy to the external circuit; or recombine and emit a photon of energy roughly equal to the energy gap. The recombination process is a limiting factor to the efficiency of solar cells; see Section 9.2.3.

Because the potential energy of the electron–hole pair equals the gap energy, the

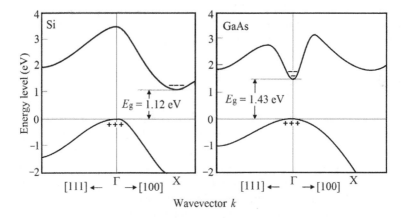

Figure 9.2 Direct and indirect semiconductors. Depending on the relative positions of the top of the valence band and the bottom of the conduction band in the wavevector space, the energy gap of a semiconductor can be direct or indirect. Direct semiconductors have a much higher absorption coefficient than that of indirect semiconductors. As shown, Si is an indirect semiconductor and GaAs is a direct semiconductor.

Figure 9.3 Absorption spectra of semiconductors commonly used for solar cells. The most commonly used material for solar cells, silicon, is an indirect semiconductor, and has a relatively low absorption coefficient, typically $10^3\,\mathrm{cm}^{-1}$. A thickness of 0.01 cm is required for efficient absorption. Direct semiconductors, such as GaAs, CuInSe$_2$ and CdTe, have absorption coefficients ranging from 10^4 to $10^5\,\mathrm{cm}^{-1}$. A thickness of a few micrometers is sufficient for an almost complete absorption.

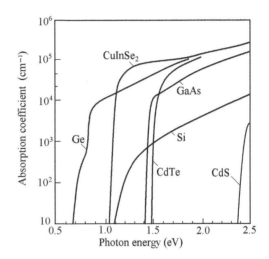

best material should have a band gap close to the center of the solar spectrum. Another factor that affects the efficiency of solar cells is the type of the energy gap. Depending on the relative positions of the top of the valence band and the bottom of the conduction band in the wavevector space, the energy gap of a semiconductor can be direct or indirect; see Fig. 9.2. For semiconductors with a direct gap, such as GaAs, CuInSe$_2$, and CdTe, a photon can directly excite an electron from the valence band to the conduction band; the absorption coefficient is high, typically greater than $1 \times 10^4\,\mathrm{cm}^{-1}$. For semiconductors with an indirect gap, such as Ge and Si, the top of valence band and the bottom of the conduction band are not aligned in the wavevector space, and the excitation must be mediated by a phonon, in other words, by lattice vibration. Therefore, the absorption coefficient is low, typically smaller than $1 \times 10^3\mathrm{cm}^{-1}$. A thicker substrate is required. Figure 9.3 shows the absorption spectra of commonly used semiconductor materials for solar cells. Table 9.1 gives the properties of frequently used solar cell materials.

Table 9.1: Properties of common solar-cell materials.

Material	Ge	CuInSe$_2$	Si	GaAs	CdTe
Type	Indirect	Direct	Indirect	Direct	Direct
Band gap (eV)	0.67	1.04	1.11	1.43	1.49
Absorption edge (μm)	1.85	1.19	1.12	0.87	0.83
Absorption coef. (cm^{-1})	5.0×10^4	1.0×10^5	1.0×10^3	1.5×10^4	3.0×10^4

9.1.1 Generating Electric Power

Figure 9.4 shows how electric current and voltage are created by radiation on a solar cell. As shown in Fig. 9.4(a), the impinging photons generate electron–hole pairs in the p-type region. The built-in electrical field intensity in the depletion layer points to the p-type region, see Fig. 8.7. The negatively charged electrons drift to the n-type region. Depending on the external circuit, there are two cases.

By connecting the two terminals, as shown in Fig. 9.4(a), electrons drifted to the n-type region flow through the external circuit and back to the p-type region. The *short-circuit current* I_{sc} equals the rate of electron–hole pair generation.

As shown in Fig. 9.4(b), if the two electrodes are not connected, the electrons stay in the n-type region, generate a forward bias voltage for the diode. Another way of looking at it is, the electrons drifted to the n-type region partially neutralize the positive charges in the depletion region, reduce the thickness of the transition region and then reduce the diffusion potential V_0. A forward diode current is created. When the forward diode current equals the drift current of the electrons generated by the photons, an equilibrium is established. The voltage on the two terminals is the *open-circuit voltage* V_{oc} of the solar cell under a given illumination.

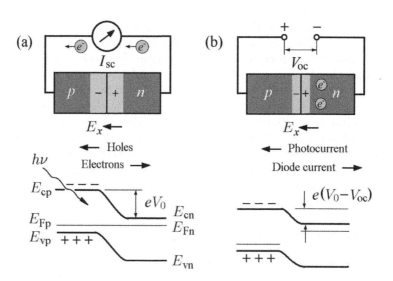

Figure 9.4 Generating voltage and current by the electron–hole pair. (a) The incoming photons generate electron–hole pairs in the p-type region. The built-in electrical field in the transition region is pointing toward the p-type region. The electrons, negatively charged, is dragged by the field into the n-type region. By connecting the two terminals, an electrical current is generated. The *short-circuit current* I_{sc} equals the rate of electron–hole pair generated by the radiation. (b) If the terminals are not connected, the electrons migrate to the n-type region are accumulated to compensate the positive charges in the depletion layer. A positive forward bias voltage is created. At equilibrium, an *open-circuit voltage* V_{oc} is established.

9.1.2 Solar Cell Equation

A solar cell can be represented by a current source connected in parallel with a *pn*-junction diode; see Fig. 9.5. The current source is the photocurrent generated by sunlight, see Section 9.2 for a detailed discussion. Therefore, we have

$$I = I_{\text{sc}} - I_0 \left(e^{eV/k_B T} - 1 \right),\tag{9.1}$$

which is the fundamental equation of solar cells, termed *solar cell equation*.

The open-circuit voltage is the voltage when the current is zero,

$$I_0 \left(e^{eV_{\text{oc}}/k_B T} - 1 \right) = I_{\text{sc}}.\tag{9.2}$$

Consequently,

$$V_{\text{oc}} = \frac{k_B T}{e} \ln \left(\frac{I_{\text{sc}}}{I_0} - 1 \right).\tag{9.3}$$

Because I_{sc} is always much bigger than I_0, Eq. 9.2 can be simplified to

$$V_{\text{oc}} = \frac{k_B T}{e} \ln \frac{I_{\text{sc}}}{I_0}.\tag{9.4}$$

9.1.3 Maximum Power and Fill Factor

The output power of a solar cell is determined by the product of voltage and current, $P = IV$. As discussed in Section 1.3.2, it is always smaller than the product of the short-circuit current I_{sc} and the open-circuit voltage V_{oc}, see Fig. 1.14. The rated power of a solar cell is the maximum power output with an influx of photons of one sun, or 1 kW/m^2, under favorable impedance-matching conditions. In general, the condition of maximum power is

$$dP = I dV + V dI = 0,\tag{9.5}$$

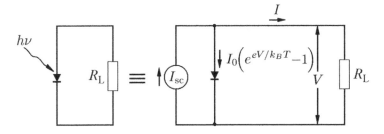

Figure 9.5 Equivalent circuit of solar cell. A solar cell can be represented by a current source connected in parallel with a *pn*-junction diode. The current source is the photocurrent generated by incoming sunlight.

in other words,

$$\frac{dI}{dV} = -\frac{I}{V}. \tag{9.6}$$

According to the solar cell equation, Eq. 9.1, the output power as a function of the output voltage V is

$$P = IV = \left[I_{sc} - I_0\left(e^{eV/k_BT} - 1\right)\right]V. \tag{9.7}$$

From Fig. 1.14, we observe that the voltage of maximum power is only slightly smaller than the open-circuit voltage. Introduce a voltage offset v, we write

$$V = V_{oc} - v. \tag{9.8}$$

Using Eq. 9.4, Eq. 9.7 can be simplified to

$$P \approx I_{sc}\left(V_{oc} - v\right)\left[1 - e^{-ev/k_BT}\right]. \tag{9.9}$$

Taking the derivative of P with respect to v, the condition of maximum power is

$$e^{ev/k_BT} = 1 + \frac{eV_{oc}}{k_BT}. \tag{9.10}$$

Again, using Eq. 9.4, Eq. 9.9 becomes

$$e^{ev/k_BT} = 1 + \ln\frac{I_{sc}}{I_0}. \tag{9.11}$$

Because $I_{sc} \gg I_0$, we find

$$v = \frac{k_BT}{e}\ln\ln\frac{I_{sc}}{I_0}. \tag{9.12}$$

Therefore, the voltage at maximum power is

$$V_{mp} = V_{oc} - v = V_{oc}\left(1 - \frac{\ln\ln(I_{sc}/I_0)}{\ln(I_{sc}/I_0)}\right), \tag{9.13}$$

and the current at maximum power is

$$I_{mp} = I_{sc}\left(1 - e^{-ev/k_BT}\right) = I_{sc}\left(1 - \frac{1}{\ln(I_{sc}/I_0)}\right). \tag{9.14}$$

After some simplification, maximum power is

$$P_{mp} = I_{mp}V_{mp} = V_{oc}I_{sc}\left(1 - \frac{1 + \ln\ln(I_{sc}/I_0)}{\ln(I_{sc}/I_0)}\right). \tag{9.15}$$

The fill factor η_f,[1] defined as

$$\eta_f \equiv \frac{I_{mp}V_{mp}}{V_{oc}I_{sc}}, \tag{9.16}$$

---

[1] In the literature, the fill factor often takes a notation FF. Since it is an efficiency multiplier in Shockley and Queisser's theory of solar cells, to maintain notational consistency, we use η_f instead.

is

$$\eta_f = 1 - \frac{1 + \ln\ln(I_{\text{sc}}/I_0)}{\ln(I_{\text{sc}}/I_0)}. \tag{9.17}$$

From Eq. 9.4, we have

$$\ln\frac{I_{\text{sc}}}{I_0} = \frac{eV_{\text{oc}}}{k_B T}. \tag{9.18}$$

The typical values are, $V_{\text{oc}} = 0.6$ V, and $k_B T/e = 0.027$ V. We have

$$\ln\frac{I_{\text{sc}}}{I_0} \cong 22.2. \tag{9.19}$$

From Eq. 9.17, we find

$$\eta_f = 1 - \frac{1 + \ln 22.2}{22.2} \cong 0.82. \tag{9.20}$$

9.2 The Shockley–Queisser Limit

In 1961, William Shockley and Hans Queisser made a thorough analysis of pn-junction solar cell, and established an upper limit for the efficiency of single-junction photovoltaic cells as a consequence of the principle of detailed balance [100]. The efficiency is defined as the ratio of power delivered to a matching load versus the incident radiation power to the solar cell. Three parameters are involved: The temperature of the Sun, $T_\odot$; the temperature of the cell, T_c; and the energy gap of the semiconductor, E_g. Actually, efficiency only depends on two dimensionless ratios:

$$x_s = \frac{E_g}{k_B T_\odot}, \tag{9.21}$$

$$x_c = \frac{E_g}{k_B T_c}. \tag{9.22}$$

Typically, $k_B T_\odot = 0.5$ eV, $k_B T_c = 0.025$ eV, and E_g is on the order of 1–2 eV. Therefore, the typical order of magnitude is $x_s \approx 2 - 4$, and $x_c \approx 40 - 80$.

The analysis of Shockley and Queisser was based on the following assumptions:

1. A single p–n-junction

2. One electron–hole pair excited per incoming photon

3. Thermal relaxation of the electron–hole pair energy in excess of the band gap

4. Illumination with unconcentrated sunlight

The above assumptions are well satisfied by the great majority of conventional solar cells, and the limit is well verified by experiments, unless one or more assumptions are explicitly removed, for example, using concentrated sunlight or using tandem solar cells. In these cases, the arguments of the Shockley–Queisser theory are still valid.

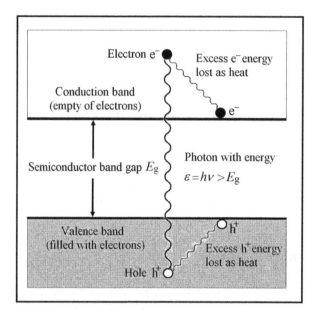

Figure 9.6 Generation of an electron–hole pair. A photon of energy greater than the band gap of the semiconductor can excite an electron from the valence band to the conduction band. The electron–hole pair energy in excess of the band gap dissipated into thermal energy of the electrons quickly (with a time scale of 10^{-11} s). The useful part of the photon energy for converting to electrical energy equals the band gap.

9.2.1 Ultimate Efficiency

Shockley and Queisser first considered the effect of band gap [100]. Assuming that solar radiation is a blackbody radiation of temperature $T_\odot$, power spectrum is (see Eq. 2.129),

$$u(\epsilon, T_\odot)\, d\epsilon = \frac{2\pi e^4}{c^2\, h^3} \frac{\epsilon^3}{\exp(\epsilon/k_B T_\odot) - 1}\, d\epsilon. \tag{9.23}$$

Equation 9.23 represents the radiation power density at the surface of the Sun. At the location of Earth, the spectral power density is diluted by a factor f defined by Eq. 2.130,

$$f = \left(\frac{r_\odot}{A_\odot}\right)^2 = \frac{\left[6.96 \times 10^8\right]^2}{\left[1.5 \times 10^{11}\right]^2} = 2.15 \times 10^{-5}. \tag{9.24}$$

Shockley and Queisser evaluated the ratio of the power density of electron–hole pairs and the incident radiation power density by assuming that the absorptivity of the semiconductor for photons with energy less than E_g is zero and those greater than E_g is unity. This simplification is also used to evaluate the rate of radiative combination of the electron–hole pair thus generated; see Section 9.2.3. For photons with $\epsilon > E_g$, the electron-pair energy is quickly thermalized to E_g; see Fig. 9.6. By replacing one of variables ϵ in Eq. 9.23 is by E_g, the power density of thus generated electron–hole pairs is

$$P_{ep} = \frac{2\pi e^4 E_g f}{c^2 h^3} \int_{E_g}^{\infty} \frac{\epsilon^2 \, d\epsilon}{\exp(\epsilon/k_B T_\odot) - 1} = \frac{2\pi e^4 f (k_B T_\odot)^4}{c^2 h^3} x_s \int_{x_s}^{\infty} \frac{x^2 \, dx}{e^x - 1}. \tag{9.25}$$

On the other hand, the incident radiation power is

$$P_s = \frac{2\pi e^4 f}{c^2 h^3} \int_0^{\infty} \frac{\epsilon^3}{\exp(\epsilon/k_B T_\odot) - 1} \, d\epsilon = \frac{2\pi e^4 (k_B T_\odot)^4 f}{c^2 h^3} \frac{\pi^4}{15}. \tag{9.26}$$

The efficiency as a function of the dimensionless variable x_s is

$$\eta_u(x_s) = \frac{P_{ep}}{P_s} = \frac{15}{\pi^4} x_s \int_{x_s}^{\infty} \frac{x^2 \, dx}{e^x - 1}. \tag{9.27}$$

The integral in Eq. 9.27 can be evaluated using a simple numerical integration program or the quickly converging expansion in Problem 9.1. The result is shown in Fig. 9.7. A

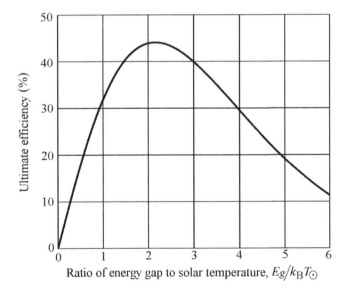

Figure 9.7 Ultimate efficiency of solar cells. Only photons with energy greater than the band gap can be absorbed, and the electron–hole pair energy in excess of the band gap is dissipated into thermal energy of the electrons, which make an ultimate limit to the efficiency of solar cells. The maximum efficiency is 0.44, at $E_g = 2.2 \, k_B T_\odot$. For $T_\odot = 5800$ K, $k_B T_\odot \approx 0.5$ eV, the optimum band gap is 1.1 eV.

qualitative explanation of the result is as follows. If the band gap is small, the range of photon absorption is large. However, most of the photon energy is dissipated as heat; see Fig. 9.6. For a large band gap, the range of spectral absorption is reduced. Therefore, it should have a maximum somewhere. The position and the value of the maximum efficiency can be obtained using a numerical program, and the results are

$$E_{g\,max} = 2.2\,k_B\,T_\odot; \quad \eta_{u\,max} = 0.44. \tag{9.28}$$

Taking $T_\odot = 5800$ K, $k_B T_\odot \approx 0.5$ eV. The optimum band gap is 1.1 eV. Shockley and Queisser called it the *ultimate efficiency*.

9.2.2   Role of Recombination Time

The ultimate efficiency determines the maximum open-circuit current of a solar cell. If the solar radiation power received by a solar cell is P_S, the power of the electron–hole pairs generated by the solar radiation is $\eta_u P_S$. It corresponds to the maximum short-circuit current of the solar cell,

$$I_{sc} = \frac{e}{E_g}\,\eta_u P_S. \tag{9.29}$$

The open-circuit voltage at the terminals of the solar cell is determined by the diode equation Eq. 9.3. Combining Eq. 9.3 with Eq. 9.29, we have the nominal power, defined as the product of the short-circuit current and the open-circuit voltage,

$$P_{no} = I_{sc}V_{oc} = \frac{\eta_u(x_s)}{x_c}\,\ln\left(\frac{I_{sc}}{I_0} - 1\right)P_S. \tag{9.30}$$

Clearly, the *reverse saturation current* of a *pn*-junction, I_0, is the limiting factor, determined by Eq. 8.47,

$$I_0 = en_i^2\left(\frac{1}{N_A}\sqrt{\frac{D_n}{\tau_n}} + \frac{1}{N_D}\sqrt{\frac{D_p}{\tau_p}}\right). \tag{9.31}$$

The reverse saturation current, Eq. 9.31, can be estimated using actual data from semiconductors. A general observation is as follows. The higher the reverse saturation current, the smaller the open-circuit voltage. By looking at Eq. 9.31, it is clear that the determining factor is the *recombination time*, τ_n and τ_p. Once an electron–hole pair is generated by absorbing a photon, the pair has a tendency to recombine by generating radiation or giving up the energy to the lattice. Shockley and Queisser found a fundamental limit due to radiative recombination of electron–hole pairs based on a detailed-balance argument.

9.2.3   Detailed-Balance Treatment

In the steady state, the electron–hole pairs in a solar cell are undergoing two major processes: the generation of the pairs by solar radiation and various recombination processes. Both quantities can be characterized by a current density. The current density of electron–hole pair generation I_{ep} is calculated by assuming that the spectral density of sunlight is a blackbody radiation with temperature $T_\odot$ but diluted by the distance between the Sun and Earth by a factor $f = 2.15 \times 10^{-5}$, defined by Eq. 9.24. From Eq. 9.25, we obtain

$$I_{ep} = \frac{eP_{ep}}{E_g} = \frac{2\pi q^5 f}{c^2 h^3} \int_{E_g}^{\infty} \frac{\epsilon^2}{\exp(\epsilon/k_B T_\odot) - 1} d\epsilon. \tag{9.32}$$

Some factors that contribute to recombination, such as those related to defects or surfaces, can be reduced or avoided. Radiative recombination, however, is a process of fundamental nature which sets a firm on the efficiency of solar cells. To evaluate the rate of radiative recombination of the electron–hole pairs, Shockley and Queisser considered first the equilibrium of a solar cell with environment at cell temperature T_c, typically 300 K, without sunlight. The principle of detailed balance requires that the rate of generation must equal the rate of recombination. This principle, manifested as Kirchhoff's law, states that emissivity equals absorptivity at any given wavelength. With respect to the simplified model of the semiconductor by Shockley and Queisser [100], for photons with energy greater than the energy gap, the emissivity is 1; otherwise it is 0 (see Fig. 9.8). The rate of radiative electron–hole recombination can also be expressed as a current density I_{c0} using an integral similar to Eq. 9.32 but at the environment temperature T_c without the diluting factor f,

$$I_{c0} = \frac{2\pi q^5}{c^2 h^3} \int_{E_g}^{\infty} \frac{\epsilon^2}{\exp(\epsilon/k_B T_c) - 1} d\epsilon. \tag{9.33}$$

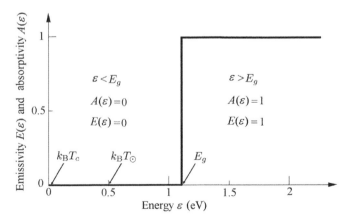

Figure 9.8 A simplified optical model of semiconductors. To evaluate the effect of radiative recombination, Shockley and Queisser used a simplified optical model of semiconductors. For photons with energy greater than the band gap of the semiconductor, the absorptivity is 1, and hence the emissivity is also 1. Otherwise the emissivity is 0.

Typically, the parameter x_c in Eq. 9.22 equals $E_g/k_B T_c \approx 40 - 80$, and thus the term 1 in the denominator of the integrand of Eq. 9.33 can be neglected. To high accuracy, the result is

$$I_{c0} = \frac{2\pi q^5 (k_B T_c)^3}{c^2 h^3} \int_{x_c}^{\infty} x^2 e^{-x} \, dx = \frac{2\pi q^5 (k_B T_c)^3}{c^2 h^3} e^{-x_c} \left[x_c^2 + 2x_c + 2 \right]. \qquad (9.34)$$

The above result is valid when the semiconductor is at equilibrium. With sunlight, excess minority carriers are generated. For example, if the bulk of a solar cell is made of p-type semiconductor, according to Eq. 8.33, with an external voltage V, the electron concentration changes from its equilibrium value n_{p0} to

$$n_p = n_{p0} \exp\left(\frac{eV}{k_B T_c} \right). \qquad (9.35)$$

Thus, in general, the current density due to radiative recombination is

$$I_c = I_{c0} \exp\left(\frac{eV}{k_B T_c} \right). \qquad (9.36)$$

By connecting the solar cell with a load resistor, a load current I arises. At a steady state, the current of electron–hole pair generation I_{ep} must equal the sum of radiative recombination current and the load current I,

$$I_{ep} = I_c + I. \qquad (9.37)$$

Define a *short-circuit current* I_{sc} as the load current at $V = 0$, or when

$$I_c = I_{c0}, \qquad (9.38)$$

according to Eq. 9.37, it is

$$I_{sc} = I_{ep} - I_{c0}. \qquad (9.39)$$

From Eq. 9.36, in general, the load current I is

$$I = I_{sc} - I_{c0} \left[\exp\left(\frac{eV}{k_B T_c} \right) - 1 \right]. \qquad (9.40)$$

The *open-circuit voltage* V_{oc} is defined as the voltage with

$$I = 0. \qquad (9.41)$$

From Eq. 9.40, it is

$$V_{oc} = \frac{k_B T_c}{e} \ln\left(\frac{I_{sc}}{I_{c0}} - 1 \right). \qquad (9.42)$$

Using Eq. 9.39, we obtain

$$V_{oc} = \frac{k_B T_c}{e} \ln\left(\frac{I_{ep}}{I_{c0}} \right). \qquad (9.43)$$

9.2.4 Nominal Efficiency

Shockley and Queisser defined the *nominal efficiency* as

$$\eta_n = \frac{V_{oc}I_{sc}}{P_s}. \tag{9.44}$$

Using Eqs. 9.43, 9.29, and 9.22, after some reduction, we find the expression

$$\eta_n = \eta_u\eta_d, \tag{9.45}$$

where the ultimate efficiency η_u is defined by Eq. 9.27, and the detailed balance efficiency η_d is defined as

$$\eta_d = \frac{eV_{oc}}{E_g} = \frac{1}{x_c}\ln\left(\frac{I_{ep}}{I_{c0}}\right), \tag{9.46}$$

More explicitly, using Eqs. 9.32 and 9.33,

$$\frac{I_{ep}}{I_{c0}} = \frac{f\left(\dfrac{x_c}{x_s}\right)^3 \displaystyle\int_{x_s}^{\infty} \dfrac{x^2\,dx}{e^x - 1}}{e^{-x_c}\left[x_c^2 + 2x_c + 2\right]}. \tag{9.47}$$

As shown in Eq. 9.46, the detailed-balance efficiency η_d is the fundamental upper limit of the ratio of the open-circuit voltage and the band gap of the semiconductor in volts. It depends on the temperature of the cell. If the temperature of the solar cell is very low, $x_c \to \infty$, the detailed-balance efficiency becomes

$$\eta_d \to \frac{1}{x_c}\ln\left(C\,e^{x_c}\right) \to \frac{x_c + \ln C}{x_c} \to 1, \tag{9.48}$$

because the expression C varies much less than the exponential. Therefore, at very low cell temperatures, if there is no other recombination mechanism other than radiative recombination, the open-circuit voltage approaches the band gap energy in volts and the nominal efficiency approaches the ultimate efficiency.

Equation 9.40 is very similar to Eq. 9.1, except the reverse saturation current density I_0, represented by Eq. 9.31, is replaced by the radiation recombination current density I_{c0}. The radiation recombination current density I_{c0} is the natural limit of the observed reverse saturation current density I_0, Eq. 9.31. In addition to radiative recombination, there are other types of recombination processes that increase the reverse saturation current density and then further reduce the efficiency of solar cells; see Section 9.3.

9.2.5 Shockley–Queisser Efficiency Limit

With a load of matched impedance, the output power of a solar cell can be maximized. As shown in Section 9.1.3, the maximum power is related to the nominal power by a form factor η_f approximately

$$\eta_{Se} = \eta_u\,\eta_d\,\eta_f. \tag{9.49}$$

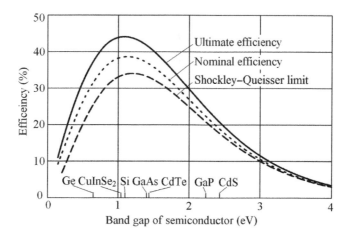

Figure 9.9 Efficiency limit of solar cells. Efficiency limit as determined by detailed balance. The solar radiation is approximated by blackbody radiation at 5800 K. The temperature of the solar cell and its surroundings is assumed as 300 K. The abscissa is the band gap of the semiconductor in electron volts. The values of several important solar cell materials are marked. The ultimate efficiency is determined by the absorption edge and thermalization of excited electrons. The nominal efficiency, which is always lower than the ultimate efficiency, is a result of radiative recombination of electrons and holes. While driving an external load at maximum power, the radiative recombination is further increased, and the efficiency is further reduced. The detailed-balance limit of efficiency is a fundamental limit for the conditions listed by Shockley and Queisser. *Source*: Adaped from Ref. [100].

The detailed-balance efficiency limit, together with the ultimate efficiency and the nominal efficiency, for $T_\odot = 5800$ K and $T_c = 300$ K, is shown in Fig. 9.9. The abscissa is the band gap of the semiconductor in electron volts. Values of several important solar cell materials are indicated. The ultimate efficiency is determined by the absorption edge and thermalization of excited electrons. The nominal efficiency, which is always lower than the ultimate efficiency, is a result of radiative recombination of electrons and holes. While driving an external load at maximum power, the radiative recombination is further increased, and the efficiency is further reduced.

In the original paper [100], Shockley and Queisser also discussed several factors that affect the efficiency limit, such as both sides of a solar cell can radiate which only one side receives sunlight, the non-blackbody behavior of the cell, nonradiative electron–hole pair recombination processes, and the difference between the solar constant and the AM-1.5 solar radiation. The efficiency is further reduced by a few percentage points.

9.2.6 Efficiency Limit for AM1.5 Radiation

Shockley and Queisser assumed that solar radiation is a blackbody radiation [100]. The actually solar radiation spectrum received on Earth is influenced by scattering and the absorption of water vapor, carbon dioxide, and so on; thus the spectrum is different

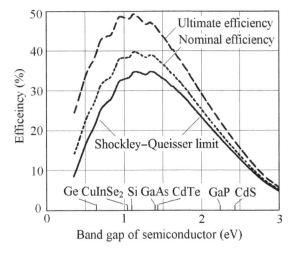

Figure 9.10 Efficiency limit of solar cells for AM1.5 solar radiation. Efficiency limit as determined by detailed balance for the AM1.5 solar radiation. The abscissa is the band gap of the semiconductor in electron volts. The values of several important solar cell materials are marked. Similar to Fig. 9.9, the ultimate efficiency, the nominal efficiency, and the detailed-balance limit of efficiency, the fundamental limit for the conditions listed by Shockley and Queisser, are shown.

from blackbody radiation; see Section 5.2. How is the efficiency limit for actual solar radiation spectrum changed from that of blackbody radiation?

The problem can be resolved using the same method as Shockley and Queisser. Instead of using the blackbody radiation spectrum, the measured solar radiation, the standard AM1.5 spectrum (see Plate 1, Section 5.2.1, and Appendix E) is used to compute the integrals in Eqs. 9.27 and 9.47. Because the strongest absorption and scattering are in the mid- to far-infrared and ultraviolet regions, visible light is basically unchanged. Because the infrared radiation with photon energy lower than the energy gap of the semiconductor does not participate in the generation of the electron–hole pair, and ultraviolet photons would lose more energy in the thermalization process, efficiency with regard to total radiation power is slightly higher. The strong absorption peaks and valleys are smoothed out by the integration process. Therefore, the final result is qualitatively identical to that from a blackbody radiation approximation. For details, see Fig. 9.10.

9.3 Nonradiative Recombination Processes

The Shockley–Queisser limit is solely based on the thermodynamics of radiative recombination of electron–hole pairs. There are several other recombination mechanisms and factors that could limit the solar-cell efficiency. Some of the factors are intrinsic and others can be mitigated or avoided by better cell design and manufacturing.

As shown in Section 9.2.2, the most serious limiting factor to solar cell efficiency is the recombination rate of electrons and holes. According to the basic equation of solar cells (Eq. 9.4), the open-circuit voltage is determined as

$$V_{oc} = \frac{k_B T}{e} \ln \frac{I_{sc}}{I_0}. \tag{9.50}$$

The dark current I_0 is determined by Eq. 8.47. The solar cells are always made of a thicker, lightly doped substrate and a thinner, highly doped film. For example, the typical silicon solar cell is made of a p–n^+ junction. Therefore, only one of the two terms dominate. Thus, Eq. 8.47 can be simplified to

$$I_0 = \frac{e n_i^2}{N} \sqrt{\frac{D}{\tau}}. \tag{9.51}$$

The open-circuit voltage is

$$V_{oc} = \text{const} + \frac{k_B T}{2e} \ln \tau. \tag{9.52}$$

Obviously, the greater the recombination time τ, the greater the open-circuit voltage V_{oc}, the efficiency will be greater.

If there are several recombination processes, the rate is additive, and thus the inverse of the recombination time is additive,

$$\frac{1}{\tau} = \frac{1}{\tau_1} + \frac{1}{\tau_2} + ... + \frac{1}{\tau_n}. \tag{9.53}$$

Each additional recombination process causes a reduction of the total recombination time, and thus, a reduction of efficiency.

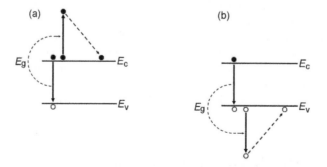

Figure 9.11 The Auger recombination process. The electron–hole pair can recombine and transfer the energy E_g into either a free electron near the conduction band edge E_c, (a), or free hole near the valance band edge, E_v, (b). Then, the excited electron quickly loses its excess energy to the lattice as phonons.

Various types of nonradiative recombination processes are discussed in details in Chapter 7 of Jacques I. Pankove's book [79]. Their effect on the efficiency of solar cells is discussed in Chapter 3 of Martin A. Green's book [34] and subsequent papers [35, 108].

9.3.1 Auger Recombination

As shown in Section 9.2.3, after an electron–hole pair is created, both the free electron and free hole quickly transfer excess energy to the lattice as phonons and stay near the band edge. The electron–hole pair can recombine and emit a photon, as shown in Section 9.2.3. An alternative process, the Auger process, is to transfer the energy E_g into either a free electron near the conduction band edge, E_c, as shown in Fig. 9.11(a), or a free hole near the valance band edge, E_v, as shown in Fig. 9.11(b). Then the excited electron quickly loses its excess energy to the lattice as phonons.

Clearly, Auger recombination is an intrinsic process which cannot be eliminated by smart design. Detailed calculations and experiments have shown that, for good-quality crystalline silicon, it is the dominant recombination process besides radiative recombination, which would further reduce the theoretical efficiency from the Shockley–Queisser limit of about 32% to about 28% [35, 108].

9.3.2 Trap-State Recombination

As shown in Section 8.1.2, the impurities in a semiconductor create states in the energy gap. The gap states are effective intermediate media for a two-step recombination process; see Fig. 9.12(a). Clearly, the higher the concentration of impurities, the more the gap states, and thus the shorter the electron–hole pair lifetime. As a general guideline, high-purity semiconductor materials are preferred.

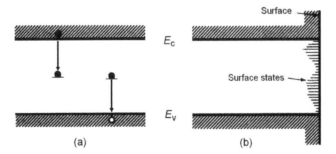

Figure 9.12 Two-step recombination processes. The electron–hole pair can recombine and transfer the energy E_g into either a free electron near the conduction band edge, E_c (a), or free hole near the valance band edge, E_v (b). Then the excited electron or hole quickly loses its excess energy to the lattice as phonons.

9.3.3   Surface-State Recombination

Surfaces of the semiconductor materials can represent a high concentration of defects, or surface states. Each surface state can become a mediator for two-step recombination; see Fig. 9.12(b). An experimentally verified effective method to reduce or eliminate surface states is *passivation*. For silicon, typical methods are either to create a Si–SiO$_2$ interface by oxidation, or to create a Si–Al$_2$O$_3$ interface by depositing a thin layer of alumina. Both SiO$_2$ and Al$_2$O$_3$ are insulators, and thus prevent the formation of direct metallic conductor on the surface. At both the top and back sides, only a small area is allowed to make ohmic contact. We will discuss this issue in section 9.5.2.

9.4   Antireflection Coatings

As discussed in Section 2.2.3, because all semiconductors have high refractive indices, according to the Fresnel formulas, reflection loss at the semiconductor–air interface is significant.

The solution, that is, antireflection coatings, were invented in early 20th century and has been applied to reduce the reflection of lenses in eyeglasses, cameras, telescopes, and microscopes. The concept of antireflection coatings is shown in Fig. 9.13. Without antireflection coatings, the reflection coefficient at the interface of two media with refractive indices n_1 and n_2 is determined by the Fresnel formula 2.76,

$$\mathcal{R} = \left(\frac{n_1 - n_2}{n_1 + n_2} \right)^2.$$ (9.54)

For example, for silicon, where $n = 3.8$, $\mathcal{R} = 0.34$, which is very high. By coating the

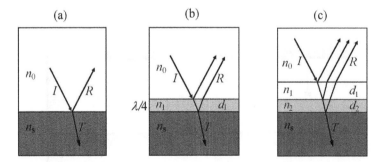

Figure 9.13   Antireflection coatings (a) At the interface of two dielectric media, the reflection coefficient is determined by the Fresnel formula. (b) By coating the interface with a film of dielectric material with an refractive index equal to the geometric mean of the indices of air and the bulk and a thickness equal to a quarter of the wavelength in that medium, the reflection can be eliminated. However, it only works for one wavelength. (c) Using multiple coatings, the wavelength range for almost zero reflection can be extended.

surface with a film of thickness equal to a quarter wavelength in that medium, the two reflected light waves should have a phase difference of $180°$. If the intensities of the two reflected light waves are equal, complete cancellation can take place. The condition of cancellation is then

$$\left(\frac{n_1 - n_2}{n_1 + n_2} \right)^2 = \left(\frac{n_2 - n_3}{n_2 + n_3} \right)^2. \tag{9.55}$$

It can happen only when $n_1/n_2 = n_2/n_3$, or

$$n_2 = \sqrt{n_1 n_3}. \tag{9.56}$$

In other words, the reflection can be completely eliminated when the refractive index of the thin film is the geometric mean of the refractive indices of air and the bulk medium. Obviously, it only works for a single wavelength. As shown in Fig. 9.13(c), by using multiple coatings, two or more reflection minima can be created, and the wavelength range for almost zero reflection can be extended.

The above argument based on interference is intuitive but not accurate. Multiple reflections take place in the film. The simple interference argument does not work for multiple antireflection coatings. The standard treatment is based on a matrix method, as presented in the next section.

9.4.1 Matrix Method

A general treatment of the optics of stratified media can be found in Born and Wolf [13] and Macleod [61]. In both books, very general cases are discussed, and the mathematics is quite complicated. Here we present a simple treatment for the case of normal incidence. For antireflection coatings on solar cells, it represents most of the essential physics and is mathematically transparent. For advanced readers, it can serve as a bridge to understand more sophisticated treatments.

Consider an antireflection coating with s films each with refractive index n_j and thickness d_j; see Fig. 9.14. Consider the electromagnetic wave of one polarization with circular frequency ω; see Fig. 2.2. Because, for a vacuum, $E_x = cB_y$ (see Eq. 2.36), we combine them as a two-dimensional vector,

$$\mathbf{F}(z) = \begin{pmatrix} F_1(z) \\ F_2(z) \end{pmatrix} = \begin{pmatrix} E_x(z) \\ cB_y(z) \end{pmatrix}. \tag{9.57}$$

Because both E_x and B_y are continuous at a boundary of two layers of dielectrics, F_1 and F_2 are also continuous at a boundary of two dielectrics. Within the jth film, following Eqs. 2.37–2.40, Maxwell's equations for $F_1(z)$ and $F_2(z)$ are

$$\frac{dF_1(z)}{dz} = ik_0 F_2(z), \tag{9.58}$$

$$\frac{dF_2(z)}{dz} = in_j^2 k_0 F_1(z), \tag{9.59}$$

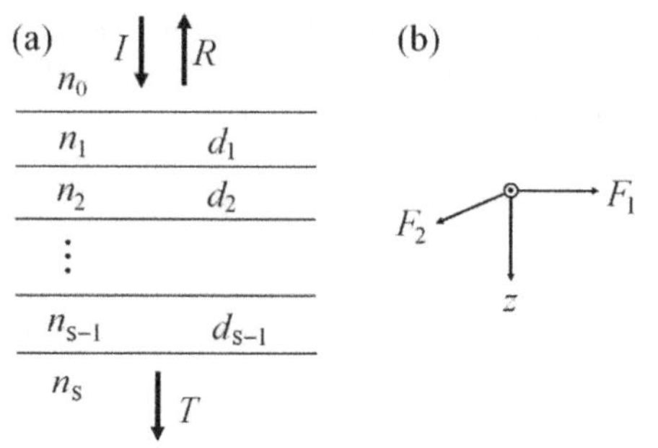

Figure 9.14   Matrix method for antireflection coatings. For each layer of antireflection coating, Maxwell's equations for a continuous medium is valid. Its effect can be represented by a 2×2 matrix. (a) The incident, reflected, and transmitted light can be represented by a two-dimensional vector for each. (b) The field vectors and the z-direction.

where $k_0 = \omega/c$ as the wavevector of the electromagnetic wave in a vacuum. Both F_1 and F_2 satisfy the following second-order differential equations:

$$\frac{d^2 F_1(z)}{dz^2} + n_j^2 \, k_0^2 \, F_1(z) = 0, \tag{9.60}$$

$$\frac{d^2 F_2(z)}{dz^2} + n_j^2 \, k_0^2 \, F_2(z) = 0, \tag{9.61}$$

There are two easily verifiable solutions: the forward-running wave,

$$\mathbf{F}(z) = \begin{pmatrix} 1 \\ n_j \end{pmatrix} F_0 \, e^{i n_j k_0 z} \tag{9.62}$$

and the backward-running wave,

$$\mathbf{F}(z) = \begin{pmatrix} 1 \\ -n_j \end{pmatrix} F_0 \, e^{-i n_j k_0 z}. \tag{9.63}$$

In general, the field is a linear combination of both waves. To obtain the general solution in a multilayer film system, we write down the solutions of Eqs. 9.58 and 9.59 with boundary conditions $F_1(0)$ and $F_2(0)$ at $z = 0$. The solutions can be easily verified as

$$F_1(z) = F_1(0) \cos k_0 n_j z + \frac{i}{n_j} F_2(0) \sin k_0 n_j z, \tag{9.64}$$

$$F_2(z) = i n_j \, F_1(0) \sin k_0 n_j z + F_2(0) \cos k_0 n_j z. \tag{9.65}$$

Equations 9.64 and 9.65 can be written conveniently in matrix form:

$$\begin{pmatrix} F_1(z) \\ F_2(z) \end{pmatrix} = \begin{pmatrix} \cos k_0 n_j z & \dfrac{i}{n_j} \sin k_0 n_j z \\ i n_j \sin k_0 n_j z & \cos k_0 n_j z \end{pmatrix} \begin{pmatrix} F_1(0) \\ F_2(0) \end{pmatrix}. \tag{9.66}$$

Introducing the two-by-two matrix

$$\mathsf{M}_j(z) = \begin{pmatrix} \cos k_0 n_j z & \dfrac{i}{n_j} \sin k_0 n_j z \\ i n_j \sin k_0 n_j z & \cos k_0 n_j z \end{pmatrix}, \tag{9.67}$$

Eq. 9.66 can be written in concise form

$$\mathbf{F}(z) = \mathsf{M}_j(z)\,\mathbf{F}(0). \tag{9.68}$$

The matrix format has some interesting properties. By direct arithmetic, it is easy to prove that

$$\mathsf{M}_j(z_1 + z_2) = \mathsf{M}_j(z_1)\,\mathsf{M}_j(z_2), \tag{9.69}$$

and then

$$\mathbf{F}(z_1 + z_2) = \mathsf{M}_j(z_1 + z_2)\,\mathbf{F}(0). \tag{9.70}$$

Using the inverse matrix

$$\mathsf{M}_j^{-1}(z) = \begin{pmatrix} \cos k_0 n_j z & -\dfrac{i}{n_j}\sin k_0 n_j z \\[2mm] -i n_j \sin k_0 n_j z & \cos k_0 n_j z \end{pmatrix}, \tag{9.71}$$

where it can be verified directly that $\mathsf{M}_j(z)\,\mathsf{M}_j^{-1}(z) = \mathsf{M}_j^{-1}(z)\,\mathsf{M}_j(z) = 1$, a reverse equation can be established,

$$\mathbf{F}(0) = \mathsf{M}_j^{-1}(z)\,\mathbf{F}(z). \tag{9.72}$$

Because at the boundaries the x- and y-components of E and B are continuous, for a series of films with thicknesses d_j,

$$\mathbf{F}(d) = \mathsf{M}_{s-1}(d_{s-1})\,\mathsf{M}_{s-2}(d_{s-2})\ldots\mathsf{M}_2(d_2)\,\mathsf{M}_1(d_1)\,\mathbf{F}(0), \tag{9.73}$$

where d is the total thickness of the antireflection films. The inverse expression is used more often,

$$\mathbf{F}(0) = \mathsf{M}_1^{-1}(d_1)\,\mathsf{M}_2^{-1}(d_2)\ldots\mathsf{M}_{s-2}^{-1}(d_{s-2})\,\mathsf{M}_{s-1}^{-1}(d_{s-1})\,\mathbf{F}(d). \tag{9.74}$$

In the following, we will treat the problems of single-layer antireflection (SLAR) coatings and double-layer antireflection (DLAR) coatings, see Fig. 9.15.

9.4.2 Single-Layer Antireflection Coating

The reflectance of the SLAR coating can be easily calculated using the matrix method. Because on the semiconductor side there is only transmitted light, the calculation starts with the field vector of transmitted light,

$$\mathbf{F}(0) = \mathsf{M}_1^{-1}(d_1)\,\mathbf{F}(d). \tag{9.75}$$

Since we are only interested in the ratio of the intensities of the incident light and the reflected light, the absolute magnitude and phase of the light waves are not important. Following Eq. 9.62, the transmitted light can be represented by

$$\mathbf{F}(d) = \begin{pmatrix} 1 \\ n_s \end{pmatrix}, \tag{9.76}$$

where n_s is the refractive index of the substrate. The field vector $\mathbf{F}(0)$ is a mixture of incident and reflected light waves. It can be separated by a projection matrix

$$\mathsf{P} = \begin{pmatrix} n_0 & -1 \\ n_0 & 1 \end{pmatrix}. \tag{9.77}$$

In fact, from Eqs. 9.62 and 9.63, at $z = 0$,

$$\mathsf{P}\,\mathbf{F}(0) = \mathsf{P}\left[\begin{pmatrix} 1 \\ n_0 \end{pmatrix} I + \begin{pmatrix} 1 \\ -n_0 \end{pmatrix} R\right] = 2n_0 \begin{pmatrix} R \\ I \end{pmatrix}. \tag{9.78}$$

Since we are interested in the ratio of R and I, the factor $2n_0$ has no effect. Combining Eqs. 9.71 and 9.75–9.77, we find

$$R = (n_0 - n_s)\cos\delta_1 - i\left(n_1 - \frac{n_0\,n_s}{n_1}\right)\sin\delta_1, \tag{9.79}$$

$$I = (n_0 + n_s)\cos\delta_1 - i\left(n_1 + \frac{n_0\,n_s}{n_1}\right)\sin\delta_1, \tag{9.80}$$

where δ_1 is the phase shift of the film,

$$\delta_1 = n_1 k_0 d_1. \tag{9.81}$$

The reflectivity is

$$\mathcal{R} = \left|\frac{R}{I}\right|^2 = \frac{n_1^2(n_0 - n_s)^2\,\cos^2\delta_1 + (n_1^2 - n_0\,n_s)^2\,\sin^2\delta_1}{n_1^2(n_0 + n_s)^2\,\cos^2\delta_1 + (n_1^2 + n_0\,n_s)^2\,\sin^2\delta_1}. \tag{9.82}$$

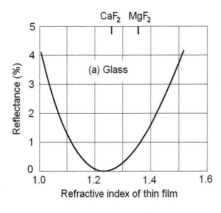
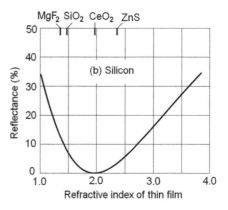

Figure 9.15 Choice of materials for SLAR coatings. The minimum reflectance for an SLAR coating is determined by the refractive index of the material; see Eq. 9.83. Two cases are shown. For glass, calcium fluoride and magnesium fluoride are the best choices. For silicon, cerium (ceric) oxide is the best choice.

Two special cases are worth noting. If the thickness of the coating is one quarter of the wavelength in that medium, that is, $\delta_1 = \pi/2$, then

$$\mathcal{R} = \left(\frac{n_1^2 - n_0\,n_s}{n_1^2 + n_0\,n_s}\right)^2. \tag{9.83}$$

When $n_1^2 = n_0\,n_s$, the reflectivity is zero. This verifies the result based on the naïve interference argument resulting in Eq. 9.56.

If the thickness of the coating is one-half of the wavelength in that medium, that is, $\delta_1 = \pi$, then

$$\mathcal{R} = \left(\frac{n_0 - n_s}{n_0 + n_s}\right)^2, \tag{9.84}$$

which coincides with the Fresnel formula 2.76, as if the antireflection coating does not exist.

9.4.3 Double-Layer Antireflection Coatings

The above treatment can be extended to double-layer antireflection (DLAR) coatings readily. The mathematical details are left as an exercise. The general result for the reflectance is

$$\mathcal{R} = \frac{R_1^2 + R_2^2}{I_1^2 + I_2^2}, \tag{9.85}$$

where

$$
\begin{aligned}
R_1 &= (n_0 - n_s)\,\cos\delta_1\,\cos\delta_2 - \left(\frac{n_0 n_2}{n_1} - \frac{n_s n_1}{n_2}\right)\sin\delta_1\,\sin\delta_2, \\[4pt]
R_2 &= \left(\frac{n_0 n_s}{n_2} - n_2\right)\cos\delta_1\,\sin\delta_2 + \left(\frac{n_0 n_s}{n_1} - n_1\right)\sin\delta_1\,\cos\delta_2, \\[4pt]
I_1 &= (n_0 + n_s)\,\cos\delta_1\,\cos\delta_2 - \left(\frac{n_0 n_2}{n_1} + \frac{n_s n_1}{n_2}\right)\sin\delta_1\,\sin\delta_2, \\[4pt]
I_2 &= \left(\frac{n_0 n_s}{n_2} + n_2\right)\cos\delta_1\,\sin\delta_2 + \left(\frac{n_0 n_s}{n_1} + n_1\right)\sin\delta_1\,\cos\delta_2,
\end{aligned}
\tag{9.86}
$$

and the phase shift of the jth film is given as

$$\delta_j = n_j k_0 d_j. \tag{9.87}$$

As a special case, if the thicknesses of both films are a quarter-wavelength, where $\delta_1 = \delta_2 = \pi/2$, all cosines are zero and all sines are unity. We have

$$\mathcal{R} = \left|\frac{R}{I}\right|^2 = \left(\frac{n_0 n_2^2 - n_s n_1^2}{n_0 n_2^2 + n_s n_1^2}\right)^2. \tag{9.88}$$

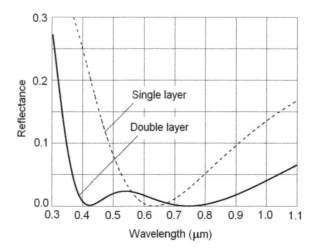

Figure 9.16 Wavelength range of antireflection coatings. Single-layer antireflection coating has one minimum-reflectance wavelength. Shown here is the reflectance for a 80-nm-thick CeO_2 SLAR film on silicon substrate. DLAR coatings can have two minimum-reflectance wavelengths. Shown here is the reflectance of a 101-nm ZnS film and a 56-nm MgF_2 film on a silicon substrate. The wavelength range of low reflectance is greatly increased.

The condition for zero reflectance is

$$n_0 n_2^2 = n_s n_1^2. \tag{9.89}$$

Therefore, the choice of materials can be broadened.

The major advantage of multilayer antireflection coatings is the wavelength range. As discussed in the previous section, even if the best material is chosen, for an SLAR coating, complete reflection cancellation can only happen for a single wavelength. For DLAR coatings, this can happen for two different wavelengths. This can be seen from the expression of reflectance $\mathcal{R}$. If the condition in Eq. 9.89 is satisfied, when $\cos \delta_1$ and $\cos \delta_2$ vanishes at two different wavelengths, reflection is eliminated at both wavelengths. Figure 9.16 shows two cases used in manufacturing. The dashed curve is the reflectance for an 80-nm-thick CeO_2 SLAR film on silicon. The solid curve is the reflectance of a DLAR cotings with a 101-nm-ZnS film and a 56-nm MgF_2 film on a silicon substrate, which provides two minimum-reflectance wavelengths. The wavelength range of low reflectance is greatly increased. See Ref. [124].

9.5   Crystalline Silicon Solar Cells

The first practical solar cell, invented in 1954, used crystalline silicon. To date, crystalline silicon solar cells still have 80–90% of market share. The material has many advantages:

1. Silicon, comprising 27% of Earth's crest, is the second most abundant element after oxygen.

2. Its band gap is almost optimum regarding the solar spectrum.

3. Chemically, silicon is very stable.

4. Silicon is nontoxic.

5. Because of the microelectronics industry, the production and processing of ultra-pure silicon are well developed.

6. After more than 50 years of research and development, the efficiency of silicon solar cells, 24.7% for research prototypes, is already close to its theoretical limit. The mass-produced modules, limited for cost reduction considerations, have reached 20% efficiency.

9.5.1 Production of Pure Silicon

The raw material for silicon, silica, is the most abundant mineral on Earth, including quartz, chalcedony, white sand, and numerous noncrystalline forms. The first step in silicon production is to reduce silica with coke to generate metallurgical-grade silicon,

$$SiO_2 + C \longrightarrow Si + CO_2. \tag{9.90}$$

The silicon thus produced is typically 98% pure. For applications in solar cells, at least 99.9999% purity is required, so-called solar-grade silicon. However, many processes can generate silicon with impurity levels less than 10^{-9}, which is needed for high-efficiency solar cells.

Two processes are commonly used. In the Siemens process, high-purity silicon rods are exposed to trichlorosilane at 1150°C. The trichlorosilane gas decomposes and deposits additional silicon onto the rods, enlarging them:

$$2HSiCl_3 \longrightarrow Si + 2HCl + SiCl_4. \tag{9.91}$$

Silicon produced from this and similar processes is called polycrystalline silicon. The byproduct, silicon tetrachloride, cannot be reused and becomes waste. The energy consumption in this process is also significant.

In 2006 Renewable Energy Corporation in Norway (REC) announced the construction of a plant based on fluidized-bed technology using silane, taking silicon tetrachloride as the starting point:

$$\begin{aligned}
3SiCl_4 + Si + 2H_2 &\longrightarrow 4HSiCl_3, \\
4HSiCl_3 &\longrightarrow 3SiCl_4 + SiH_4, \\
SiH_4 &\longrightarrow Si + 2H_2.
\end{aligned} \tag{9.92}$$

The purification process takes place at the silane (SiH$_4$) stage. According to REC, the energy consumption of this new process is significantly reduced from the Siemens process. Also, using the almost free hydroelectric power in Norway, REC is expecting to reduce the cost of solar-grade pure silicon to less than \$20 per kilogram.

To prepare for solar cell production, pure silicon can go through two alternative processes. For single-crystal silicon solar cells, the Czochraski process or the float-zone process is applied to generate single-crystal silicon ingots. Alternatively, the pure silicon can be melted in an oven to produce polycrystalline ingots.

9.5.2   Solar Cell Design and Processing

The efficiency of solar cells has gradually improved since the invention of the silicon solar cell in 1954. Recently, the 25% efficiency of laboratory prototypes of monocrystalline silicon solar cells has approached its theoretical limit; see Plate 8.

There are several silicon solar cell designs. Here we present one by the University of New South Wales in Australia, the passivated emitter, rear locally diffused (PERL) cells (Fig. 9.17), which has achieved up to 24.7% efficiency under standard global solar spectra; see Refs. [124] and [36].

The design of the PERL solar cell is shown in Fig. 9.17. In addition to using a rather thick wafer (370- or 400-μm) high-quality monocrystalline silicon, it has several features that enhance efficiency.

Passivation and Metal Contacts

To reduce surface recombination, both sides of the wafer are passivated with a layer of silicon dioxide. Because SiO$_2$ is an insulator, metal contacts must be made with small holes in the SiO$_2$ film. By locally diffusing boron to the rear contact areas, the effective

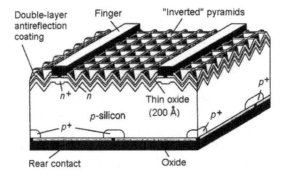

Figure 9.17 Typical high-efficiency silicon solar cell. The front side of the relatively thick silicon wafer is textured to capture light. A double-layer antireflection film is applied. The back side is passivated with silicon dioxide to reduce surface recombination. The rear contacts are enabled through highly doped $p+$ regions.

recombination rate is reduced by suppressing minority-carrier concentration in these regions. The doping process uses BBr_3 as the source dopant. The contact resistance is also markedly reduced. On the front side, heavy phosphorous doping is made using liquid PBr_3 as the dopant carrier. The width of the front metal contact is reduced to increase the area of silicon surface.

Textured Front Surface

The front surface is textured to become a two-dimensional array of inverted pyramids. The photons entering the substrate can be trapped by the textured top surface.

Double-Layer Antireflection Coating

As presented in Section 9.4, DLAR coatings can significantly improve overall efficiency. A ZnS and MgF_2 DLAR coating is evaporated onto the cells.

9.5.3 Module Fabrication

Single silicon solar cells are fragile and vulnerable to the elements. In all applications, silicon solar cells are framed and protected to become solar modules. Figure 9.18 shows a cross section of a typical solar module. To manufacture such a module, a piece of low-iron window glass, two sheets of ethylene vinyl acetate (EVA) film (with typical thickness of 0.5 mm), a rectangular array of solar cells, and a back plate, either a Mylar film or a sheet of metal, are stacked together in a heated press machine. The EVA is softened at about 150°C, then tightly bonded with the solar cells and other components. Finally, the glass module is framed by a metal structure with a protection gasket.

The monocrystalline solar module and polycrystalline solar module can be identified by the look. Figure 9.19 shows the two types of solar nodules. Figure 9.19(a) is a monocrystalline solar module. The solar cells are cut from a cylindrical single crystal

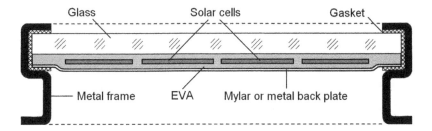

Figure 9.18 Cross section of typical solar module. A complete solar module is made of a piece of low-iron window glass, two sheets of EVA film, an array of solar cells, and a back plate. Those components are bonded together in a heated press.

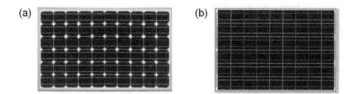

Figure 9.19 Monocrystalline solar module and polycrystalline solar module. (a) The monocrystalline solar cells are cut from a cylindrical single crystal. To save material and space, the solar cells are cut to an octagonal piece. There is always some wasted space because of the cut corners. (b) The polycrystalline solar cells are cut from a rectangular ingot. The solar cells are usually perfect squares. There is no wasted space.

to become an octagonal piece. Figure 9.19(b) is a module made of polycrystalline solar cells which are cut from a rectangular ingot.

9.6   Thin-Film Solar Cells

Silicon as a solar cell material has many advantages. However, it also has a disadvantage. As shown in Section 9.1, silicon is an indirect semiconductor. The absorption coefficient near its band edge is low. Therefore, a fairly thick substrate is required. The wafer is cut from a single crystal or an polycrystal ingot. The minimum thickness to maintain reasonable absorption and mechanical strength is 0.1 mm. The cost of the material and mechanical processing is substantial. The direct semiconductors often have an absorption coefficient one or two orders of magnitude higher than silicon; see Fig. 9.3. A thickness of a few micrometers is sufficient. In addition to a high absorption coefficient near the band gap which is close to 1 eV, there are many other factors which determine the practicality of making a solar cell. To date, besides silicon, solar cells based on two types of semiconductor materials have reached the status of mass production, namely, cadmium telluride (CdTe) and copper indium gallium selenide, $Cu(In_xGa_{1-x})Se_2$, often called CIGS. However, the material cost for these is still high. Amorphous-silicon thin-film solar cells, in spite of their relatively low efficiency, are mass produced for applications where a low efficiency is tolerable.

9.6.1   CdTe Solar Cells

Due to its high absorption coefficient and ease in making a p-type material, cadmium telluride (CdTe) is currently the most popular material for thin-film solar cells [16, 19, 87]. Another advantage is its compatibility with CdS, a wide-band-gap semiconductor for which it is easy to generate an n-type film. Because the absorption edge of CdS is 2.4 eV, it is transparent to the bulk of solar radiation. The typical structure of a CdTe solar cell is shown in Fig. 9.20. As shown, the solar cell is sandwiched between two sheets of window glass. The solar cell is made of a 5-μm film of CdTe, covered with a 100-nm CdS thin film, to form a pn-junction. To the sunny side is a film of TCO, to

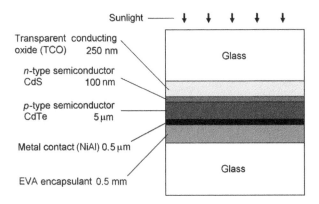

Figure 9.20 Typical structure of CdTe thin film solar cell. The *pn*-junction made of CdTe and CdS is sandwiched between two glass plates. See Refs. [16], [19], and [87].

allow radiation to come in and serves as a conductor. To the back side is a metal film for electric contact. A 0.5-mm EVA film is added for mechanical protection. The best efficiency of CdTe solar cells is 16.5%, and expected to reach 20%.

One question often asked is about the toxicity of cadmium. According to a recent study, because the quantity of cadmium is very small and is completely sealed by glass, the environment impact is negligible.

The largest manufacturer of CdTe solar cells is First Solar, with headquarters in Tempa, Arizona. For several years since 2002, First Solar was the largest manufacturer of solar cells in the world. Only in late 2011 it was dethorned by Suntech. In September 2009, First Solar signed a contract with China to built a 2 GW solar field in Ordos, Inner Mongolia. It is by far the largest solar field under development to date.

9.6.2 CIGS Solar Cells

The CuInSe$_2$/CdS system was discovered in 1974 as a photovoltaic light detector [113]. By mixing CuInSe$_2$ with CuGaSe$_2$ to form Cu(In$_x$Ga$_{1-x}$)Se$_2$, abbreviated as CIGS, the band gap can be optimized for solar cells. In 1975, a CIGS solar cell has shown an efficiency comparable to the silicon solar cells at that time [99]. In 2000s, the efficiency of CIGS thin-film solar cells has reached 19.9%, comparable to polycrystalline silicon solar cells [19, 87, 101], see Plate 8. As shown in Fig. 9.3, the absorption coefficient of CuInSe$_2$ near its band gap is about 100 times greater than that of silicon. Therefore, even if the semiconductor film were as thin as 2 μm, more than 90% of the near-infrared and visible light would be absorbed.

The typical structure of a CIGS solar cell is shown in Fig. 9.21. Similar to the CdTe solar cells, a 50-nm *n*-type CdS film is used to form a *pn*-junction. Again, because the quantity of cadmium used here is very small, and sandwiched between two glass plates, the environment impact is negligible.

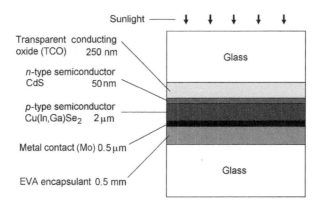

Figure 9.21 Typical structure of CIGS thin-film solar cell. A *pn*-junction of Cu(InGa)Se$_2$ and CdS is sandwiched between two glass plates. See Refs. [19], [87], [99], [101], and [113].

The CIGS solar cell can be produced with a wet process, without requiring a vacuum. Therefore, the manufacturing cost can be low. Another advantage of CIGS solar cells is that interconnections can be made on the same structure, similar to an integrated circuit, thus a higher voltage single cell, for example 12 V, can be made without requiring external connections. Figure 9.22 is an experimental 5-V CIGS solar cell, made from 10 individual CIGS solar cells in a single glass envelop. The boundaries between adjacent solar calls are apparent.

9.6.3   Amorphous Silicon Thin-Film Solar Cells

Because the cost of some key materials in CdTe and CIGS thin-film solar cells, notably tellurium and indium, is high, silicon thin-film solar cells, in spite of their low efficiency, have been mass produced for many years. For applications where efficiency is not crucial, such as hand-held calculators and utility-scale solar fields in deserts, silicon thin-film solar cells provide an advantage. Especially, silicon thin-film solar cells can

Figure 9.22 CIGS solar cell integrated circuit. An experimental solar cell made of 10 individual CIGS cells. The nominal voltage of each CIGS solar cell is 0.5 V. By connecting 10 individual CIGS solar cells in series internally, a 5-V solar cell is built. Because the connections are integrated, the device is compact and rugged. Photo taken by the author.

be manufactured on flexible substrates.

A major disadvantage of silicon is its low absorption coefficient. However, by substantially doping amorphous silicon with hydrogen, up to 10%, its absorption coefficient can be made as high as 10^5 cm^{-1}, with the band gap shifted from 1.1 eV to 1.75 eV, similar to that of CdTe. In the literature, this material is often abbreviated as a-Si:H [19]. Because of high defect density, the recombination rate in high. The efficiency of the best experimental a-Si:H solar cells is around 10%, and for the mass-produced solar cells, it is around 5%.

9.7 Tandem Solar Cells

As discussed in Section 9.2.1, over the entire solar spectrum, the highest efficiency comes from photons with energy just above the band gap of the semiconductor material. For photons with energy lower than the band gap, the semiconductor is transparent. There is no energy conversion. For photons with energy much higher than the band gap, the energy of the electron–hole pair quickly relaxes to that of the energy gap. The excess photon energy over the band gap is lost. Therefore, by stacking two or more solar cells in tandem, the efficiency can be much higher than the Shockley–Queisser limit.

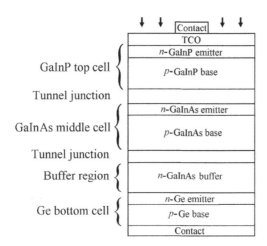

Figure 9.23 Multijunction tandem solar cell. Through the TCO layer, the sunlight falls on the top cell first. The semiconductor material of the top cell has a large band gap, in this case GaInP, 1.9 eV. It is transparent to photons of energy smaller than 1.9 eV. Photons with energy greater than 1.9 eV will generate an electron–hole pair with energy about 1.9 eV. The middle cell, made of GaInAs, has band gap 1.35 eV. Photos with energy between 1.35 and 1.9 eV will generate an electron–hole pair of about 1.35 eV. The bottom cell is made of Ge, with a band gap of 0.67 eV. For photons with energy greater than 0.67 eV and smaller than 1.35 eV, an electron–hole pair of about 0.67 eV is generated. The voltage is additive, and thus it can generate much more power than the single cells, often exceeding the Shockley–Queisser limit.

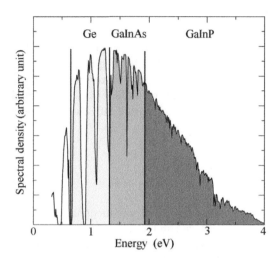

Figure 9.24 Working principle of multijunction tandem solar cells. The solar spectrum is divided into three sections. Each section generates a voltage. The tandem cell can be considered as a solar battery made of three cells. The voltage is additive, and thus it could generate much more power than the single cells, often exceeding the Shockley–Queisser limit.

Figure 9.23 shows the schematic of a three-junction tandem cell. The top cell is made of GaInP, with a band gap of 1.9 eV. The photons with energy greater than 1.9 eV will generate an electron–hole pair with energy about 1.9 eV. For photons with energy smaller than 1.9 eV, that GaInP layer is transparent. The middle cell, made of GaInAs, with a band gap of 1.35 eV, would absorb photons with energy between 1.35 and 1.9 eV, and generate an electron–hole pair of about 1.35 eV. The GaInAs film is transparent to photons with energy smaller than 1.35 eV. Those photons then go to a layer of Ge, with a band gap of 0.67 eV. For photons with energy greater than 0.67 eV and smaller than 1.35 eV, an electron–hole pair is generated in the bottom cell. The current should be continuous, but the voltage is additive. Therefore, for solar radiation with a rich spectrum, the tandem cell can generate much more power than the single cells, thus often exceeding the Shockley–Queisser limit, see Fig. 9.24. Recently, a group at Spectrolab has demonstrated a tandem solar cell of efficiency exceeding 40%. See [48].

Much of the material used in tandem cells is expensive. The major application of tandem solar cells is with concentrated solar radiation. By concentrating solar radiation 100 folds or more, the area of the solar cell is less than 1% without concentration. Economically it can be even better than crystaline silicon solar cells.

Problems

9.1. The ultimate efficiency of Shockley and Queisser can be easily computed by expanding the denominator of Eq. 9.27,

$$\eta_u(x_s) = \frac{15}{\pi^4} x_s \int_{x_s}^{\infty} \frac{x^2 \, dx}{e^x - 1}$$

$$= \frac{15}{\pi^4} x_s \sum_{n=1}^{\infty} \int_{x_s}^{\infty} e^{-nx} x^2 \, dx. \tag{9.93}$$

Prove that

$$\eta_u(x_s) = \frac{15}{\pi^4} x_s \sum_{n=1}^{\infty} e^{-nx_s} \left[\frac{x_s^2}{n} + \frac{2x_s}{n^2} + \frac{2}{n^3} \right]. \tag{9.94}$$

9.2. If the temperature of the solar cell can be maintained at room temperature, what is the effect of concentration of the sunlight by 100 times on the efficiency of solar cells? For blackbody radiation, work out the enhancement for a silicon solar cell.
Hint: Instead of using the usual geometric factor $f = 2.15 \times 10^{-5}$, using a larger number, for example, $100f$ in Eq. 9.47.

9.3. The absorption coefficients of some frequently used solar cell materials at the center of the visible-light spectrum are listed in Table 9.1. To achieve a total absorption of 95%, what is the required thickness?

9.4. A typical solar cell is made from a 10-cm × 10-cm piece of single-crystal silicon. Reverse saturation current I_0 is 3.7×10^{-11} A/cm^2. For $V = 0, 0.05, 0.1, \ldots, 0.55, 0.6$V, calculate the forward current for the solar cell at room temperature.

9.5. For a typical silicon solar cell, the acceptor concentration in the p-region is $N_A = 1 \times 10^{16}$/cm^3, and the donor concentration in the n-region is $N_D = 1 \times 10^{19}$/cm^3. Assuming that the built-in potential is 0.5 V, calculate the pn-junction capacitance of the 10-cm × 10-cm solar cell. (The permittivity of free space is $\varepsilon_0 = 8.85 \times 10^{-14}$F/cm, and the relative permittivity of silicon is $\varepsilon_r = 11.8$.)

Hint: the donor concentration is too high, thus the thickness of the n-type region is negligible. Only the acceptor concentration is needed for calculation. The dielectric constant (permittivity) of silicon is the product of the permittivity of free space and the relative permittivity of silicon.

9.6. For an interface of air and glass with refractive index n, show that the transmittance of each interface is

$$\tau = \frac{4n}{(1+n)^2}. \tag{9.95}$$

Verify the validity of the relation by showing that if $n=1$, $\tau = 1$.

9.7. If a solar device is covered with N sheets of glass with refractive index n, show that the transmittance of the entire cover is

$$\tau = \frac{(4n)^{2N}}{(1+n)^{4N}}. \tag{9.96}$$

9.8. A solar cell of 100 square centimeter area has a reversed-bias dark current of 2×10^{-9} A, and a short-circuit current at one sun (1 kW/m^2) of 3.5 A. At room temperature, what is the open-circuit voltage? What is the optimum load resistance (V_{mp}/I_{mp})? What is the maximum power output?

9.9. A typical silicon solar cell has the following parameters:

Band gap of silicon is 1.1 eV.

The p-type material has an acceptor concentration $N_A = 1 \times 10^{16}$ cm^3, with hole diffusion coefficient $D_p = 40$ cm^2/s and lifetime $\tau_p = 5\,\mu s$.

The n-type material has a donor concentration $N_D = 10^{19}$ cm^3, with electron diffusion coefficient $D_n = 40$ cm^2/s and lifetime $\tau_n = 1\mu s$.

The intrinsic carrier concentration in silicon is $n_i = 1.5 \times 10^{10}$ cm^3.

For a 1-cm $\times$ 1-cm solar cell, calculate the following:

1. Reverse saturation current I_0

2. Short-circuit current under one sun solar radiation

3. Open-circuit voltage at room temperature

Chapter 10

Solar Photochemistry

In the previous chapter, we discussed semiconductor solar cells, where photons from the Sun generate electron–hole pairs, and then the energy in the electron–hole pair is converted into electric energy. Solar photochemistry follows a different route: photons from the Sun causes a molecule to go from its ground state to an excited state. The energy stored in the excited molecule can be converted into either electrical energy, or permanent chemical energy. The most important example of solar photochemistry is photosynthesis, the conversion of solar energy into chemical energy stored in organic material, such as glucose.

10.1   Physics of Photosynthesis

To date, most of the energy human society uses originates from photosynthesis. The energy stored in food and traditional fuels, such as firewood, hey, and vegetable and animal oil, comes directly or indirectly from photosynthesis. Fossil fuel is the remains of ancient organisms, that is, stored product of photosynthesis. As the energy source of all life on Earth, photosynthesis is a natural process that has evolved in the last one billion years through natural selection. The study of photosynthesis will inspire us to create high-efficiency systems to harvest solar energy. For more details on photosynthesis, see Blankenship [10] and Voet [110].

Photosynthesis is arguably the most important chemical reaction on Earth. As evidence, nine Nobel Prizes in Chemistry have awarded for research of photosynthesis:

1915: Richard Wilstätter, for chlorophyll purification and structure, carotenoids, anthocyanins.

1930: Hans Fischer, for haemin synthesis, chlorophyll chemistry.

1937: Paul Karrer, for carotenoid structure, flavins, vitamin B_2.

1938: Richard Kuhn, for carotenoids, vitamins.

1961: Melvin Calvin, for carbon dioxide assimilation.

1965: Robert Burns Woodward, for total synthesis of vitamin B_{12}, chlorophyll, and other natural products.

1978: Peter Mitchell, for oxidative and photosynthetic phosphorylation, chemiosmotic theory.

1988: Hartmut Michel, Robert Huber, and Johannes Deisenhofer, for X-ray structure of the bacterial photosynthetic reaction center.

1992: Rudolph Marcus, for electron transfer theory and application to the primary photosynthetic charge separation.

1997: Paul D. Boyer and John E. Walker, for elucidation of enzymatic mechanism underlying the synthesis of adenosine triphosphate (ATP).

A typical chemical equation for photosynthesis is the generation of glucose or fructose, $C_6H_{12}O_6$, from carbon dioxide, water, and energy from solar radiation:

$$6CO_2 + 6H_2O + 29.79\,eV \longrightarrow C_6H_{12}O_6 + 6O_2. \tag{10.1}$$

To understand the elementary process, reaction 10.1 is often written as

$$CO_2 + H_2O + 4.965\,eV \longrightarrow \frac{1}{6}(\text{glucose}) + O_2. \tag{10.2}$$

Because the typical energy of a photon from sunlight is 1–3 eV, the above process is inevitably a multiphoton process.

10.1.1 Chlorophyll

Although there have been many different types of photosynthetic processes, only very few survived the natural selection process to support life on Earth. Photosynthesis relies on chlorophyll, the green pigment on plants. The word chlorophyll is derived from two Greek words, *cholos* ("green") and *phyllon* ("leaf").

The most common chlorophyll in plants and algae is chlorophyll *a*. The chemical structure is shown in Fig. 10.1. It is a squarish planar molecule about 1 nm on each side. A Mg atom at the center of the molecule is coordinated with four nitrogen atoms. Each nitrogen atom is part of a pyrrole ring. At the external sites groups such as CH_3

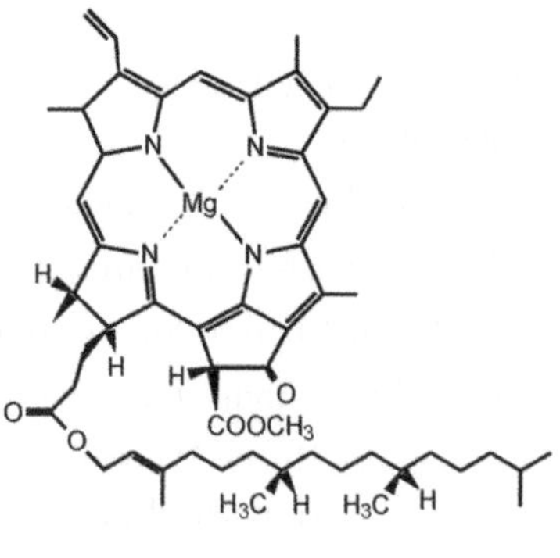

Figure 10.1 Chlorophyll. Chemical structure of the most common chlorophyll, chlorophyll *a*. At the center of the squarish molecule is a Mg atom. Each nitrogen atom is part of a pyrrole ring. A long hydrocarbon tail is attached through an oxygen site. At the external sites, different groups are bonded.

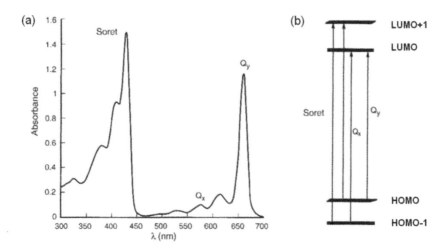

Figure 10.2 Absorption spectra of chlorophyll a. (a) The absorption peaks of chlorophyll *a* are in the red, yellow, and blue through the near-ultraviolet ranges. It is transparent to green light, which gives its characteristic green color. (b) Energy-level diagram of chlorophyll. The absorption peak in the red near 660 nm corresponds to the transition from the HOMO to the LUMO. The absorption peak in the yellow, around 570 nm, corresponds to the transition from one level below the HOMO to the LUMO. The peak in the violet, near 430 nm, corresponds to the transition from energy levels below the HOMO to energy levels above the LUMO. In a very short period of time, all those excitations are relaxed to the molecular state with one LUMO, which is about 1.88 eV above the ground state.

and C_2H_5 are bonded. Many other types of chlorophylls share the basic structure but with different groups at the external positions.

The absorption spectrum of chlorophyll *a* and its interpretation are shown in Fig. 10.2. It has three major groups of absorption peaks, centered at 662 nm (red), 578 nm (yellow), and 430 nm (blue). The green region is transparent, which gives rise to its characteristic green color. The energy diagram of chlorophyll *a* is shown in Fig. 10.2(b). In addition to the highest occupied molecular orbital (HOMO) and the lowest unoccupied molecular orbital (LUMO), two additional energy levels, one below the HOMO and one above the LUMO, are required to explain the absorption spectrum in the visible region. The absorption peak in the red, near 660 nm, corresponds to the transition from the HOMO to the LUMO. The absorption peak in yellow, around 570 nm, corresponds to the transition from one level below the HOMO to the LUMO. The peak in the violet, near 430 nm, corresponds to the transition from energy levels below the HOMO to energy levels above the LUMO. In a very short period of time, all those excitations are relaxed to the molecular state with one LUMO, which is about 1.88 eV above the ground state.

The energy stored in the excited chlorophyll molecule is transferred to an energy storage molecule, adenosine triphosphate (ATP). Then, ATP drives the process to synthesize sugar from carbon dioxide and water.

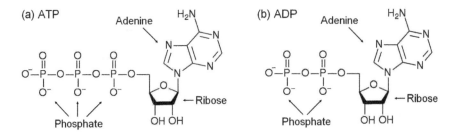

Figure 10.3　ATP and ADP. Both molecules contain an adenine, a ribose, and two or three phosphate groups. It takes energy to attach a phosphate group to ADP to form ATP. By detaching a phosphate group from ATP to recover ADP, energy is released. ATP is the universal "rechargeable battery" in biological systems.

10.1.2　ATP: Universal Energy Currency of Life

The structures of ATP and a related molecule adenosine diphosphate (ADP) are shown in Fig. 10.3. Both contain a nitrogenous base called *adenine*, a five-carbon sugar called *ribose*, and two or three *phosphate* groups. (Adenine is also one of the four nitrogenous bases which are the building blocks of DNA, the genetic code material.) These molecules are the universal rechargeable batteries in biological systems. ATP is the charged battery and ADP is the discharged battery. It takes some energy to attach a phosphate group to ADP to form ATP. By detaching a phosphate group from ATP to recover ADP, energy is released:

$$\text{ATP}^{4-} + \text{H}_2\text{O} \longrightarrow \text{ADP}^{3-} + \text{HPO}_4^{2-} + \text{H}^+ + 0.539\,\text{eV}. \tag{10.3}$$

The molecules ADP and ATP were isolated in 1929 from muscle tissues. In 1940, Fritz Lipmann (who won the Nobel Prize for Physiology and Medicine, 1953) proposed that ATP is the universal energy currency in cells. For example, when a human being is doing an aerobic exercise, glucose is oxidized by the oxygen in the blood into carbon dioxide and water, at the same time releasing energy. The energy is temporarily stored as ATP. Then the energized ATP drives the contraction of muscles.

The central role of ATP in photosynthesis was proposed by Daniel Arnon in the 1950s. His idea was met with a lot of skepticism. Then, a colleague in the same university, Berkeley, Melvin Calvin, did a series of experiments and discovered the process of photosynthesis and verified Arnon's hypothesis.

The molecule ATP plays such an important role in the energetics of life that further elucidation of its synthesis mechanism resulted in the 1997 Nobel Prize in Chemistry for Paul D. Boyer and John E. Walker and was named "The Molecule of the Year" by *Science* magazine in 1998.

10.1.3　NADPH and NADP$^+$

Besides transferring energy, reduction of CO_2 into carbon hydride is also required in the synthesis of glucose. The reduction agent in this process is NADPH, which can

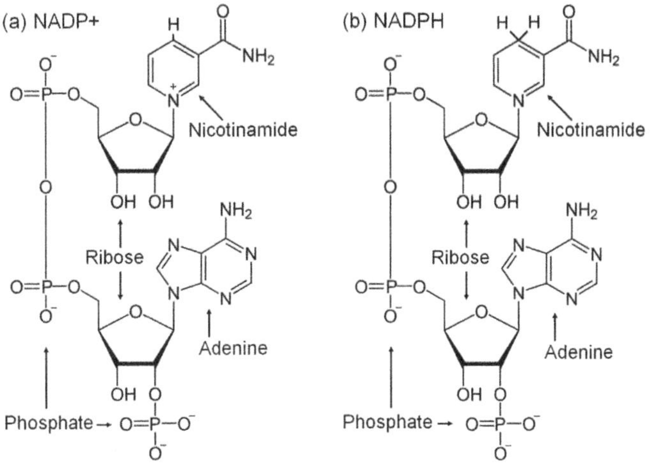

Figure 10.4 NADPH and NADP$^+$. The reduction agent in living systems. Both molecules contain a nicotinamide group, an adenine group, two ribose groups, and three phosphate groups. NADPH has two hydrogen atoms at a site on the nicotinamide. During reduction, it releases a hydrogen atom with an electron and becomes NADP$^+$.

release a hydrogen atom with an electron and become NADP$^+$. Both molecules contain a nicotinamide group, an adenine group, two ribose groups, and three phosphate groups (Fig. 10.4).

10.1.4 Calvin Cycle

During 1940s and 1950s, Melvin Calvin and his colleagues performed a series of experiments using a carbon isotope and worked out the pathway of photosynthesis. By adding $^{14}CO_2$ to a liquid containing the green alga *Chlorella pyrenoidosa*, using two-dimensional paper chromatograms, the subsequent molecules of the photosynthesis process were traced by the radioactivity of the ^{14}C atoms. The details of the process, the *Calvin cycle*, is rather complicated. Interested readers are recommended to check the books of Blankenship [10] or Chapter 24 of Voet [110]. Here we highlight some key points of the Calvin cycle with respect to energy transfer processes.

Figure 10.5 shows key steps in the Calvin cycle. The most significant discovery of Calvin's experiments is that the assimilation of carbon from CO_2 results in the generation of two identical three-carbon molecules (3-phosphoglycerate) by inserting a carbon atom and then cracking a five-carbon molecule (ribulose-1,5-biphosphate). The first step of the Calvin cycle is called *carbon fixation*. The next step is to reduce the carboxyl group to form triose phosphate by NADPH. The entire process of generating glucose requires that carbon fixation be repeated six times. In each loop, the five-carbon molecule (ribulose-1,5-biphosphate) must be regenerated to prepare for the next carbon assimilation process. It takes 9–10 photons to complete the process of fixing each carbon atom into the final product, for example, glucose.

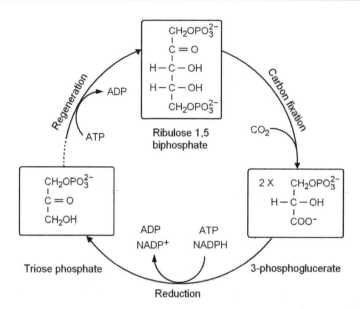

Figure 10.5 Key steps in the Calvin cycle. The Calvin cycle has three major steps. The first step, *carbon fixation*, is the insertion of a CO_2 molecule into a molecule with five-carbon atoms and two phosphate groups. The resulting molecule is split, generating two identical molecules with three carbon atoms each. In a subsequent step, the molecule is reduced by the action of NADPH. The process repeats six times to generate a glucose molecule. In each loop, the five-carbon molecule (ribulose-1,5-biphosphate) must be regenerated to prepare for the next carbon fixation step.

10.1.5 C4 Plants versus C3 Plants

In the conventional Calvin cycle, there is an alternative reaction in the carbon fixation step: An oxygen molecule can take the role of the CO_2 and generate two different products, one with three carbon atoms and another with two carbon atoms. The process using oxygen, *photoperspiration*, reduces the efficiency.

In some plants, such as maize (corn) and sugarcane, a better process takes place in the carbon fixation step to circumvent photorespiration by a CO_2 pumping mechanism which generates two identical molecular fragments of four carbon atoms each. Because the initial products of carbon fixation is a 4-carbon-atom molecule instead of a three-carbon-atom molecule, such variation of the Calvin cycle is called a C4 cycle and plants such as maize and sugarcane are called C4 plants. Especially under high temperature and high insolation environments, the photosynthesis efficiency of C4 plants is significantly higher than the great majority of plants which have a C3 cycle [10].

10.1.6 Chloroplast

In plants, photosynthesis actions and ingredients are contained in disklike units called *chloroplasts*, see Fig. 10.6(a). The chloroplast, with a typical size of 5 μm, has a well-

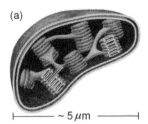
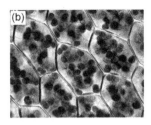

Figure 10.6 Chloroplast. (a) Chloroplast is the site of photosynthesis in plants. With a typical size of 5 μm, it has a well-defined structure to facilitate the flow of water, CO_2, and products. (b) A typical leaf cell contains 20–60 chloroplasts.

defined structure to facilitate the flow of water, CO_2, and products. Typically, each leaf cell contains 20–60 chloroplasts; see Fig. 10.6(b). A 1-mm^2 section in a typical corn leaf can have as many as a half million chloroplasts.

10.1.7 Efficiency of Photosynthesis

From an engineering point of view, the efficiency of photosynthesis is a critical parameter. By efficiency we mean the ratio of the chemical energy of the products of photosynthesis and the solar energy falling on the leaves.

Figure 10.7 shows the results of a study by Bolton and Hall [12]. The first loss is due to the wavelength range. The chlorophyll only absorbs less than one-half of the solar radiation, which is red, orange, and blue. The rest has no effect. The second loss is relaxation. As shown in Fig. 10.2, the excited molecule quickly relaxes to the state with only one LUMO, which is about 1.8 eV. The energy in the excited state of chlorophyll must be temporarily stored as usable chemical energy, the energy in ATP and NADPH, which is about 0.54 eV for each molecule. Sixty-eight percent of the energy is lost. The Calvin cycle is also not 100% efficient:. 35–45% of energy is lost. As a result, the net efficiency is about 5%.

Although the numerical value of the efficiency looks miserable, because of the enormous area of ground covered by plants, the total production of chemical energy by photosynthesis over the world each year is 3×10^{21} J, which is six times the total global consumption of energy in 2008.

To compare photosynthesis with other solar energy utilization processes, one often-used measure is *power density* in watts per square meter, defined as the chemical energy produced each year on a square meter of land divided by the number of seconds in a year. Note that the final useful product, for example, sugar or biodiesel, is only a small part of the total chemical product of photosynthesis. A large part of the products of photosynthesis, such as roots, branches, and leaves, is not useful. Table 10.1 is based on the data provided by a report published by United Nations Development Programme [90]. For comparison, data on total solar radiation and the power density of typical solar

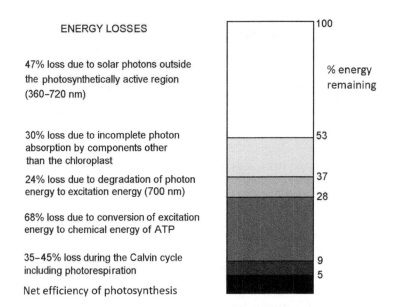

Figure 10.7 Efficiency of photosynthesis. The net efficiency of photosynthesis is about 5%. Notice that the largest percentage loss is due to the conversion of excitation energy in chlorophyll into ATP, that is, charging the biological rechargeable battery; and the limited wavelength range of the absorption spectrum of chlorophyll [12].

cells are also included. The average insolation is 1500 h/year. The average efficiency of crystalline silicon solar cell is 15%. The energy density of biomass is still much lower than that of typical solar cells. However, the cost of plants is much lower than solar cells.

Table 10.1: Power density of photosynthesis

Item	Energy Density	Power Density
Parameter	(MJ/year/m^2)	(W/m^2)
Average solar radiation	5400	171
Average silicon solar cell	810	25.6
Wood (commercial forestry)	3–8	0.095–0.25
Rapeseed (northwest Europe)	5–9	0.16–0.29
Sugarcane (Brazil, Zambia)	40–50	1.27–1.58

Source: *World Energy Assessment: Energy and
the Challenge of Sustainability*, UNDP, 2000 [90].

10.2 Artificial Photosynthesis

For many decades, scientists have been trying to mimic the elegant process of photosynthesis to convert sunlight into fuel which can be stored and applied to, for example, transportation. The most studied approach is to use sunlight to split water into hydrogen and oxygen,

$$H_2O + 2.46\,eV \longrightarrow H_2 + \frac{1}{2}O_2. \tag{10.4}$$

Hydrogen can be used directly as a clean fuel. Once hydrogen is generated, by combining with carbon dioxide, liquid fuel could be generated. Therefore, if this process can be demonstrated, it should be a true revolution.

The current status and difficulties of this approach have been summarized in a review paper [7]. Direct cleavage of water into hydrogen and oxygen by sunlight is still a lofty dream. One experimentally verified approach to generate hydrogen and oxygen in significant quantity is to generate electricity using solar cells and then split water through electrolysis. Because of the high cost and low efficiency, it is not competitive with other means of energy storage, such as rechargeable batteries (see Chapter 12).

10.3 Genetically Engineered Algae

Although artificial photosynthesis is progressing slowly, an alternative approach using biotechnology seems extremely promising. The focus is on algae. As a source of biological fuel, algae have several advantages. First, they live in water and thus do not occupy arable land or require irrigation. Second, algae can have very high oil content, up to 50%. Third, the waste disposal problem could be minimal. The yield of oil per unit area per year for algae could be many times greater than even the most efficient land-based oil-producing plant, the oil palm. Recently, the use of genetically altered algae to produce liquid fuel has enjoyed much attention. Through gene modification combined with directed selection, new species or variations of algae could be created which will grow faster, contain more fuel, and be easy to harvest. For details, see a report by U.S. Department of Energy [25].

10.4 Dye-Sensitized Solar Cells

The principles of photosynthesis have inspired the invention of a novel type of solar cell, the *dye-sensitized solar cell* [32, 33, 77]. It has several advantages over the common crystalline silicon solar cell. The cost of materials and processing is greatly reduced because most of the process is by liquid-phase deposition instead of in a vacuum. In addition, it can be made on lightweight flexible substrates. To date, nearly 10% overall conversion efficiency from AM1.5 solar radiation to electrical power is achieved.

The sensitization of semiconductors to light of wavelength longer than that corresponding to the band gap has been used in photography and photo-photochemistry.

The silver halides used in photography have band gaps on the order of 2.7–3.2 eV, and are not sensitive to most of the visible spectrum. Panchromatic films were made by adding dyes to sensitize silver halides, making them responsive to visible light.

The typical structure of a dye-sensitized solar cell is shown in Fig. 10.8. The most used semiconductor titanium dioxide (TiO_2), has many advantages for sensitized photochemistry and photophotochemistry: It is a low-cost, widely available, nontoxic and biocompatible material. As such it is also used in health care products as well as domestic applications such as paint pigmentation. The band gap, 3.05 eV, corresponding to a wavelength of 400 nm, lies in the near-ultraviolet region, which is too high for the solar spectrum. A dye is needed to mitigate this problem.

The ideal sensitizer for a single junction photovoltaic cell converting standard global AM1.5 sunlight to electricity should absorb all light below a threshold wavelength of about 920 nm. In addition, it must also carry attachment groups such as carboxylate or phosphonate to firmly graft it to the semiconductor oxide surface. Upon excitation it should inject electrons into the solid with a quantum yield of nearly unity. The energy level of the excited state should be well matched to the lower bound of the conduction band of the oxide to minimize energetic losses during the electron transfer reaction. Its redox potential should be sufficiently high that it can be regenerated via electron donation from the redox electrolyte or the hole conductor. Finally, it should be stable enough to sustain about 20 years of exposure to natural light. Much of the research in dye chemistry is devoted to the identification and synthesis of dyes matching these requirements while retaining stability in the photoelectrochemical environment. The

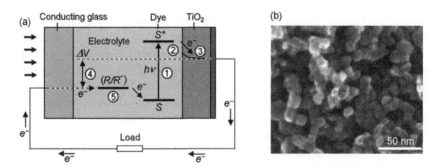

Figure 10.8 Structure of dye-sensitized solar cell. (a) The cell is built on top of a glass substrate with a conducting film. A nanostructured TiO_2 film of grain size about 15 nm and thickness about 10 μm is deposited on top of that conducting film. Dye molecules with a strong absorption band in the visible region are deposited on the surface of TiO_2 nanoparticles. The counter electrode is a film of transparent conducting oxide. The area between the cathode and the anode is filled with an electrode, typically a solution of lithium iodide. (b) a microscopic image of the TiO_2 film. The process of generating an electrical power is as follows: (1) Absorption of a photon by the dye to elevate an electron to the excited state, typically a LUMO. (2) Transfer of the electron to the TiO_2 film. (3) The electron relaxes to be at the bottom of the conduction band of TiO_2. (4) A photovoltage is generated by the cell, corresponding to the difference between the Fermi level in the semiconductor and the Nernst potential of the redox couple in the electrolyte. *Source:* Adapted from Refs. [33] and [77].

attachment group of the dye ensures that it spontaneously assembles as a molecular layer upon exposing the oxide film to a dye solution.

One of the most studied and used dyes is the N3 ruthenium complex shown in Fig. 10.9. The strong absorption in the visible region makes the dye a deep brown-black color, thus the name "black dye." The dyes have an excellent chance of converting a photon into an electron, originally around 80% but improving to almost perfect conversion in more recent dyes. The overall efficiency is about 90%, with the "lost" 10% being largely accounted for by the optical losses in the top electrode. The spectral response for a dye-sensitized solar cell using an N3 ruthenium dye is shown in Fig. 10.9(b). The photocurrent response of a bare TiO_2 film is also shown for comparison.

The four-step process of generating electrical power is as follows (see Fig. 10.8(a)).

1. A photon is absorbed by the dye to elevate an electron to the excited state, typically a LUMO.

2. The electron is transferred to the TiO_2 film.

3. The electron relaxes to be at the bottom of the conduction band of TiO_2.

4. A photovoltage is generated by the cell, corresponding to the difference between the Fermi level in the semiconductor and the Nernst potential of the redox couple in the electrolyte.

Nevertheless, dye-sensitized solar cells have some disadvantages. First, the efficiency is about one-half of that of the crystalline silicon solar cells. Second, the necessity of a liquid-phase electrolyte made the solar cell mechanically weak. Third, the long-term stability of the organic materials needs to be improved.

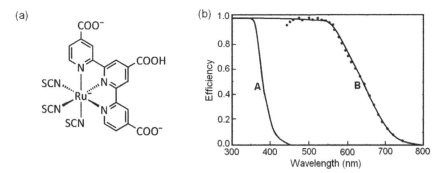

Figure 10.9 The N3 ruthenium dye and photocurrent spectrum. (a) Chemical structure of the N3 ruthenium complex used as a charge transfer sensitizer in dye-sensitized solar cells. (b) Photocurrent action spectra obtained with the dye as sensitizer, curve B. The photocurrent response of a bare TiO_2 films, A, is also shown for comparison. *Source*: Adapted from Refs. [33] and [77].

10.5   Bilayer Organic Solar Cells

Another approach to mitigate the high cost of crystalline silicon solar cells is to use organic semiconductors, or semiconducting polymers, to replace the expensive purified silicon. Because of its high absorption coefficient in the visible region, a very thin film of organic material is sufficient. These polymers can be deposited by screen printing, inkjet printing, and spraying, as these materials are often soluble in a solvent. Furthermore, these deposition techniques can take place at low temperature, which allows devices to be fabricated on plastic substrates for flexible devices.

The basic structure of a bilayer organic solar cell is shown in Fig. 10.10(a). There are two layers of polymer films: a film of an absorbing polymer, the *electron donor*, and a film of *electron acceptor*. The double layer is sandwiched between the anode, a TCO film, and a metal back contact, the cathode. The process of generating a photocurrent has four steps, see Fig. 10.10(b). In the first step, a photon is absorbed by the polymer, the electron donor. An exciton, an electron–hole pair, is generated. In the second step, the exciton diffuses inside the absorbing polymer (the donor) toward the interface to the acceptor. In the third step, the electron transfers to the acceptor. Finally, the electron is collected by the cathode, or the back contact. Through the external electric circuit, the electron goes back to the anode (TCO) and eliminates the hole.

In the first successful bilayer organic solar cell, copper phthaocyanine (CuPc) is used for the absorbing polymer [105]. The chemical structure and absorption spectrum are shown in Fig. 10.11. It is a solid with dark blue color, as the red, yellow, green, and violet radiations are heavily absorbed. The absorption coefficient in some ranges is more than 10^6 cm^{-1}. Therefore, a very thin film of the absorbing polymer, typically

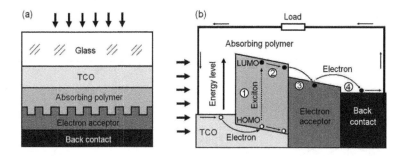

Figure 10.10 Bilayer organic solar cell. (a) A cross sectional view. Solar radiation comes from the top. Through the glass and a transparent conducting oxide (TCO) film, light is absorbed by the absorbing polymer film, or the *electron donor*. The electron thus generated transfers to the *electron acceptor*, and then to a metal back contact or the cathode. (b) The working process. (1) A photon generates an exciton, typically an electron in LUMO and leaves a hole in a HOMO. (2) The exciton diffuses towards the acceptor. (3) The exciton dissociates into a free electron and a hole. (4) The electron moves to the cathode, then drives the external circuit. *Source: Source:* Adapted from Refs. [105, 39].

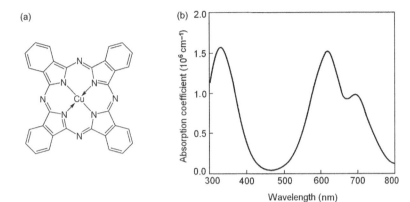

Figure 10.11 CuPc and its absorption spectrum. (a) Chemical structure of CuPc. (b) Absorption spectra of a solid thin film of CuPc. The red, yellow, green, and violet radiations are heavily absorbed. The material shows a dark blue color. *Source*: Adapted from Ref. [39].

around 100 nm, is used. A larger thickness is conversely a disadvantage because of the short diffusion length; see below.

The second process, exciton diffusion, deserves much attention. In contrast to semiconductor solar cells, the *diffusion lengths* of excitons in organic polymers are very short, typically 5–10 nm. Therefore, the lifetime of the excitons are very short. If the polymer is too thick, the excitons generated by photoexcitation might not reach the donor–acceptor interface and then disappear. In order to increase the probability for excitons to reach the acceptor, a nonplanar interface is often used; see Fig. 10.10(a).

In the third process, the exciton dissociates into a free electron and a free hole. The material for the acceptor should facilitate the dissociation and the final transfer of the carrier to the back contact, the cathode. In the first successful demonstration of a bilayer organic solar cell, a perylene tetracarboxylic derivative is used [105]. Later, C_{60} (buckminsterfullerene) and its derivatives are often used [15, 39].

Similar to dye-sensitized solar cells, the bilayer organic solar cells have disadvantages. Its efficiency is less than one-half that of the crystalline silicon solar cells. The long-term stability of the organic materials still needs to be improved.

Problems

10.1. Assume that (1) 50% of the leaf surface is stuffed with chlorophyll, (2) 30% of the solar photon hitting the chlorophyll generates one ATP, and (3) it takes 10 ATP to generate one-sixth of glucose, a unit of CH_2O. What is the efficiency of that photosynthesis process?

Hint: Use the blackbody radiation formula to estimate the average photon energy of sunlight. Note that

$$\int_0^\infty \frac{x^2\,dx}{e^x - 1} = 2\,\zeta(3) = 2.404. \tag{10.5}$$

10.2. The crown of a typical sugar maple tree is roughly a sphere of 5 m radius. If the photosynthesis process of the leaves is devoted to making syrup, on a sunny summer day, how many kilograms of condensed maple syrup (60% sugar in weight) can this tree produce?

Hint: 1 eV equals 96.5 kJ/mol. Use the experimental value of photosynthesis efficiency (5%) to estimate the solar radiation required to produce 1 kg of syrup.

Chapter 11

Solar Thermal Energy

11.1 Early Solar Thermal Applications

One of the earliest documentations of solar thermal energy applications is in *Code of Zhōu Regulations* (Zhōu Lǐ), a government document on the organization and laws of the West Zhōu dynasty (11th century *B.C.* to 771 *B.C.*). An entry in that book says, "The fire-maker uses a solar igniter (yángsuì) to start a fire using sunlight." Mòzǐ, a philosopher and physicist living in the Zhōu dynasty (468 *B.C.* to 376 *B.C.*), expounded the imaging properties of concave mirrors. A good quantitative understanding had been achieved at that time. To date, six such solar igniters have been found from various West Zhōu dynasty tombs. Sixteen from the East Zhōu dynasty (770 *B.C.* to 221 *B.C.*) were also unearthed. Figure 11.1 shows an example, dated about 1000 *B.C.*, discovered in 1995 in Fúfōng county of Shǎnxī province. Its diameter is 90.5 mm. Its radius of curvature is 207.5 mm, and the focal length is 103.75 mm. The original mirror was rusty. By making a mold from the original and cast in bronze, after polishing, the replica can ignite a straw using sunlight in a few seconds. Similar solar igniters have been found in other Bronze Age cultures over the world.

According to a Greek legend, in 212 *B.C.*, Archimedes used mirrors to focus sunlight on ships of an invading Roman fleet at Syracuse and destroyed the fleet. The use of burning mirrors for military purposes was a favorite theme of Middle Age and Renaissance scholars. However, later experiments in the 17th through the 19th centuries

Figure 11.1 A 3000-years-old solar igniter. A bronze solar igniter (yángsuì) of the early Zhōu dynasty, dated 1000 *B.C.*, discovered in 1995 from a Zhōu dynasty tomb in Fúfōng county of Shǎnxī province. With a radius of curvature 207.5 mm, its focal length is 103.75 mm. A replica of the original can ignite a piece of straw using sunlight in a few seconds.

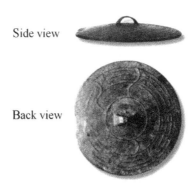

Side view

Back view

showed that even with modern technology and large mirrors the focused sunlight is not intense enough to burn ships at a reasonable distance. The story is probably a myth.

In 1767, French-Swiss scientist Horace Benedict de Saussure designed and built the first solar heat trapper that could be used for cooking [17]. Figure 11.2 shows the device. It is made of two wooden boxes, a small one inside a large one, with insulation (cork) in between. The inside of the small box is painted black. The top of the small box is covered by three separate sheets of glass, with air between adjacent glass sheets. By facing the top of the box toward the Sun, and moving the box to keep the glass perpendicular to sunlight, in a few hours, the temperature inside the small box reaches above 100°C. Therefore, it is a hot box heated by the Sun. To identify the source of the heat, de Saussure carried a hot box to the top of Mt. Cramont. He found that although the air temperature there is about 5–10°C lower than on the plain, the interior of the box could reach the boiling point of water as well. He attributed this effect to the clearness of the air on top of the mountain, where solar radiation is stronger.

Horace de Saussure's experiment is a demonstration of the *greenhouse effect*. It motivated Joseph Fourier to explain the equilibrium temperature of Earth by infrared absorption of Earth's atmosphere. Fourier explained his theory of the greenhouse effect by making an analogy to de Saussure's hot box where the glass sheets keep heat inside.

However, de Saussure's hot box was too slow to heat up, and the temperature could not become high enough for cooking, for example, 150°. This was probably the reason it did not become a popular product. The first mass-produced solar thermal device is the solar oven invented by W. Adams in the 1870s in Bombay, India [1]. He added a modest solar energy concentration device to de Saussure's hot box, as shown in Fig. 11.3. Eight glass mirrors (A) form an octagonlike reflector. Concentrated sunlight floods into the wooden box covered with glass (B), which contains a pot (C). The

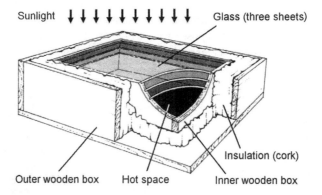

Figure 11.2 Hot box of Horace de Saussure. In 1767, Horace de Saussure designed and built the first solar cooking device, the hot box. Using three sheets of glass, the sunlight could come in but heat could not escape, making the temperature inside the box rise to the boiling point of water. Adapted from Butti and Perlin [17], courtesy of John Perlin.

Figure 11.3 Adams solar oven. In 1878, W. Adams invented a solar oven in Bombay, India. Eight mirrors made of silvered glass (A) form an octagonal reflector. Sunlight is concentrated and floods into a wooden box covered with glass (B), which contains a pot (C). The box can be rotated by hand to align with sunlight. The temperature in the box could exceed 200°C [1].

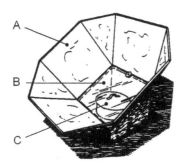

mirrors and the glass are inclined so that the rays of the Sun fall perpendicular to the box. When the Sun moves, the box can be rotated by hand to align with sunlight. In a paper published in 1878 in *Scientific American* [1], Adams reported, "The rations of seven soldiers, consisting of meat and vegetables, are thoroughly cooked by it in two hours, in January, the coldest month of the year in Bombay, and the men declare the food to be cooked much better than in the ordinary manner." His solar oven was then mass produced in India. In the United States, the Adams solar oven is mass-produced as a hobby and education device for teenage students.

A solar oven currently in widespread use in third world countries is shown in Fig. 11.4. It is a parabolic reflector made of cast iron and plated with chromium. Two steel ribs support a steel holder for a pot. The pot is positioned near the focal point of the parabolic dish. By manually aligning the parabolic dish with the sunlight, the pot can receive significant solar heat. This type of solar oven is mass produced in Eastern China and more than 10,000 units are sold annually in Tibet.

In the 20th century, the solar water heater was invented and improved, and became increasingly popular [17]. A brief history of the solar water heater is presented in Chapter 1. On the other hand, concentrated solar thermal electricity is a bright spot in solar power generation. In the following sections, we present the physics of solar thermal energy applications.

Figure 11.4 Cast-iron solar oven. Cast-iron parabolic reflector supported by a tripod with a joint which can be turned in two axes. Two steel ribs support a pot holder. By manually aligning the axis of the parabolic dish with the sunlight, the pot can always receive the maximum solar heat.

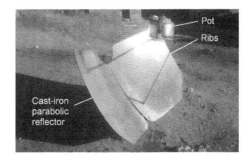

11.2   Solar Heat Collectors

For all solar thermal applications, the first step is to convert solar radiation energy into heat. From both materials and mechanical structure points of view, the key requirement is to absorb as much sunlight as possible and to lose as little heat energy as possible. Three methods are used: selective absorption surface, vacuum to block heat conduction and convection, and focused sunlight to change the ratio of the absorbing surface area and the emitting surface area.

11.2.1   Selective Absorption Surface

In early solar thermal applications, such as water heaters in the early 20th century, the absorbing surface was painted black. As a blackbody, it absorbs the maximum amount of solar radiation. When the body becomes hot, it lost energy by radiation.

The blackbody radiation spectra of the Sun and a hot body, for example, a pot of boiling water or a solar thermal absorber at $400°$, could be separable on the energy scale; see Table 2.4 and Fig. 2.5. It becomes more intuitive to make a plot of the relative spectral power density on a logarithmic scale of wavelength (see Problem 2.3),

$$u(\lambda, T) = \frac{2\pi hc^2}{\lambda^5 \left[e^{hc/\lambda k_B T} - 1\right]}$$

$$= \frac{3.75 \times 10^8}{\lambda^5 \left[e^{\lambda_T/\lambda} - 1\right]} \left(\frac{W}{m^2}\right) \left(\frac{1}{\mu m}\right),$$

$$(11.1)$$

where $\lambda_T = hc/k_B T$, and the power density is expressed in watts per square meter and the spectrum is expressed in terms of wavelength in micrometers. Figure 11.5 shows the relative spectral power density of AM1.5 solar radiation and from hot bodies. As shown, the peak of solar radiation is around 0.5 μm. On the other hand, the radiation from a hot body at $400°C$ is 4 μm and that of boiling water is 8 μm. The blackbody radiation spectrum from the hot liquid is well separated from that of solar radiation. Therefore, by designing a material having a high absorptivity for wavelength shorter than 2 μm and a low emissivity for longer wavelength, the absorption of solar radiation can be maximized and radiation loss can be minimized.

In the solar energy literature, the property of the surface is usually denoted by its *reflectance* at different wavelengths, $R(\lambda)$. The absorption surface for solar thermal applications is opaque. Energy conservation requires that the sum of reflectivity and absorptivity be

$$R(\lambda) + A(\lambda) = 1; \qquad (11.2)$$

see Eq. 5.2. According to Kirchhoff's law, the emissivity of a surface equals its absorptivity (Eq. 5.3),

$$E(\lambda) = A(\lambda). \qquad (11.3)$$

Therefore, the requirement of an ideal selective absorbing surface is low reflectivity for wavelengths shorter than 2 μm and a high reflectivity for longer wavelengths.

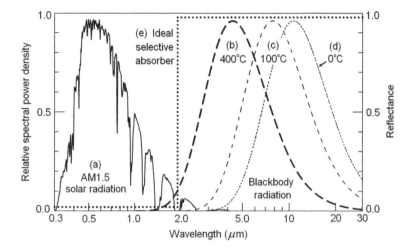

Figure 11.5 Spectral power density of solar radiation and hot bodies. (a) The spectral density of solar radiation concentrates to wavelengths shorter than 2 μm. (b)–(d) The spectral density of bodies on Earth. Even at 400°, it is concentrated in wavelengths greater than 2 μm. (e) The ideal selective absorber is a perfect blackbody for $\lambda < 2\,\mu$m and a perfect miror for $\lambda > 2\,\mu$m.

Two dimensionless quantities are defined to characterize the global performance of theselective absorption surface as follows. The solar absorptivity α is defined as

$$\alpha = \frac{\int_0^\infty u_\odot(\lambda)\left[1 - R(\lambda)\right] d\lambda}{\int_0^\infty u_\odot(\lambda)\, d\lambda}, \tag{11.4}$$

where the solar radiation power density $u_\odot(\lambda)$ is the AM1.5 direct normal spectral power density. The thermal emissivity ε is defined as

$$\varepsilon = \frac{\int_0^\infty u(\lambda, T)\left[1 - R(\lambda)\right] d\lambda}{\int_0^\infty u(\lambda, T) d\lambda}, \tag{11.5}$$

where $u(\lambda, T)$ is the blackbody radiation of the solar device at temperature T defined in Eq. 11.1. The photothermal conversion efficiency can be calculated using the formula,

$$\eta = \alpha - \varepsilon \frac{\sigma T^4}{C I_\odot}, \tag{11.6}$$

where σ is the Stefan–Boltzmann constant, T is the temperature of the solar heat collector, C is the concentration factor, and $I_\odot$ is the solar radiation power density, with a typical value of 1 kW/m^2 on a sunny day under normal incidence conditions.

Since the concept of *selective absorption surface* was proposed in the 1950s, it became an intensive international research project, especially in the United States, Israel, and Australia. For a review with an extensive list of references, see Ref. [46]. For a systematic treatment of theory and processes, see Ref. [52]. The ideal selective absorption curve is the dotted line in Fig. 11.5, with $\alpha = 1$ and $\varepsilon = 0$. Nevertheless, very good approximations have been achieved. There is still much research interest in this, especially for solar thermal electricity applications, where the temperature of the fluid should be as high as possible. According to basic thermodynamics, the maximum efficiency is the Carnot efficiency (Eq. 6.19),

$$\eta_c = 1 - \frac{T_L}{T_H}. \tag{11.7}$$

For example, if the fluid heated by solar radiation is 40°C higher than ambient temperature at 20°C (293 K), the maximum efficiency of a heat engine is 12%.

The requirements for a selective absorption coating are as follows:

1. High solar absorptivity, ideally $\alpha = 90\%$ to 97%

2. Low thermal emissivity, ideally $\varepsilon = 3\%$ to 10%

3. Durable at working temperature, for example, 30 years at 400°

4. Stable in air

5. Low-cost large-scale manufacturing

There are several types of selective absorption surfaces:

1. Intrinsic selective absorbers, such as semiconductors

2. Metal mirror coated with a thin absorption layer

3. Thin film with a window of transmittivity on a transparent cover.

4. Multilayer interference film

Because of the high stability, high contrast, and relative ease of mass production, a metal mirror coated with a thin absorption layer is the most popular type. The mirror can be made of any metal of high reflectivity for infrared radiation, such as copper, aluminum, nickel, and stainless steel. The absorption film is made of metal oxides, and named "dark mirrors."

Table 11.1 lists several well-studied and commercialized dark mirrors. the first three cases, black nickel, black chrome, and black copper, are produced with a liquid-phase process and are stable in air up to a few hundred degrees Celsius. The $Al–AlN_x$ system must be processed in a vacuum chamber, and is stable in a vacuum up to 500°C. It is particularly convenient for vacuum tube thermal absorbing systems because the deposition of aluminum and AlN can be made in the same processing chamber. The

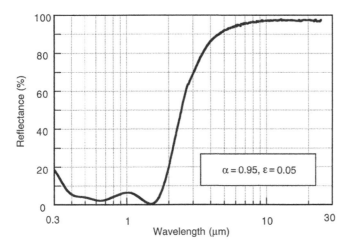

Figure 11.6 Reflectance curve for cermet selective-absorbing surface. The measured reflectance spectrum of a surface with AlN cermet composite on stainless steel base. It approaches the ideal behavior shown in Fig. 11.5.

first step, aluminum deposition, is performed in a vacuum. Then, by controlled bleeding of nitrogen, good-quality AlN$_x$ film can be formed. Usually, such selective coating is working under a vacuum, it does not deteriorate.

For high-temperature systems, for example, solar thermal electricity system working at 400–500°C, especially with a requirement of stability in air, a cermet coating is preferred. A cermet is a composite material composed of ceramic (cer) and metallic (met) materials which could have the properties of both a ceramic, such as high-temperature resistance and hardness, and a metal, such as the ability to undergo plastic deformation. Therefore, properly designed cermet coatings can withstand high temperature and are stable in air [47]. Figure 11.6 shows the reflectivity spectrum of a surface with an AlN cermet composite on a stainless steel base. As shown, it approaches the ideal behavior illustrated in Fig. 11.5.

Table 11.1: Selective absorbing surfaces

System	Mirror	Absorbent	α	ϵ
Black nickel	Ni or steel	NiS–ZnS	0.88–0.96	0.03–0.10
Black chrome	Cr	Cr_2O_3	0.97	0.09
Black copper	Cu	Cu_2O–CuO	0.97–0.98	0.02
Aluminum nitride	Al	Al_3N_4	0.97	0.10

Source: Ref. [46].

11.2.2 Flat-Plate Collectors

The early solar water heaters use flat-panel solar heat collectors. It is relatively easy to build and rugged enough to withstand the elements. Figure 11.7 shows a schematic. It has a copper plate with copper pipes soldered on, painted black. Usually one or two sheets of glass are installed for top-side heat insulation. Water can be heated up to 60° or 80°. It is a rugged device which can last for many decades.

The standard treatment of the flat-panel solar heat collectors is based on the Hottel–Whillier model [42]. Details can be found in Duffie and Beckman [23, 24], Lunde [59], and other publications [49, 51, 88].

Most solar thermal collectors are covered with glass. The normal-incidence transmittance of the glass cover is (see Chapter 9),

$$\tau = \frac{(4n)^{2N}}{(1+n)^{4N}}, \tag{11.8}$$

where N is the number of sheets and n is the refractive index of glass, typically $n = 1.5$ (see Section 9.4).

Consider a solar collector plate with total area A and absorptivity α. If the portion of effective absorption area is F and the solar power density is P_0, the input power to the plate is

$$Q_I = FA\tau\alpha P_0. \tag{11.9}$$

Because of the solar power, the temperature of the plate is elevated from ambient temperature T_a to T_p. If the temperature difference is not too great, the heat loss is proportional to the temperature difference. In all practical cases, the area of the panel, FA, is much greater than the area of the edge. For clarity and brevity, the edge effect is neglected. The heat loss is also proportional to the area of the hot plate,

$$Q_L = U_L FA (T_p - T_a). \tag{11.10}$$

Here, U_L is the *combined heat loss coefficient*. The efficiency of the solar thermal energy collector is

$$\eta = \frac{Q_I - Q_L}{P_0 A} = F\left(\tau\alpha - \frac{1}{P_0}U_L(T_p - T_a)\right)^+. \tag{11.11}$$

Figure 11.7 Flat-plate solar heat collector. It is essentially a de Saussure hot box, Fig. 11.2, hosting a copper plate with copper pipes soldered on, painted black. Usually one or two sheets of glass are installed for top-side heat insulation. Water can be heated up to 60° or 80° by sunlight. It is a quite rugged device which could last for many decades.

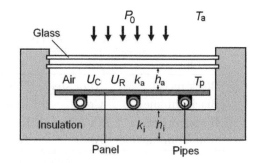

The plus sign indicates that only a positive value of the expression is taken. In other words, if the expression is negative, the value is taken as zero.

Therefore, the problem of the efficiency of the solar heat collector is reduced to the evaluation of the heat loss coefficient U_L. For flat-panel collectors, the problem is nontrivial. Many factors contribute to the loss through the front cover, and some of them are quite difficult to quantify:

1. *Conduction through the back-side insulation* is easy to quantify, as the corresponding loss factor equals k_i/h_i, the thermal conductivity and thickness of the insulation material.

2. *Conduction through the air space* is also easy to quantify, as the corresponding loss factor equals k_a/h_a, the thermal conductivity and thickness of air.

3. *Convection in the air space* is difficult to quantify. It has a complicated dependence on spacing and the tilt angle.

4. *Convection outside the glass cover* depends not only the temperature but also the wind speed of ambient air.

5. *Radiation.* The thermal radiation of the panel first reaches the glass panel. It essentially absorbs most of the radiation. The glass panel, being heated, radiates again to the ambient.

The transmittivity spectrum of common window glass is shown in Fig. 11.8. It is transparent for visible and near-infrared radiation, but opaque for ultraviolet and far-infrared radiation. The blackbody radiation from the hot plate is almost completely absorbed by the glass.

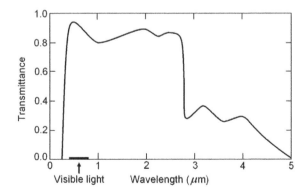

Figure 11.8 Transmittance of window glass. The window glass is transparent for the visible and near-infrared radiation, but opaque for ultraviolet and far-infrared radiation. The blackbody radiation from the hot plate is almost completely absorbed by the glass. Plotted using the data from American Institute of Physics Handbook.

Table 11.2: Typical parameters of flat-plate solar heat collectors

Parameter	Description	Symbol	Unit	Value
Cover glass	Refractive index	n	—	1.50
Panel	Absorptance	α	—	0.95
Insulation	Thermal conductivity	k_i	W/m²·K	0.02
Insulation	Thickness	d_i	meter	0.05
Air space	Thermal conductivity	k_a	W/m²·K	0.024
Air space	Thickness	d_a	m	0.025

The *combined heat loss coefficient* can be estimated as

$$U_L = \frac{k_a}{h_a} + \frac{k_i}{h_i} + U_C + U_R. \tag{11.12}$$

Where k_a and k_i are the thermal conductivities of the air and insulation material, respectively, h_a and h_i are the thicknesses of the air and insulation wall, respectively (see Fig. 11.7), U_C is the convection heat loss coefficient, and U_R is the radiation heat loss coefficient. The first two terms can be estimated using the typical parameters listed in Table 11.2,

$$\frac{k_a}{h_a} + \frac{k_i}{h_i} \approx 1.36 \, (\text{W/m}^2 \cdot \text{K}). \tag{11.13}$$

However, the last two terms are usually much larger than the first two. An effective way to reduce convection and radiation loss is to increase the number of sheets of glass. According to the calculations of Duffie and Beckman [23, 24], under normal conditions (wind speed 5.0 m/s, average plate temperature 60°C, slope 45°, ambient temperature minus 20–40°C), the top loss coefficient is 6.9 W/m²·K for one cover, 3.5 W/m²·K for two covers, and 2.4 W/m²·K for three covers. However, more covers result in more loss of transmittance. From Eq. 11.8, the transmittance of one sheet of glass is 0.92. It is reduced to 0.85 for two sheets, and 0.782 for three sheets. Therefore, using three covers does not represent an advantage.

11.2.3 All-Glass Vacuum-Tube Collectors

As shown in the previous section, the single most important factor affecting the efficiency of solar heat collectors is the heat loss through the top cover. As early as 1911, William L. R. Emmet invented vacuum tube heat collectors (U.S. Patent 980,505) which could in principle completely resolve the problem of top-cover heat loss. It took 80 years to make them suitable for mass production.

Figure 11.9 shows a modern evacuated-tube solar thermal collector. It is made of two concentric glass tubes sealed at one end. The space in between is evacuated to a medium high vacuum. A metal spacer is placed between the tubes to support the tubes and as a source for getter, typically a mixture of barium and titanium. After being sealed, the getter support is heated from outside using microwave power to evaporate the getter onto the inner surface of the glass tubes. A high vacuum thus can be maintained. Good-quality evacuated tubes should have a vacuum better than 10^{-4} Pa, or 10^{-6} Torr. A selective absorption coating is applied on the *outer surface* of the *inner glass tube*. Therefore, the selective coating is always under high vacuum. It has a significant advantage over the flat-panel heat collector as a selective coating stable in air is not required and the coating could stay intact virtually forever. Furthermore, an antireflection film can be applied on top of the selective absorption coating even if the film is not stable in air.

Another advantage over the flat-panel heat collector is that the materials are inexpensive and abundant and can be mass produced at very low cost. To date, 200 million evacuated tubes are produced annually in China.

An important consideration is that the area ratio F is much less than 1 for vacuum tube collectors, because the diameter of the inner tube determines the absorption area, and when the tubes are installed on a system, there should be spacing between adjacent outer tubes. Typically the space is 20 mm. The typical area ratio $F = 47/(58 + 20) \approx 0.6$. The smaller factor F actually is not a serious disadvantage. First, in residential applications, there is always more roof space than needed. Second, An important consideration is the *cost* of the solar thermal energy collector. Because there is empty space between the tubes, there is no additional cost. Third, because of the empty space, there is no additional weight on the roof. Finally, if the sunlight is not perpendicular to the plane of the tubes, up to an angle of incidence $\theta = \arccos(F)$, maximum power can be maintained.

11.2.4 Thermosiphon Solar Heat Collectors

Evacuated-tube solar energy collectors are used primarily for direct-flow solar water heaters, where the usable water goes directly into the tubes. It has very high effi-

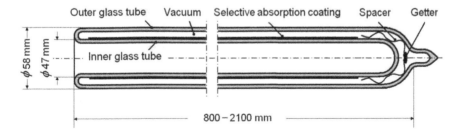

Figure 11.9 Evacuated-tube solar thermal collector. It is made of two concentric glass tubes sealed at one end. The space in between is evacuated to better than 10^{-4} Pa, or 10^{-6} Torr. A metal spacer is placed as a source for the getter, typically a mixture of barium and titanium. After it is sealed, the getter is evaporated onto the inner surface of the glass tubes. A high vacuum thus can be maintained. A selective absorption coating is applied on the outer surface of the inner glass tube.

ciency. However, the hot water could be contaminated by the system, and the pressure comes directly from gravitation. For systems requiring pressurized hot water and more stringent sanitation, thermosiphon solar heat collectors are used.

Figure 11.10(a) shows a cross-section of a thermosiphon solar heat collector. At the center is a sealed metal tube typically made of copper. A small amount of volatile liquid is in the metal tube, typically water. The metal tube is connected with metal fins covered with selective absorption coatings. The metal tube is mounted on a metal flange, typically stainless steel. A glass–metal joint is formed between the flange and the glass tube. A vacuum is drawn in the glass tube. Figure 11.10(b) is a photo of the device. The tube must be installed at a tilted position with the evaporator at the top. With sunlight falling on the metal fins, the liquid in the metal tube is evaporated, then condensed at the top, which is thermally connected to a heat load.

A key technical problem is formation of the glass–metal joint. A widely used technology is a metal gasket of relatively low melting point, such as tin, lead, or aluminum. By heating the joint under pressure at a temperature lower but close to the melting point of the metal gasket, a good joint can be formed. The high vacuum in the tube constantly exerts pressure on the glass–metal joint; therefore, the probability of a leakage is small. Compared with all-glass vacuum tubes, thermosiphon solar heat collectors have several advantages. First, because there is no running water in the tubes, it can withstand bitter cold without breaking the glass. Second, because the thermal mass of the tube is much smaller than the water in the all-glass tubes, the start-up time is much

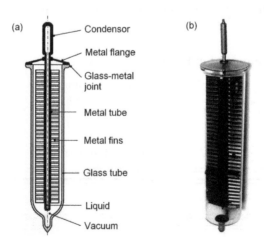

Figure 11.10 Thermosiphon solar heat collector. (a) Structure of a thermosiphon solar heat collector. At the center is a sealed metal tube, typically made of copper. A small amount of volatile liquid is filled in the metal tube, typically water. The metal tube is connected with metal fins, covered with selective absorption coatings. The metal tube is mounted on a metal flange, typically stainless steel. A glass-metal joint is formed between the flange and the glass tube. A vacuum is drawn in the glass tube. (b) Photograph of the collector. With sunlight falling on the metal fins, the liquid in the metal tube is evaporated, then condensed at the top, which is thermally connected to a heat load.

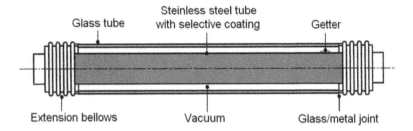

Figure 11.11 High-pressure vacuum tube collector. At the core is a stainless steel tube coated with a selective absorption film. Both ends are fitted with extension bellows. Through a glass–metal joint, each side is attached to an end of the glass tube. A vacuum is drawn between the stainless tube and the glass tube. A getter is included to maintain a good vacuum.

shorter. Third, even if one of the glass tubes is broken, for example, by hail impact, there is no water leakage. Fourth, because hot water does not flow in the tubes, the tank can be pressurized and run high-standard clean water. Lastly, because the liquid returns to the bottom of the siphon tube by gravity, there is a *thermal diode effect* — the heat only flows from the collector to the tank and cannot be reversed. However, because of the metal structure and the glass–metal joint, the cost is much higher than the all-glass tubes. Therefore, it is used in high-end solar water heaters.

11.2.5 High-Pressure Vacuum Tube Collectors

For solar thermal applications, the working fluid inside the tube is not hot water. It is either oil at high temperature (300°C or more) or superheated steam at high pressure (10–100 atm). The glass tube would not withstand such temperature and pressure. The inner tube must be made of a strong metal, typically stainless steel. The outer tube must be transparent, made of glass. Therefore, there is a problem of mismatch of thermal expansion coefficients, and a metal–glass joint is required. The typical structure of such a solar heat collector is shown in Fig. 11.11. At the core is a stainless steel tube coated with a selective absorption film. Both ends are fitted with extension bellows. Through a glass-metal joint, each side is attached to an end of the glass tube. A vacuum is drawn between the stainless steel tube and the glass tube. A getter is included to maintain a good vacuum.

11.3 Solar Water Heaters

Currently, in the United States and Europe, solar water heaters using flat-panel solar thermal collectors similar to the Day-and-Night solar water heater are still very common. However, worldwide, great majority of solar water heaters are direct-flow type using all-glass evacuated-tube solar heat collectors. Usually, it is installed on the roof

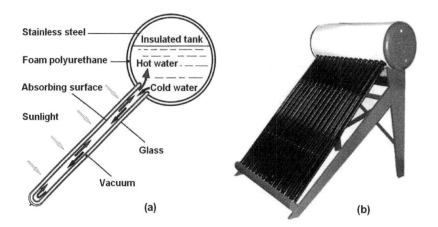

Figure 11.12 Solar water heater with thermosiphon collectors. (a) Design of the solar water heater. The evaporators of the thermosiphon solar heat collector are in thermal contact with the water in the tank through copper blocks. (b) A system on a roof after more than 10 years of operation. No degradation is observed. Photo taken by the author.

of a single-family house or an apartment building. The hot water flows simply by gravitation. Because of low manufacturing cost as a result of large-scale mass production, whenever such a system can be used, the investment can be recouped in a few years without government subsidy. After the initial cost is paid off, the system can work properly for 20–30 years with no maintenance. In China alone, more than 10 million such systems have been installed. For an example, see Fig. 11.12.

11.4   Solar Thermal Power Systems

A more important area of solar thermal application is electric power generation. According to the second law of thermodynamics, the upper limit of the efficiency of converting heat to mechanical power is the Carnot efficiency (Eq. 6.19),

$$\eta_c = 1 - \frac{T_L}{T_H},$$ (11.14)

where T_L is the temperature of the cold reservoir and T_H is the temperature of the hot source. In order to improve efficiency, the temperature of the hot source should be as high as possible. Since T_L cannot be lower than the atmospheric temperature, on average 300 K, in order to achieve an efficiency of 50%, T_H should at least be 600 K, or 327°C. Without concentration, the temperature can hardly reach 150°C.

11.4.1 Parabolic Trough Concentrator

Before the popularity of solar photovoltaics, much of solar electricity was generated by concentrated solar thermal power plants is from parabolic trough systems. The structure of the system is shown in Plate 11. Parabolic mirrors are mounted on an axis to track the Sun. At the focal line of the parabolic mirrors is a linear collector, typically a high-temperature evacuated tube.

To date, the largest solar power plants are the Solar Energy Generating Systems (SEGS) in the California Mojave Desert. The nine SEGS power plants built between 1984 and 1990 have a total capacity of 354 MW. The total area of the 232,500 parabolic mirrors is 6.5 km^2, with a total length of 370 km. The sunlight bounces off the mirrors and is directed to a central tube filled with synthetic oil which heats to over 400°C. The reflected light focused at the central tube is 71–80 times more intense than ordinary sunlight. The synthetic oil transfers its heat to water, which boils and drives the Rankine cycle steam turbine, thereby generating electricity. Synthetic oil is used to carry the heat (instead of water) to keep the pressure within manageable parameters.

Figure 11.13 is an aerial photo of the SEGS. The axis of the parabolic mirrors is north–south, which turns from east to west every day. The most expensive parts are the curved mirrors. Wind damage occurs frequently, and about 3000 mirrors are replaced each year. The mirrors are periodically cleaned by a special machine. To avoid shadows from neighboring mirrors, the distance between mirrors is about twice the width of the mirrors. The cost of electricity from the SEGS is still not competitive with coal-burning power plants.

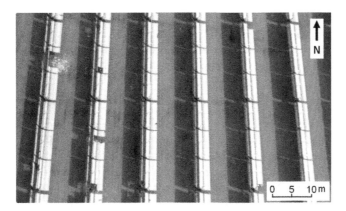

Figure 11.13 Aerial photo of SEGS system. The axes of the parabolic mirrors are north–south, which turn from east to west every day. About 3000 mirrors are replaced each year because of wind damage. Several damaged mirrors are shown in this photo. To avoid shadows from neighboring mirrors, the distance between adjacent mirrors is about twice the width of the mirrors.

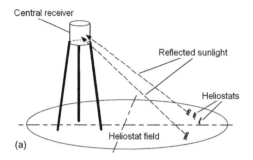

Figure 11.14 Solar power plant with central receiver. To achieve high power at a centralized receiver, hundreds of *heliostats* are used, that is, mirrors mounted on a two-dimensional shaft to follow the position of the Sun by a centralized computer. The temperature of the central receiver can reach 500°C or more.

11.4.2   Central Receiver with Heliostats

To achieve high power at a centralized receiver, hundreds of *heliostats*, mirrors mounted on a two-dimensional shaft to track the position of the Sun by a centralized computer distributed on a typically circular or oval field; see Fig. 11.14. Because very high temperatures can be reached (e.g. 565°C), superheated steam or molten salt (e.g. 60% $NaNO_3$ + 40% KNO_3) is usually used as the working material. Finally, it drives a standard steam turbine to generate electrical power [91].

The first pilot project, Solar One, in the Mojave Desert, was completed in 1981, and was operational from 1982 to 1986. It has 1818 mirrors, each 40 m^2. Oil or water was used as the working fluid. An efficiency of 6% was demonstrated. A photo of Solar One is shown in Fig. 11.13(b). In 1995, Solar One was upgraded to Solar Two. Molten salt, 60% sodium nitride and 40% potassium nitride were used as the working material. Ten megawatts of power is demonstrated, and its efficiency was improved to 16%. On November 25, 2009, the Solar Two tower was demolished.

11.4.3   Paraboloidal Dish Concentrator with Stirling Engine

The third type of concentration solar power plant is the paraboloidal dish concentrator with Stirling engine. According to a report released by Sandia National Laboratory in February 2008, the Stirling engine system has shown a solar-to-grid energy conversion efficiency of 31.25%, the highest of any solar-to-electricity conversions, recorded on a perfectly clear and cold New Mexico winter day [54].

The Stirling engine was invented in 1816 by Robert Stirling, a priest studying mechanical engineering, built the first one in his home machine shop. A schematic diagram of the Stirling engine is shown in Fig. 11.15. This engine is different from the two popular heat engines, the steam engine and the internal combustion engine. Similar to the steam engine, it uses an *external heat source*. However, instead of constantly

Figure 11.15 Stirling engine. Invented in 1816 by Robert Stirling, the engine without a valve is the simplest heat engine. A fixed amount of gas, typically hydrogen, is used as the working medium. It can be driven by a single heat source, with an efficiency close to the Carnot limit. It is the ideal heat engine driven by concentrated sunlight.

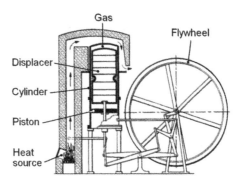

evaporating water into steam and then discarding it, the Stirling engine uses a fixed body of gas in a closed cylinder. It is the simplest heat engine: There are no valves. It can approach the Carnot efficiency. It can be operated by any type of single heat source; thus concentrated sunlight is perfect. In Fig. 11.15, the heat is provided by burning wood or coal. A piston is tightly fit in a cylinder which drives a flywheel through a crankshaft. A gas displacer, loosely fit in the cylinder, is driven by the crankshaft with a phase shift in the motion of the piston. The working media, the gas, is always contained in the cylinder. The details of its working cycles are shown in Fig. 11.16. During step (a), the gas is cold, the displacer is in the innermost position, and the piston pushes into the cylinder. Then the hot source heats up the gas in step (b). During step (c), the expanding hot gas pushes the piston outward. Finally, step (d), the displacer returns to the innermost position, and the gas is cooled by the surrounding.

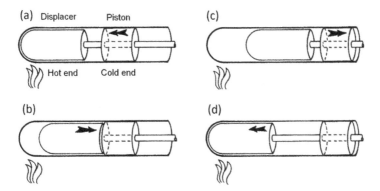

Figure 11.16 Working principle of Stirling engine. (a) The gas is cold, the displacer is in the innermost position, and the piston pushes into the cylinder. (b) The hot source heats up the gas. (c) The expanding hot gas pushes the piston outwards. (d) The displacer returns to the innermost position and the gas is cooled by the surrounding.

In order to achieve effective operation, the gas must have high thermal conductivity. The most frequently used gas is hydrogen. However, the diffusion coefficient of hydrogen in steel is very high. Therefore, either a special material with low diffusion coefficient for hydrogen is used to construct the cylinder or the hydrogen is periodically supplemented.

The Stirling engine is not suitable for vehicle applications because of its large volume and the requirement of an effective cooling mechanism.

Problems

11.1. Based on the algebraic definition of a parabola, $y = x^2/4f$, prove that a parabola is the locus of equal distance from a focus to a directrix. See Fig. 11.17.

11.2. Using the geometric definition in Problem 11.1, prove that all light rays parallel to the y-axis would be reflected by a parabolic surface to the focus. See Fig. 11.17.

11.3. A vacuum solar heat collector has an internal diameter of 45 mm and a length of 1800 mm and is filled with water of 20°C. The axis of the tube is perpendicular to the sunlight. Assuming the efficiency is 90%, with full sunlight, how long does it take to boil the water inside the tube?

11.4. A solar hot water system consists of 24 vacuum tubes with outer diameter 58 mm and inner diameter of 47 mm, and length of 1800 mm. It is connected to an insulated

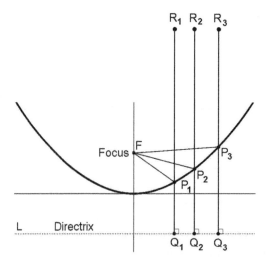

Figure 11.17 Focusing property of a parabola.

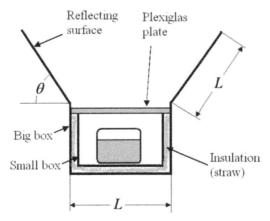

Figure 11.18 Kyoto box. A simple solar oven used extensively in Africa.

tank containing 200 L of water. On a sunny day, with the sunlight perpendicular to the plane of vacuum tubes, and the efficiency is 90%, how long it takes to heat up the water by 20°C? (see Figure 1.32, Figure 11.11, and Figure 11.17).

11.5. Kyoto box is a simple solar oven used extensively in Africa; see Fig. 11.18. Assuming a square box with four reflectors of $L = 75$ cm, on a sunny day with direct sunlight from the zenith, what is the optimum angle θ? What is the total solar power received by the box? If the efficiency is 70%, how long it will take to heat 1 gallon (3.785 L) of water from 25°C to the boiling point?

Chapter 12

Physical Energy Storage

Because solar energy is intermittent, if it is the main supply, energy storage is a necessity. In general, there are two types of energy storage: utility-scale massive energy storage and the application-related, distributed energy storage. For utility-scale energy storage, the most effective method is using a reversible hydroelectric plant, which stores mechanical energy as potential energy of water in a high-level reservoir. The principle of such energy storage mechanics has been discussed in Section 1.5.1. We will present more details in Section 12.1. For distributed energy storage, thermal energy storage is efficient and economical. The most versatile energy storage system is rechargeable battery, based on electrochemistry. We will discuss it in Chapter 13.

12.1 Pumped Hydro Storage

Pumped hydro storage (PHS) is based on pumping water from a lower reservoir to another at a higher elevation at low-demand period. When demand hits the peak, the collected water is discharged to the lower reservoir through a turbine to re-produce electricity. The storage capacity is determined by the height in water discharge and volume. Typically, PHS can turned on within a minute when necessary to provide reliable power for many hours [89].

An example of such PHS system is the Nant de Drance plant in Switzerland, officially inaugurated in September 2022, see Fig. 12.1. In that cross-section view, (A) is the upper reservoir, (B) is the vertical water tunnel, (C) is the machine room containing a turbine-pump, (D) is the horizontal water tunnel, and (E) is the lower reservoir. When there is surplus energy, the machine can function as a pump to move water from the lower reservoir to the upper reservoir. When electricity is needed, the machine can function as a turbine to generate electricity. The round-trip efficiency is about 80%.

There are several key parameters for the design of the PHS system. The maximum volume of water in the reservoir V, the average height difference of the two reservoirs H, and the power of the turbine, P. The maximum gravitational energy is

$$E = VHg, \tag{12.1}$$

where G is the gravitational acceleration, 9.8 m/s^2. If the power of the turbine is P,

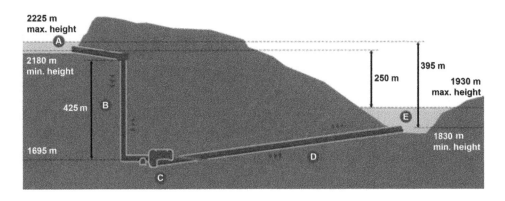

Figure 12.1 The Nant de Drance pumped hydro storage plant. (A) is the upper reservoir, (B) is the vertical water tunnel, (C) is the turbine-pump, (D) is the horizontal water tunnel, and (E) is the lower reservoir. It can maintain a 900 MW output for about 22 h.

the maximum time the system can support is

$$t = \frac{E}{P}. \tag{12.2}$$

The design parameters of the Nant de Drance pumped hydro PHS plant is as follows. The volume of the reservoir is 25 million cubic meters. The average difference of height, or the *head*, is 300 meters. The total amount of stored energy is

$$E = 25 \times 10^9 \times 9.8 \times 300 \cong 7.35 \times 10^{13}\,\mathrm{J}. \tag{12.3}$$

Because one GWh is 3.6 $\times 10^{12}$ J, the total stored energy is approximately 20 GWh. The station has six 150 MW turbines. The total power is 900 MW. Therefore, the storage system can maintain output for about 22 hours.

12.2 Sensible Heat Energy Storage

Storage of energy as heat content of matter is inexpensive and easy to implement. It can be applied to space heating and cooling as well as for power generation. Two types of thermal energy are used: sensible thermal energy, essentially proportional to temperature difference, and phase transition thermal energy, such as the latent heat during freezing and melting, which could maintain a fixed temperature with energy content much greater than sensible thermal energy. Phase-change materials (PCM) are well suited for the storage of solar energy.

Sensible heat energy storage utilizes the heat capacity and the change in temperature of the material during the process of charging or discharging—the temperature of the storage material rises when energy is absorbed and drops when energy is withdrawn. One of the most attractive features of sensible heat storage systems is that charging and discharging operations can be expected to be completely reversible for an unlimited number of cycles, that is, over the lifespan of the storage.

In sensible heat energy storage, the thermodynamic process of the material is almost always *isobaric*, or under constant pressure, typically atmospheric pressure. A solid or liquid is usually used. The specific heat of gas is too low and not practical for thermal energy storage. The heat Q delivered by the material from initial temperature T_1 to final temperature T_2 is

$$Q = M \int_{T_1}^{T_2} c_p \, dT, \tag{12.4}$$

where M is the mass, c_p is the isobaric specific heat. In most applications, the density and specific heat can be treated as a constant. Equation 12.4 can be simplified to

$$Q = Mc_p \left(T_2 - T_1 \right). \tag{12.5}$$

The quantity of material required for the storage tank and the heat losses are approximately proportional to the surface area of the tank. The storage capacity is proportional to the volume of the tank. Larger tanks have a smaller surface area–volume ratio and therefore are less expensive and have less heat losses per unit energy stored.

An important issue in thermal energy storage is thermal conduction, or temperature equalization in the medium. In liquids, heat conduction has two major paths: conduction and convection. Temperature in a liquid medium can become equalized much faster than in a solid. Therefore, liquid is preferred whenever applicable. Table 12.1 shows some thermal properties of commonly used liquids for sensible heat thermal energy storage.

Table 12.1: Thermal properties of commonly used materials

Materials	Density ρ	Heat capacity c_p	Product ρc_p	Temperature range ΔT
	(10^3 kg/m^3)	(10^3J / kg·K)	$(10^6 \text{ J / m}^3\text{·K})$	$(°C)$
Water	1.00	4.19	4.19	0 to 100
Ethanol	0.78	2.46	1.92	−117 to 79
Glycerine	1.26	2.42	3.05	17 to 290
Canola Oil	0.91	1.80	1.64	−10 to 204
Synthetic Oil	0.91	1.80	1.64	−10 to 400

Source: American Institute of Physics Handbook, 3rd Ed.,
American Institute of Physics, New York, 1972.

12.2.1   Water

As shown in Table 12.1, water has the largest heat capacity both per unit volume and per unit weight, and it is free. Therefore, it is logical to use water as the material for sensible heat storage. A typical case is the hot-water tank used in most homes. The tank is typically insulated by foam polyurethane, which has a thermal conductivity κ = 0.02 W/mK and density ρ = 30 kg/m^3.

   If the entire tank is filled with water, the volume is

$$V = \frac{1}{4}\pi D^2 L \qquad (12.6)$$

and the heat capacity is

$$C_p = \rho c_p V. \qquad (12.7)$$

The total surface area of the tank is

$$A = \frac{1}{2}\pi D^2 + \pi D L. \qquad (12.8)$$

The rate of heat loss is

$$\frac{dQ}{dt} = \frac{\kappa A}{\tau}\left(T_w - T_a\right), \qquad (12.9)$$

where $T_w - T_a$ is the difference of water temperature T_w and ambient temperature T_a. The rate of temperature loss is

$$\frac{dT}{dt} = \frac{2\kappa(D + 2L)}{\tau \rho c_p D L}\left(T_w - T_a\right). \qquad (12.10)$$

   The rate of temperature drop through the tank skin is proportional to the total surface area and inversely proportional to the volume. If the tank is too thin or too flat, then the heat loss is high. Therefore, for a tank of fixed volume, there should be an optimal ratio L/D to minimize the heat loss. Intuitively, the condition should be

Figure 12.2   Water in an insulating tank. Calculation of energy storage behavior of an insulated water tank. The energy loss is proportional to the total surface area, and the energy content is proportional to the volume. The rate of temperature drop is minimized when the diameter D equals the length L. With a tank of linear dimension about 1 m with a 5-cm-thick foam polyurethane insulation, it take 8 h for the water temperature to drop by 1°C.

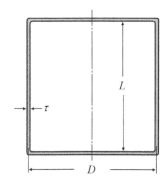

$L \approx D$. In Problem 12.1, one can show that the intuition is correct: The optimum condition is $L = D$, and Eq. 12.10 becomes

$$\frac{dT}{dt} = \frac{6\kappa}{\tau \rho c_p D} (T_w - T_a). \tag{12.11}$$

It verifies another qualitative argument: The larger the dimension, the better the tank could preserve temperature.

Here is a numerical example. If $D = L = 1$ m, $\tau = 5$ cm $= 0.05$ m, $T_w = 80°C$, and $T_a = 20°C$, the rate of temperature drop is

$$\frac{dT}{dt} = \frac{6 \times 0.02}{0.05 \times 4.19 \times 10^6 \times 1} \times 60 = 3.4 \times 10^{-5} \,°C/s = 0.124 \,°C/h. \tag{12.12}$$

The temperature takes 8 h to drop by 1°C. Such an energy storage unit is extensively used in hot-water systems.

12.2.2   Solid Sensible Heat Storage Materials

In contrast to water and other liquids, solid materials can provide a larger temperature range and can be installed without a container. However, thermal conductivity becomes a significant parameter. Table 12.2 shows the thermal properties of typical solid materials. For many items, such as soil and rock, the values are only approximate or

Table 12.2: Thermal properties of solid materials

Material	Density, ρ	Heat Capacity, c_p	Product, ρc_p	Thermal Conductivity, k
	$(10^3$ kg / m$^3)$	$(10^3$ J / kg·K$)$	$(10^6$ J / m^3·K$)$	(W/m·K)
Aluminum	2.7	0.89	2.42	204
Cast iIron	7.90	0.837	3.54	29.3
Copper	8.95	0.38	3.45	385
Earth (wet)	1.7	2.1	3.57	2.5
Earth (dry)	1.26	0.795	1.00	0.25
Limestone	2.5	0.91	2.27	1.3
Marble	2.6	0.80	2.08	$2.07 - 2.94$
Granite	3.0	0.79	2.37	3.5
Bricks	1.7	0.84	1.47	0.69
Concrete	2.24	1.13	1.41	$0.9 - 1.3$
Wood (oak)	0.48	2.0	0.96	0.16

Source: American Institute of Physics Handbook, 3rd ed.,
American Institute of Physics, New York, 1972; and Ref. [30].

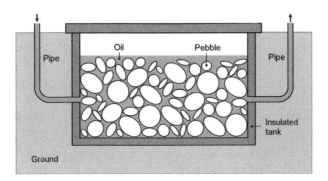

Figure 12.3   Rock-bed thermal energy storage system. Using a mixture of synthetic oil and pebbles, a thermal energy storage system at high temperature (e.g. 400°C) can be built with reasonable cost. The heat conduction is mainly through convection of oil, and the pebbles provide heat capacity.

an average, because those materials vary widely. For example, the thermal parameters of soil could vary by one order of magnitude depending on the water content.

As shown in Table 12.2, materials with high thermal conductivities usually have low heat capacity. To use solid materials with high heat capacities, a long temperature equalizing time is expected.

12.2.3   Synthetic Oil in Packed Beds

Because the temperature range of water is limited, in order to store sensible heat at higher temperature, for example, in solar power generation systems, synthetic oil should be used. However, synthetic oil is expensive. A compromised solution is to use a mixture of synthesized oil and inexpensive solid materials, such as pebbles. Figure 12.3 shows such a thermal energy storage system schematically. A thermal energy storage system at high temperature (e.g. 400°C) can be built with limited cost. The heat conduction is mainly through convection of oil, and the pebbles provide heat capacity.

12.3   Phase Transition Thermal Storage

In sensible heat thermal energy storage systems, the process of charging or discharging of energy is related to a change of temperature, and the temperature is related to the amount of heat energy content. The storage density is limited by the heat capacity of the material. Using phase-change materials (PCMs), a considerably higher thermal energy storage density can be achieved that is able to absorb or release large quantities of energy ("latent heat") at a constant temperature by undergoing a change of phase. Theoretically, three types of phase changes can be applied: solid–gas, liquid–gas and solid–liquid. The first two phase changes are generally not employed for energy storage

Table 12.3: Commonly used phase-change materials

Materials	Transition Temperature, °C	Density, ρ $(10^3\mathrm{kg/m^3})$	Latent Heat, h $(10^6\mathrm{J/m^3})$
Water-ice	0	1.00	335
Paraffin wax	58–60	0.90	180–200
Animal fat	20–50	0.90	120–210
$CaCl_2\ (6{\to}2)H_2O$	29	1.71	190.8
$Na_2SO_4\ (10{\to}0)H_2O$	32.4	1.46	251
$Ba(OH)_2\ (8{\to}0)H_2O$	72	2.18	301
$MgCl_2\ (6{\to}4)H_2O$	117	1.57	172

Source: Refs. [28] and [30].

in spite of their high latent heats, since gases occupy large volumes. Large changes in volume make the system large, complex, and impractical. Solid–liquid transformations involve only a small change in volume (often only a few percent) and are therefore appropriate for phase-change energy storage.

In general, the heat absorbed or released during the phase transition is

$$\Delta Q = H_2 - H_1, \tag{12.13}$$

where H is the enthalpy before and after the transition. The latent heat h, or specific enthalpy, is defined by

$$H = hV, \tag{12.14}$$

where V is the volume of the material.

Table 12.3 shows the thermal properties of several commonly used PCMs. A good PCM should provide energy storage at the desired temperature, have a large latent heat and a small volume change, and be non-flammable, noncorrosive, nontoxic, and inexpensive.

In general, PCMs are more expensive than sensible heat systems. They undergo solidification and therefore cannot generally be used as heat transfer media in a solar collector or the load. A separate heat transport medium must be employed with a heat exchanger in between. Many PCMs have poor thermal conductivity and therefore require large amount of heat exchange liquid. Others are corrosive and require special containers. These increase the system cost.

12.3.1 Water–Ice Systems

As shown in Table 12.3, the latent heat of the freezing of water or melting of ice is one of the highest. The water–ice system has already used in industry to save energy in air-conditioning systems. Figure 12.4 is a photo of a water–ice energy storage system,

Figure 12.4 Ice Bear energy storage system. An insulated tank is filled with water and many copper heat exchange coils. During the night, the refrigerator uses inexpensive electricity and cool air to make ice from the water. As shown in Chapter 6, the lower the ambient temperature, the higher COP. Therefore, to make a well-defined mass of ice during the night, the electricity cost is much lower than in the hot daytime. *Source*: Courtesy of Ice Energy, Inc.

named Ice Bear, designed and manufactured by Ice Energy, Inc. The system has a large insulated tank filled with water and a lot of copper heat exchange coils. During the night, the refrigerator uses inexpensive electricity and cool air to make ice from the water. As we shown in Chapter 6, the lower the ambient temperature, the higher the coefficient of performance (COP). Therefore, to make a well-defined mass of ice during the night, the electricity cost is much lower than in the hot daytime. During the hot daytime, the system uses the ice to cool the building. With this system, the efficiency of energy storage can be better than 90%. The overall energy savings can be as high as 30%.

Below is a numerical example of a water–ice system. Using the insulated container in Fig. 12.2, the total latent heat is

$$Q = \Delta H = H_2 - H_1, \tag{12.15}$$

Using Eq. 12.6,

$$\Delta H = \frac{1}{4}\pi h D^2 L. \tag{12.16}$$

For a tank of $D = L = 1$ m, the total enthalpy of phase transition is

$$\Delta H = 335 \times \pi \times 4 \times 10^6 \text{ J} = 2.63 \times 10^8 \text{ J}. \tag{12.17}$$

If the ambient temperature is $20°$C, for such a tank, according to Eq. 12.9, the rate of heat loss is

$$\frac{dQ}{dt} = \frac{3 \times \pi \times 0.02}{2 \times 0.05} \times 20 = 1.89 \text{ W}. \tag{12.18}$$

If at the beginning the tank is full of ice, then it can keep the temperature at $0°$C for $2.63 \times 10^8/1.89 = 1.39 \times 10^8$ s, or 4.4 years. Therefore, using a moderate means, the energy storage is efficient.

12.3.2 Paraffin Wax and Other Organic Materials

Paraffin wax is a byproduct of petroleum refining. The melting point of paraffin wax ranges from 50 to 90°C. Currently, paraffin wax only has a few commercially valuable applications, such as candles and floor wax. For such applications, only those with melting temperature between 58° and 60° are usable. But the supply is abundant. The melting temperature of paraffin matches the range needed for space heating and domestic hot water. It is also nontoxic and noncorrosive. One problem is its low thermal conductivity. This can be mitigated with encapsulation.

Other organic materials have similar properties as paraffin wax. An example is animal fat. Lard and chicken fat are considered harmful to human health because they can increase blood triglyceride and cause obesity. In some sense they are wastes of the food-processing industry. Animal fat is nontoxic and noncorrosive, thus it can be safely utilized for energy storage in residential environments.

12.3.3 Salt Hydrates

Many inorganic salts crystallize with a well-defined number of water molecules to become salt hydrates. Heating a salt hydrate can change its hydrate state. For example, hydrated sodium sulfate (Glauber's salt) undergoes the transition at 32.4°C

$$Na_2SO_410H_2O + \Delta Q \longrightarrow Na_2SO_4 + 10H_2O. \tag{12.19}$$

In general, the transition is

$$\text{Salt } mH_2O + \Delta Q \longrightarrow \text{Salt } nH_2O + (m - n)H_2O. \tag{12.20}$$

Thus, at the melting point the hydrate crystals break up into anhydrous salt and water or into a lower hydrate and water. The latent heat could be quite large, thus the storage density could be very high. If the water released is sufficient, a water solution of the (partially) dehydrated salt is formed.

These salt hydrates can be used in solar-operated space-heating or hot-water systems to provide uniform temperature over a longer period of time.

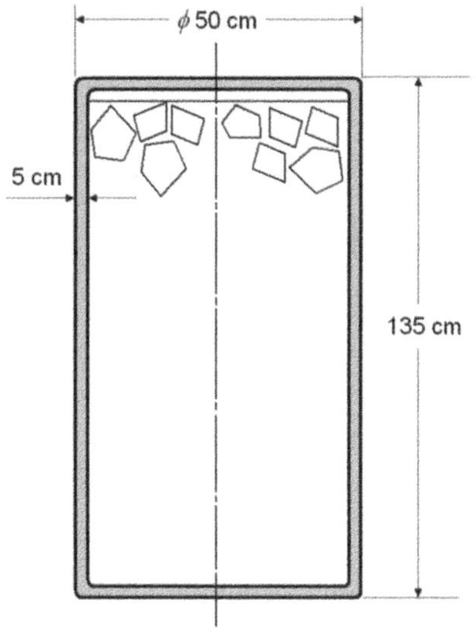

Figure 12.5 A typical domestic hot water tank.

Problems

12.1. A typical domestic hot water tank has a dimension of diameter $D = 50$ cm and height $H = 135$ cm, with a $\tau = 5$ cm thick insulation made of rigid foam polyimide, see Figure 12.5. If the difference in the external and internal temperatures is 45° C, what is the energy loss of this tank in watts? (The thermal conductivity of rigid foam polyimide is $k = 0.026$ W/(m°C).)

12.2. By storing hot water in that tank at 65°C and the environment temperature is 20°C, how long it takes to cool the water temperature down by 1°C?

12.3. By storing ice at 0°C in that insulated tank, with external temperature of 25°C, how long does it take for all the ice to melt? (The latent heat of ice is 335 kJ/L.)

Chapter 13

Rechargeable Batteries

In the last several decades, the technology of rechargeable batteries has observed a phenomenal expansion. Although the century-old lead-acid rechargeable battery has been constantly improved and is still in widespread use, new types of rechargeable batteries, especially based on the intercalation of alkali metal ions, are experiencing an explosive growth. Among them, the lithium-ion battery has already dominated the field of portable electronics, electrical vehicles, and stationary energy storage. Because of the tight resource of lithium and cobalt, intensive research and development in sodium ion rechargeable batteries have been conducted, and it will become significant in large-scale stationary energy storage. In this chapter, we present the basic theory of rechargeable batteries, electrochemistry, and various types of rechargeable batteries, especially lithium-ion batteries and sodium-ion batteries.

13.1 An Electrochemistry Primer

The field of electrochemistry was established in the middle of 19th century. As an old discipline of chemistry, Chapter 15 of the classic textbook *General Chemistry* by Linus Pauling [80] is still a good reading. For more details, see the two-volume textbook *Modern Electrochemistry* by J. O'M. Bockris and A. K. N. Reddy [11].

13.1.1 Basic Terms and Definitions

The basic concepts and terminologies in electrochemistry, established in the middle of the 19th century, are still followed to date. In the field of rechargeable batteries, there is a set of well-recognized set of terms and definitions. To facilitate reading literature in electrochemistry and rechargeable batteries, here are some explanations.

Cell and battery

According to the conventions of the battery industry, cell is the basic electrochemical unit converting electrochemical energy to electrical energy. A battery contains one or more electrochemical cells connected in series or parallel to provide electrical power. Therefore, there are one-cell batteries and multiple-cell batteries.

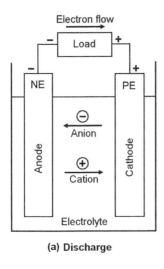
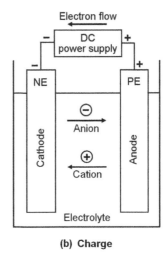

Figure 13.1 Electrochemistry of rechargeable batteries. (a) Discharging process. By connecting the cell to a load, electrons flow from the negative electrode (NE) to the positive electrode (PE). The circuit is completed in the electrolyte by the flow of cations to the cathode, here is PE; and anions to the anode, here is NE. (b) Charging process. An external DC power supply forces electrons to flow into the cathode, here is NE. The circuit is completed in the electrolyte by the flow of ions.

An electrochemical cell is a container for the electrolyte and electrodes, being inert to avoid side reactions and to ensure a good sealing. There are two electrodes connecting to an external electrical circuit: a cathode and an anode, see Fig. 13.1.

Electrolyte

The electrolyte is an ionic conductor to facilitate the movement of ions. In most cases it is a liquid. Because lithium and sodium are extremely reactive, aqueous electrolytes are not suitable. Nonaqueous solvents are required.

Oxidation and reduction

Historically, the term oxidation originated from the reaction of metals with oxygen to become metal oxides. For example, when iron is exposed to air, an iron atom can lose electrons to form rust, FeO. In the process of oxidation, an iron atom loses two electrons to become a positive ion, Fe^{2+} or Fe(II). When a positive iron ion receives electrons to become neutral, the process is called reduction. In general, oxidation means losing electron(s), and reduction means gaining electron(s).

Cathode and anode

The definitions of cathode and anode could be lengthy and confusing. Here is a succinct definition. The cathode is where electrons from the external circuit flow into, and the anode is where electrons flow out to the external circuit. That definition explains the term cathode ray: in a vacuum cell, electrons flow from the external source through the cathode (i.e., the electron gun) and emit into vacuum. Anodizing a metal surface means to oxidize the metal surface to make an anodic oxide for protection, for example forming Al_2O_3 to prevent aluminum from corrosion.

Because electron has a negative charge, the formal direction of electrical current is the opposite of the direction of electron flow. Nevertheless, for a physics point of view, thinking about the flow of electrons is more intuitive.

Cation and anion

In an electrochemical cell, a cation is defined as the ion moves to the cathode, combing with the electrons from the external circuit and being reduced. Therefore, cation means positive ion. On the other hand, an anion moves to the anode, gives up electron(s), that flow to the external circuit. Anion means negative ion.

In other words, cathode is where cations are reduced into neutral atoms or molecules by combining with electrons from external circuit; and anode is where neutral atoms or molecules are oxidized into cations by giving up electron(s) to external circuit.

Positive electrode (PE) and negative electrode (NE)

To identify the electrodes of rechargeable batteries, the terms PE and NE are more intuitive and consistent, see Fig. 13.1. By using a multimeter to check the voltage, the PE is always positive and NE is always negative. Nevertheless, in terms of electron flow, the identification of cathode and anode are reversed. During discharging, external electrons flow into PE. Therefore, PE is cathode. During charging, external electrons flow into NE. Therefore, NE is cathode. In the literature, the terms cathode and anode are often used to identify the electrodes of a battery. The convention is, cathode and anode are defined for the process of discharging. Therefore, cathode means PE and anode means NE.

13.1.2   Oxidation State

In electrochemistry, especially regarding to rechargeable batteries, a number *oxidation state*, or oxidation number, is frequently used. Typically it is an integer.

When exposed to air, an iron atom can lose two electrons to oxygen to become FeO. The iron atom become a positive ion, Fe^{2+} or Fe(II). The oxidation state of that iron ion is two. Iron has another oxidation state, Fe^{3+} or Fe(III). The oxidation state of that iron ion is three. The corresponding iron oxide is Fe_2O_3.

Strictly speaking, the concept of oxidation state can only be applied to pure ionic bonds, such a NaCl. In an aqueous solution, NaCl completely dissociates into Na^+ and Cl^-. Many bonds are only partially ionic. The oxidation state can be a non-integer. For pure covalent molecules, such as N_2 or O_2, the concept of oxidation state does not apply. In the case of rechargeable batteries using intercalation of metal ions, the concept of oxidation state in integer numbers is always valid.

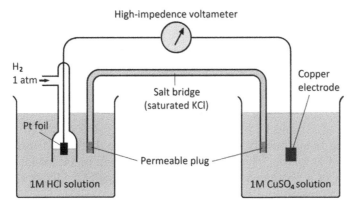

Figure 13.2 Setup for measuring standard potential. To measure the standard potential of Cu^{2+} versus Cu. The reference electrode is a Pt foil with the presence of H_2 gas.

13.1.3 Standard Oxidation-Reduction Potentials

The tendency of an atom or a molecule to become oxidized or reduced in aqueous solutions is quantitively measured by the standard oxidation-reduction potential, or simply standard potential [80, 11]. To make the physical quantity well-defined, a zero point must be defined. The reference point is hydrogen, see Table 13.1. An experimental

Table 13.1: Standard potentials at 25 °C

Oxidizing Agent			Reducing Agent	Potential (V)
F_2	$+$	$2e^-$ $\rightarrow$	$2F^-$	2.87
Cl_2	$+$	$2e^-$ $\rightarrow$	$2Cl^-$	1.36
Br_2	$+$	$2e^-$ $\rightarrow$	$2Br^-$	1.07
Ag^+	$+$	e^- $\rightarrow$	Ag	0.80
Cu^{2+}	$+$	$2e^-$ $\rightarrow$	Cu	0.15
$2H^+$	$+$	$2e^-$ $\rightarrow$	H_2	0.00
Fe^{2+}	$+$	$2e^-$ $\rightarrow$	Fe	-0.45
Cr^{3+}	$+$	$3e^-$ $\rightarrow$	Cr	-0.74
Mn^{2+}	$+$	$2e^-$ $\rightarrow$	Fe	-1.19
Al^{3+}	$+$	$3e^-$ $\rightarrow$	Al	-1.66
Mg^{2+}	$+$	$2e^-$ $\rightarrow$	Mg	-2.37
Na^+	$+$	e^- $\rightarrow$	Na	-2.71
Ca^{2+}	$+$	$2e^-$ $\rightarrow$	Ca	-2.87
K^+	$+$	e^- $\rightarrow$	K	-2.93
Li^+	$+$	e^- $\rightarrow$	Li	-3.04

setup for measuring standard potentials is shown in Fig. 13.2. The reference electrode is platinum, with the presence of gas-phase hydrogen. Showing here is to measure the standard potential of Cu^{2+} versus Cu, with $CuSO_4$ solution as the electrolyte.

13.2 Lithium-Ion Batteries

The schematics of a typical Li-ion rechargeable battery are shown in Fig. 13.3. In present time, the Li-ion batteries are manufactured in a discharged state. The NE is typically made of graphite. The PE is made of transition-metal oxides with lithium, such as $LiCoO_2$, or a polyanion compound, such as $LiFePO_4$. Out of the production line, all lithium atoms are in PE.

Before use, the Li-ion battery is charged using an external DC power supply with a voltage higher than nominal voltage of the battery, see Fig. 13.3(a). Lithium ions move from PE to NE, intercalate into the spaces between graphene sheets.

While discharging, a load is connected to the battery terminals, see Fig. 13.3(b). Lithium ions move from NE through the electrolyte then intercalated into PE, releasing electrical energy. The cycles of charging and discharging are reversable.

13.2.1 Benefit to Humankind

The Nobel Prize in Chemistry 2019 was awarded to three key inventors and developers of the lithium-ion battery, with a statement [76]:

> Lithium-ion batteries have brought the greatest benefit to humankind, as they enabled the development of laptop computers, mobile phones, electrical vehicles and the storage of energy generated by solar and wind power.

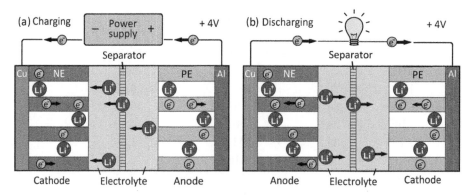

Figure 13.3 Electrochemical processes in a Li ion cell. (a) During charging, lithium ions are forced by the external voltage to leave the positive electrode (PE) and intercalated into the negative electrode (NE), typically graphite. (b) During discharging, lithium ions leave NE, drift through the electrolyte, pass the microporous separation film and intercalate to the material in PE.

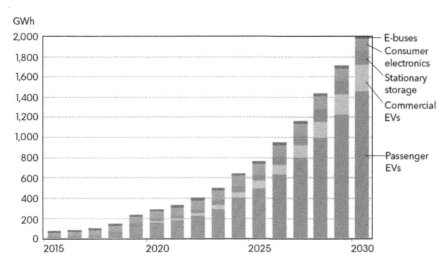

Figure 13.4 Production and demand for lithium-ion batteries. As reported by Bloomberg, in 2022, the production and demand for Li-ion batteries reached 400 GWh, a five-fold increase from 2016. The forecast for 2030 is 2000 GWh, another five-fold increase from 2022.

The above statement from the Royal Swedish Academy of Sciences is materially supported by the statistics of the rapid expansion of the production and application of the lithium-ion batteries. In Section 1.4.4, especially in Fig. 1.20, the production and application of the Li-ion batteries up to 2016 are presented. In 2016, the annual production and sales of Li-ion batteries were about 80 GWh. The installation to electric vehicles was less than one-half. According to a recent study by Bloomberg, see Fig. 13.4, in 2022, the total production of Li-ion batteries already exceeds 400 GWh, a five-fold increase. The generally accepted forecast is, in 2030, the demand of Li-ion batteries will be 2000 GWh, another fivefold increase from 2022. The implementation of Li-ion batteries in passenger electric vehicles will reach 1450 GWh, about three quarters of the total demand. The impact on human society is enormous.

Table 13.2: Comparison of rechargeable batteriess

Type	Voltage (V)	Energy Density (Wh/L)	Specific Energy (Wh/kg)	Lifetime (Cycles)
Lead–acid	2.1	70	30	300
NiMH	1.4	240	75	800
$LiCoO_2$	3.7	400	150	1000
$LiMn_2O_4$	4.0	265	120	1000
$LiFePO_4$	3.3	220	100	3000

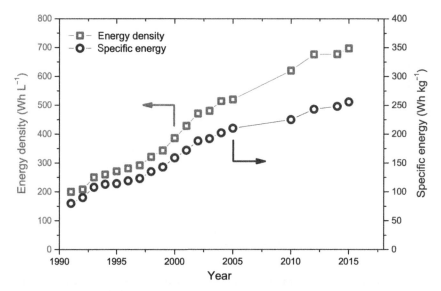

Figure 13.5 Energy density and specific energy of Li-ion batteries. Over the last two decades, the two most important technology parameters of the Li-ion batteries, energy density and specific energy, were significantly improved [86].

The progress in production and implementation is supported by a continuous improvement of the quality and continuous reduction of manufacturing cost of the Li-ion batteries, see a review article [86]. Figure 13.5 shows the progress of two critical parameters of Li-ion batteries, energy density and specific energy. Energy density of a rechargeable battery refers to the maximum amount of stored energy per unit volume. The unit of stored energy is usually in watt-hour, which equals to 3.6 kJ. In early 1990s, the typical value was 200 Wh/L In 2015, it was improved to about 700 Wh/L . Specific energy of a rechargeable battery refers to the maximum amount of stored energy per unit weight. In weight-sensitive applications such as vehicles, specific energy is more significant. In early 1990s, the specific energy was about 75 Wh/kg. In 2015, it was improved to 250 Wh/kg. In comparison, the energy density and specific energy of other rechargeable batteries are shown in Table 13.2.

Concurrent with the constant improvement of quality, the manufacturing cost of Li-ion batteries has been in a constant decline during the last decades. In 2010, the manufacturing cost of a kWh of Li-ion battery is $1,191. In 2020, it was dropped to $137, almost a 10-folds reduction. As the technology of Li-ion batteries is constantly improving, the manufacturing cost is also in constant decline.

13.2.2 Intercalation of Metal Ions

The working principle of the Li-ion battery is the intercalation of lithium ions into the chemical structures of the electrodes. In the abstract of the classic 1978 paper [115],

Table 13.3: Intercalation voltage of electrode materials

Type	Material	Structure	Voltage (V vs Li)	Capacity (Ah/kg)
Cathode	$LiCoO_2$	Layered	3.9	140
	$LiNi_{\frac{1}{3}}Mn_{\frac{1}{3}}Co_{\frac{1}{3}}O_2$	Layered	3.8	200
	$LiNi_{0.8}Co_{0.15}Al_{0.05}O_2$	Layered	3.8	200
	$LiFePO_4$	Olivine	3.45	160
Anode	Graphite (LiC_6)	Layered	0.1	360

Whittingham made a inspiring description of the concept of intercalation:

> The phenomenon of intercalation has most commonly been applied to the calendar, such as the quadrennial intercalation of February 29 into the calendar year. In chemical terms, intercalation has been associated with the insertion of guest species between the layers of a crystalline host lattice and is a special case of a topochemical reaction. Just as in the calendar usage, the process of intercalation should be reversible through appropriate chemical or thermal treatment.

Using two intercalation electrodes, one capable of accepting metal ions and the other capable of releasing metal ions, Whittingham formulated a new principle of rechargeable batteries. Metal ions "rock" between two electrodes during charging and discharging. The new battery is nicknamed "rocking-chair rechargeable battery" [114, 115].

Lithium has long received much attention as a promising candidate for intercalation, from the combination of its two unique properties: (1) it is the most electronegative metal ($\sim$ −3.04 V, see Table 13.1), and (2) it is the lightest metal (atomic weight of 7). The former confers upon it a negative potential that translates into high cell voltages, and the latter makes it an anode of high specific capacity.

When a lithium ion is intercalated to a host material, it releases a bonding energy in eV, expressed as a voltage. Different materials offer different bonding voltages, see Table 13.3. As shown, the intercalation voltages of cathode materials and anode materials are vastly different. The intercalation voltages of cathode materials are in the range of 3.5 V and 4.1 V. For typical anode materials, the voltage is close to zero. Therefore, when a lithium ion moves from the anode to the cathode, an electrical energy of roughly 4 eV is generated, that creates a battery voltage of roughly 4 V.

In addition to the intercalation voltage, the practical capacity is also an important parameter. It sets the theoretical limit of the capacity of the rechargeable battery. Table 13.3 also shows the practical capacity of some important electrode materials. Descriptions of those materials will be presented in the following subsections.

13.2.3 The Cathode Materials

Since the first demonstration of the intercalation-based rechargeable batteries with TiS_2 [114, 115], a large number of cathode materials have been studied. For a summary of various cathode materials, see the review articles by Whittingham [116, 117].

In the first successful commercial Li-ion battery, the cathode material was lithium cobalt oxide, Li_xCoC_2, discovered by John Goodenough and coworkers in 1980 [71]. Comparing with the original material, TaS_2, the nominal voltage increased from 2V to 4V. It also shows a high energy density and specific energy. The crystallographic structure of $LiCoO_2$ in a ball-and-stick model is shown in Fig. 13.6. As shown, The solid consists of layers of Li^+ sandwiched between Co^{3+} and O^{2-} sheets, and the Co^{3+} sheets are sandwiched between two O^{2-} sheets.

The discovery of the cathode material Li_xCoO_2 was based on two revolutionary concepts [71]. First, the synthetic route chosen was to prepare the cathode material in the *discharged* state rather than in the charged state such as TiS_2, because the fully charged electrode material could be thermodynamically unstable. Second, normally, the oxidation number of cobalt is 2 or 3. John Goodenough considered a rarely reported oxidation state, Co^{4+}, which can offer a high energy density. However, the pure CoO_2 is thermodynamically unstable. By synthesizing $LiCoO_2$ as the starting point, the advantage of Co^{4+} can be utilized to offer a high specific energy.

The Goodenough team prepared the battery in a completely discharged state. The initial cathode material $LiCoO_2$ contains a maximum number of lithium atoms. By introducing an index of charging, $0 < x < 1$, the charging process is

$$6C + xe^- + xLi^+ \Longrightarrow Li_xC_6 \quad \text{(anode).} \tag{13.1}$$

$$LiCoO_2 - xe^- - xLi^+ \Longrightarrow Li_{1-x}CoO_2 \quad \text{(cathode).} \tag{13.2}$$

Here $x = 0$ means completely discharged, and $x = 1$ is a hypothetical limit of charging

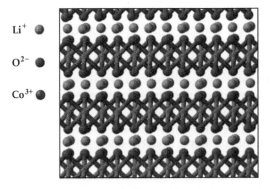

Figure 13.6 Ball-and-stick model of lithium cobalt oxide. The solid consists of layers of Li^+ sandwiched between anodic sheets of CoO_2^-. The Co^{3+} sheets are sandwiched between two sheets of O^{2-} anions. Adapted from Sigma-Aldrich product page, 2013.

Li⁺ ●

O²⁻ ●

Co³⁺ ●

defined by the chemical structure. The discharging process is:

$$\text{Li}_x\text{C}_6 - x\text{e}^- - x\text{Li}^+ \Longrightarrow 6\text{C} \qquad \text{(anode)}. \qquad (13.3)$$

$$\text{Li}_{1-x}\text{CoO}_2 + x\text{e}^- + x\text{Li}^+ \Longrightarrow \text{LiCoO}_2 \quad \text{(cathode)}. \qquad (13.4)$$

The theoretical capacity of the LiCoO_2 cell is relatively low at around 130 Ah/kg, because while $x > 0.5$, the crystallographic structure of LiCoO_2 is seriously disturbed. Practically, it operates in the range $0 < x < 0.5$ [71]. However, owing to continuous improvements, the energy density of commercial cells has almost doubled since their introduction in 1991 from 250 Wh/L to over 400 Wh/L.

Nevertheless, the scarcity and high cost of cobalt motivated researchers to look for alternatives. In the last decades, research indicated that replacing Co with a mixture of Ni, Mn, and Co not only reduced the cost, but also improved the performance. Currently, various Ni, Mn, and Co mixtures are the mainstream of cathode materials.

The mixed-metal cathodes

In the years following the first commercialized Li-ion batteries using LiCoO_2 in 1991, it was found that LiNiO_2 and LiMnO_2 have the same structure as LiCoO_2 and could form cathodes of Li-ion batteries using much less expensive materials. However, they have certain disadvantages. The focus of research and development of cathode materials is in mixed metals, having a general formula as $\text{LiNi}_{1-y-z}\text{Mn}_y\text{Co}_z\text{O}_2$ [116]. The structures of those mixed-metal cathode materials are the same as in Fig. 13.6. A large number of such mixed-metal cathode materials were synthesized and tested. Those new materials offer different compromises between cost and performance. The compositions and market shares of different materials in 2022 are shown in Fig. 13.7 [2].

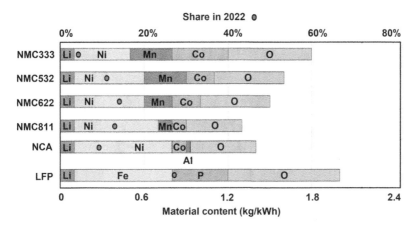

Figure 13.7 Share of different cathode materials of Li-ion batteries in 2022. Commercial name and material composition of five mixed-metal cathode materials in 2022 [2]. The red dots show the market share with the scale on the top. The share of iron phosphate cathode is also shown.

Figure 13.8 A ball-and-stick model of the phosphate polyanion. The four oxygen atoms occupy the four vertices of an tetrahedron. The phosphorous atom is at the center of the tetrahedron. Because there are eight negative charges and five positive charges, the net charge number of PO_4^{3-} is minus three.

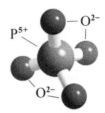

The iron phosphate cathodes

Another important cathode material is lithium iron phosphate, $Li_x FePO_4$, based on the polyanion PO_4^{3-}, shown in Fig. 13.8. The four oxygen atoms are at the four vertices of a tetrahedron with the phosphorous atom at the center. Because there are eight negative charges and five positive charges, the net charge number of PO_4^{3-} is minus three. Iron has two oxidation states, Fe^{2+} and Fe^{3+}. Because the lithium ion Li^+ is tiny, for any number of $0 < x < 1$, the crystallographic structure remains almost intact. The batteries with lithium iron phosphate cathodes can endure thousands of charging and discharging cycles with negligible deterioration.

The experiments of using $LiFePO_4$ as cathode material of the Li-ion batteries were first reported by the Goodenough group in 1997 [78]. The results showed it is an excellent candidate for Li-ion batteries because it is inexpensive, nontoxic, and environmentally benign. Especially, the crystallographic structures of $FePO_4$ and $LiFePO_4$ are almost identical, see Fig. 13.9. The tetrahedrons are the phosphoric acid anions with the oxygen anion at the vertices. The extraction and insertion of lithium ions

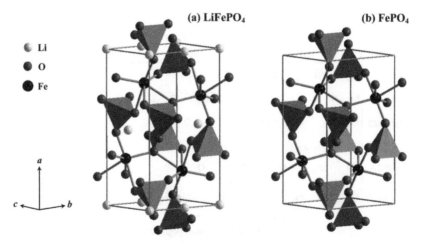

Figure 13.9 Lithium iron phosphate and iron phosphate. The tetrahedrons are phosphate anions with four oxygen atoms at the vertices and a phosphorous atom at the center. The crystallographic structures of $FePO_4$ and $LiFePO_4$ are almost identical, enabling smooth operations.

run smoothly. The extraction of lithium from $LiFePO_4$ to charge the cathodes may be written as

$$LiFePO_4 - xLi^+ - xe^- \Longrightarrow xFePO_4 + (1 - x)LiFePO_4 \quad \text{(cathode)}. \qquad (13.5)$$

And the reaction for the insertion of lithium into $FePO_4$ on discharge is

$$FePO_4 + xLi^+ + xe^- \Longrightarrow xLiFePO_4 + (1 - x)FePO_4 \quad \text{(cathode)}. \qquad (13.6)$$

Experiments showed excellent reversibility because of the similarity of the crystallographic structures of $LiFePO_4$ and $FePO_4$.

The nominal voltage of the lithium iron phosphate system is about 3.25 V, lower than that of the lithium cobalt oxide system. However, the low cost, environment friendly elements, and good safely made it the most utilized cathode material. In 2022, the iron phosphate system takes about 35% of all Li-ion batteries, see Fig. 13.7. Details of this system can be found in a 2014 review of Whittingham [117].

13.2.4 The Anode Materials

In the early stage of the Li-ion battery development, metallic lithium was used as the anode. The fire and explosion due to the formation of dendrites during charging caused a suspension of its commercialization. In 1980s, a group led by Akira Yoshino explored carbon-based material as the anode. At first, the use of graphite encountered difficulties. In the first product, the material was petroleum coke, a slightly disorder version of graphite. By improving the composition of the electrolyte, the use of graphite as anode material became practical. Since 1994, almost all Li-ion batteries use graphite as the anode material, see a 2021 review for details [123].

The maximum of lithium ion to be intercalated into graphite is the sixth of the number of carbon atoms. The generally accepted formula is LiC_6, see Fig. 13.10. As shown, each lithium ion is surrounded by 12 carbon atoms, and each carbon atom is connected with two lithium ions on each side of a graphene sheet.

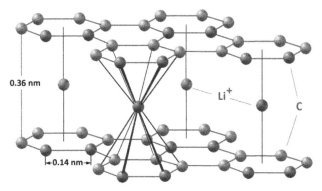

Figure 13.10 Structure of LiC_6. each lithium ion is surrounded by 12 carbon atoms, and each carbon atom is connected with two lithium ions on each side of a graphene sheet.

Table 13.4: Properties of organic solvents for Li-ion batteries

Solvent	T_m (°C)	T_b (°C)	η (cP)	ϵ_r	ρ (g/ml)
EC	36.4	248	1.90	89.8	1.32
PC	−48.8	242	2.53	64.9	1.20
DMC	4.6	91	0.59	3.11	1.06
DEC	−74.3	126	0.75	2.8	0.97
EMC	−5.3	110	0.65	2.9	1.06

For several decades, graphite has been proven to be a safe and inexpensive anode and commercialized. Nevertheless, it has a low theoretical specific capacity due to the limit of one Li atom per six C atoms. To achieve higher energy density, several alternative anode materials have been studied. For highest specific capacity, metallic lithium or its alloy is still a good choice. Studies to prevent the safety issue of dendrite reached good progress [120]. Silicon can contain a lot more lithium atoms than carbon. Use pure silicon or mixing carbon with silicon has been commercialized. The use of TiO_2 and similar materials also achieved good progress [5, 75].

13.2.5 Electrolytes

The electrolyte is an indispensable component in every electrochemical device. It physically segregates two electrodes from direct electron transfer while allowing working ions to transport both charges and masses across the cell so that the cell reactions can proceed sustainably. Because of the reactive nature of lithium, the electrolyte for the Li-ion batteries must made of non-aqueous solvents. Figure 13.11 shows the name, abbreviations, and chemical structures of five popular solvents [69, 118, 119].

Table 13.4 shows the properties of the important solvents for the Li-ion batteries. An ideal electrolyte solvent should meet the following minimal criteria: (1) It should

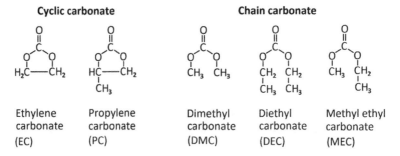

Figure 13.11 Organic solvents in Li-ion batteries. The most widely used organic solvents for the electrolytes in Li-ion batteries. Those solvents are also used in Na-ion batteries.

Figure 13.12 Structure of lithium hexafluorophosphate. The most frequently used lithium salt in the electrolyte of Li-ion batteries is lithium hexafluorophosphate. It has a rather bulky anion PF_6^-, seldom participate in forming the ionic current. The Li cation is the dominating current carrier.

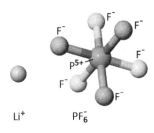

be able to dissolve salts to sufficient concentration. In other words, it should have a high dielectric constant (ϵ_r). (2) It should be fluid (low viscosity η), so that facile ion transport can occur. (3) It should remain inert to all cell components. In other words, it should be inert to the cathode and the anode. (4) It should remain liquid in a wide temperature range. In other words, it should have a low melting point T_m and a high boiling point T_b. (5) It should also be safe, nontoxic, and economical.

Among them, EC has been an essential ingredient of the electrolyte solvent. It has a high dielectric constant and a reasonable viscosity. Nevertheless, its melting point is rather high. It is often mixed with a linear carbonate as the state-of-the-art lithium-ion electrolytes and was adopted by the researchers and manufacturers.

Another essential component of the electrolyte is a lithium salt to provide cations. An ideal salt for Li-ion batteries should meet the following minimal requirements: (1) It should be able to completely dissolve and dissociate in the nonaqueous solvents, and the lithium cation should have a high mobility. (2) The anion should be stable at the cathode. (3) The anion should be inert to electrolyte solvents. (4) Both the anion and the cation should remain inert. Among the numerous lithium salts for lithium ion batteries, lithium hexafluorophosphate $LiPF_6$ was eventually commercialized. A ball-and-stick model of $LiPF_6$ is shown in Fig. 13.12. The success of $LiPF_6$ was not achieved by any single outstanding property but by the combination of a series of properties with concomitant compromises and restrictions [69, 118, 119].

13.2.6 The Separator

Separators play a key role in all batteries. Their main function is to keep the PE and NE apart to prevent electrical short circuits and at the same time allow rapid transport of ionic charge carriers that are needed to complete the circuit during the passage of current in an electrochemical cell [4].

The major requirement for an effective separator include the following (1) It is an electronic insulator. (2) It has a minimal electrolyte (ionic) resistance. (3) It has mechanical and dimensional stability. (4) It has sufficient physical strength to allow easy handling. (5) It has chemical resistance to degradation by electrolyte, impurities, and electrode reactants and products. (6) It is effective in preventing migration of particles or colloidal or soluble species between the two electrodes. (7) It is readily wetted by the electrolyte. (8) It is uniform in thickness and other properties.

Figure 13.13   An electron microscopy image of a separator for Li-ion battery.   Electron microscope image of a single-layer separator for Li-ion batteries. It is made of polypropylene by Celgrad. As shown, the film has a reasonably high percentage of pores, between 30% and 40%, and the pores are uniformly distributed.

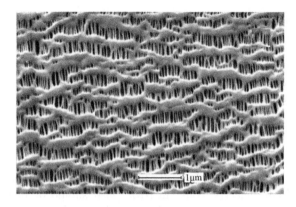

The separator can be a single sheet of polyethylene (PE) or polypropylene (PP) of thickness less than 25 μm. The film is highly porous with a porosity of around 40%. To improve the robustness, multiple-layer separators are manufactured. One design is to support two layers of ceramic material on both sides of a PE or PP film [4].

Figure 13.13 shows an electron microscopy image of a typical single-layer separator, manufactured by Celgrad. As shown, the pores are uniformly distributed [4].

13.2.7   Packaging

The battery elements are packed together tightly to contain as many layers as possible in a limited space. Figure 13.14 shows a typical way of stacking many layers in a tight space. The cathode current collectors and anode current collectors, made of metal foils, are coated with cathode and anode materials of well-considered thicknesses on

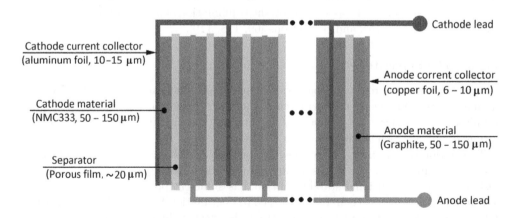

Figure 13.14   Stacking of electrodes, separators, and current collectors.   The electrodes, separators, and current collectors are stacked to contain a maximum area in a limited space.

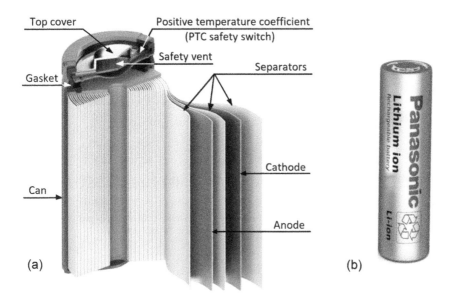

Figure 13.15 Cylindrical lithium-ion battery. The battery components, cathode, separator, and anode, are produced as sheets, then wind into a cylinder. The cathode and the anode are connected to the bottom and top, respectively. There is a PTC safety switch and a safety vent.

both sides except the outermost components. Because the battery is winded from a single composite sheet, no external connections for the layers are required.

Figure 13.15 shows a cylindrical Li-ion battery. It is winded from a single sheet of battery elements. The commonly used sizes are listed in Table 13.5. Safety is a primary concern. As shown in Fig. 13.15, there are two built-in safety devices. First one is the positive temperature coefficient safety switch, abbreviated as PCT. It is a circular-shaped grommet installed between the top cover and the cathode lead, made

Table 13.5: Cylindrical lithium ion batteries

Type	14500	14650	18500	18650
Height (mm)	50	65	50	65
Diameter (mm)	14	14	18	18
Volume (ml)	7.7	10	12.7	16.5
Mass (g)	19	26	31	42
Capacity (Ah)	0.65	0.90	1.10	1.80
Specific energy (Wh/kg)	126	128	131	155
Energy density (Wh/L)	312	333	320	410

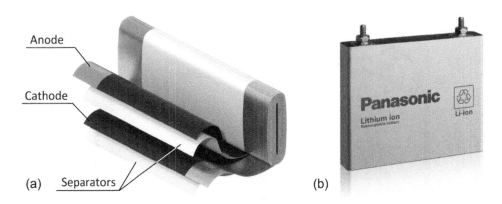

Figure 13.16 Prismatic lithium-ion battery. (A) The sheets of cathodes, separators, and anodes are winded into a rectangular stack. (B) a photo of a prismatic Li-ion battery.

of a material having very low electrical resistance at room temperature and very high electrical resistance at an elevated temperature, with a typical transition temperature around 100°C. If the temperature is elevated, it turns off the battery. Second one is a safety vent to prevent explosion.

The second popular type is the prismatic Li-ion batteries. A schematics if shown in Fig 13.16. Rather than winded to fit in a circular can, the battery elements are fit into a rectangular container. Table 13.6 shows several commercially available sizes.

Currently, both cylindrical and prismatic types of Li-ion batteries are used in portable electronics, hand-held tools, and electrical vehicles. For each electrical vehicle, 5000–9000 cylindrical Li-ion cells are needed. Safety is a serious concern. If one cylindrical cell is on fire, the neighboring cells can be ignited. To prevent accidents, fire-retarding separation boards are installed around each cylindrical cell.

Table 13.6: Prismatic lithium-ion batteries

Type	103/34/50	260/61/78	280/95/151	460/89/128
Height (mm)	50	78	151	128
Width (mm)	34.1	61	95	89
Thickness (mm)	10.3	16	28	46
Volume (mL)	18.4	76	136	465
Mass (g)	38	185	870	1108
Capacity (Ah)	1.5	7.0	35	40
Specific energy (Wh/kg)	146	145	145	156
Energy density (Wh/L)	301	345	344	372

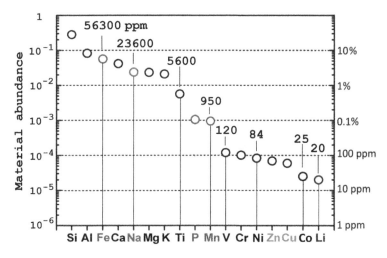

Figure 13.17 Elemental abundance in the earth's crust. The two key metal elements in the Li-ion batteries, lithium and cobalt, are among the rarest of all elements. The abundance of sodium and iron are thousands of times greater than lithium and cobalt. *Source*: Adapted from [121].

13.2.8 Mineral Resource of Lithium

As Li-ion batteries become increasingly important, the problem of the mineral resource of lithium is of great interest. Figure 13.17 shows the abundance of several relevant elements on the earth's crest [121]. Figure 13.18 shows the violent gyration of market price of lithium [40]. Lithium carbonate is from the brine in highland lakes; and

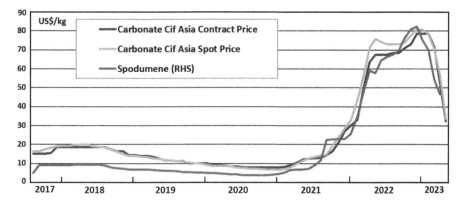

Figure 13.18 Variation of the price of lithium carbonate. The market price of lithium, showing violent gyration in the last years. Speculation was the underlying cause of the price fluctuation [40]. The two forms of mineral lithium, lithium carbonate is from the brine in highland lakes. Spodumene is a mineral with typical composition, Li_2O-Al_2O_3-$nSiO_2$, $n = 10 \sim 40$.

spodumene is a mineral with typical composition Li_2O-Al_2O_3-$nSiO_2$, $n = 10 \sim 40$.

As shown in Fig. 13.17, sodium and iron are at least one thousand times more abundant than lithium and cobalt, another essential metal for Li-ion batteries. Rechargeable batteries made with sodium and iron will be of great advantage.

13.3 Sodium-Ion Batteries

On the periodic table, sodium is the closest to lithium. The standard potential of Na versus Na^+, is -2.71 V, is only slightly smaller than that of lithium, -3.04 V. Therefore, sodium is the second ideal metal for intercalation-based rechargeable batteries. There are some differences between these systems. Na^+ ions ($r = 0.102$ nm) are larger compared to Li^+ ions ($r = 0.076$ nm), which affects the phase stability, transport properties, and interphase formation. The theoretical energy density of the Na-ion battery is lower than the Li-ion battery. Nevertheless, experiments indicated that the Na-ion batteries have several unexpected advantages. First, the Na-ion batteries are much safer than the Li-ion batteries, and can work under a much lower environmental temperature. Furthermore, as lithium easily reacts with aluminum, copper is used as the anode current collector. Sodium does not react with aluminum, which can be used for electrodes, and the cost of the entire battery is further reduced.

From economical point of view, Na-ion batteries have great potential advantages. The price of lithium carbonate in 2002 was about $37 per kilogram, whereas the price of sodium carbonate was $0.15 per kilogram, more than 200-fold less. In addition, for Na-ion batteries, cobalt is not a necessary metal. Aluminum can be used for cathode and anode. Since the 1970s, the research on Na-ion rechargeable batteries has been as intensive as for Li-ion batteries. Nevertheless, after Sony commercialized Li-ion batteries in 1991, research and development of Na-ion batteries were largely abandoned.

In the 21st century, the high price of lithium and the discovery of high-density anode material for Na-ion batteries, the research for practical Na-ion batteries exploded [45], as shown in Fig. 13.19. As shown, the number of research papers published on Na-

Figure 13.19 Number of research papers on Ni-ion battery. Since the 1970s, the research on Na-ion rechargeable batteries were as intensive as Li-ion rechargeable batteries. After Sony commercialized Li-ion batteries in 1991, Ni-ion battery research was almost abandoned. In the 21st century, the high price of lithium and the discovery of high-density anode materiel for Na-ion batteries, the research for Na-ion batteries exploded [45].

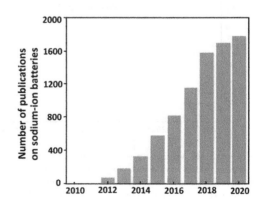

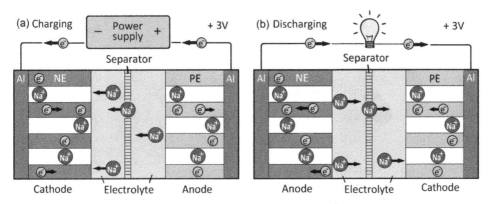

Figure 13.20 Sodium ion rechargeable battery. (a) During charging, Na ions are forced by the external power supply to leave PE and intercalated into NE, typically based on hard carbon. (b) During discharging, Na ions leave NE, drift through the electrolyte, and intercalate to PE.

ion rechargeable battery grows from almost zero in 2010 to more than 1600 annually in 2018, and it is still growing. In 2015, Faradion, a technology start-up located in United Kingdom, India, and Canada, started to commercialize Na-ion batteries for electric bicycles. In 2018, Natrion, based in Binghamton, NY, and Urbana Champaigne, IL, commercialized a Na-ion battery for mass energy storage. Na-ion batteries are considered the next-generation electrochemical energy storage system of choice.

A schematics of an Na-ion rechargeable battery is shown in Fig. 13.20. Except the Li ions are replaced by the Na ions, the structure and functions are identical to the Li-ion batteries, Fig. 13.3. The battery is manufactured in the fully discharged state. All sodium ions are in the PE.

Before use, the Na-ion battery is charged using an external DC power supply with a voltage higher than nominal voltage of the battery, see Fig. 13.20(a). Sodium ions move from PE to NE, and intercalate into the material of the NE.

While discharging, a load is connected to the battery terminals, see Fig. 13.20(b). Sodium ions move from NE through the electrolyte then intercalate into PE, releasing electrical energy. The cycles of charging and discharging are reversable.

From a fundamental point of view, the Na^+ ion is much heavier and much larger than the Li^+ ion. Because of such a difference, the cathode material and the anode material for Na-ion batteries are different from those for Li-ion batteries.

13.3.1 The Cathode Materials

It was natural that the successful cathode materials for the Li-ion batteries were tested for the Na-ion batteries, especially the layered transition-metal oxides such as $Na(NiMnCo)O_2$ and the iron phosphate $NaFePO_4$. Nevertheless, because of the difference in ion radius, those materials did not work as good as the lithium counterparts. The favored cathode materials for the Na-ion batteries is the Prussian blue (PB) and

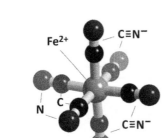 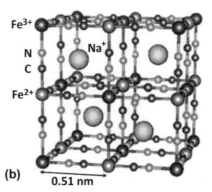

Figure 13.21 Ferrocyanide anion and Prussian blue. (a) An iron cation is bonded with six cyanide anions. If the oxidation number of iron is 2, the charge state of that ferrocynide anion is −4. If the oxidation number of iron is 3, its charge state becomes −3. Because the bonds of the cyanide group with iron are very strong, the substance is not toxic. (b) The structure of Prussian blue. The sodium ion can move almost freely in the cubic lattice of the iron-based framework.

Prussian blue analogues (PBA) [83, 93, 38].

Prussian blue was discovered in the early 18th century as the first synthesized pigment and has been used by artists and wall painters since then. The raw material is sodium ferrocyanide (or more formally, sodium hexacyanoferrate). In the early 18th century, it was produced by heating dried animal blood, iron filings, and soda together, then washing the mixture with warm water. After evaporation, the yellow crystal is produced with a name yellow blood salt, $Na_4Fe(CN)_6$. By mixing sodium ferrocyanide with ferric chloride in water, a dark blue pigment is produced,

$$Na_4Fe(CN)_6 + FeCl_3 \Longrightarrow NaFe[Fe(CN)_6] + 3NaCl. \qquad (13.7)$$

The key factor in the application of PB and its analogy to Na-ion battery is the ferrocyanide anion $Fe(CN)_6^{4-}$. A ball-and-stick diagram is shown in Fig. 13.21(a). Because iron has two oxidation states, 2+ and 3+, the anion can have two charging states, 3− and 4−. Figure 13.21(a) shows the state of the cathode material from the production line. The oxidation state of the iron atom is Fe^{2+}. During charging, a Na^+ ion is extracted, and the state of the iron atom becomes Fe^{3+}. The crystallographic structure stays unchanged. During discharging, a Na^+ ion moves into the PB structure, the oxidation state of the iron atom returns to Fe^{2+}.

PB was the cathode material in the first large-scale Na-ion battery products [38]. It has the advantages of excellent electrochemical stability and low cost. It can be prepared from abundant and non-toxic elements by simple and low-cost co-precipitation synthesis. Because of the open-framework lattice, Na^+ ion can move freely, offering high-power capabilities and long cycle life. The element Fe can be replaced by other transition metals, such as Mn, Ni, and Co, to make PB analogs.

13.3.2 The Anode Materials

Because the radius of the Na ion is much bigger than that of the Li ion, and the bonding of Na ion with graphite is weak, it is hard to intercalate to the spaces of adjacent graphene sheets of graphite. If the anode material is still made of carbon, the pores of the carbon-based material must be made bigger. In 2000, Stevens and Dahn reported the successful fabrication of high capacity anode materials based on hard carbon, produced by pyrolyzing glucose at 1000 and 1150°C [102].

Carbon has no liquid phase. It sublimates at about 4000 K. By pyrolyzing organic materials, such as sugar, glucose, starch, cork, coconut shell, apricot shell, bamboo, hard wood, or corn kernels at a temperature between 1000°C and 2000°C, the carbon skeleton of the material is left as a black matter, known as hard carbon. Hundreds of research papers showed it as a good material for the anodes in Na-ion batteries. Hard carbon anodes have the characteristics of high capacity, low cost, and wide availability, making them the most likely anode materials for Na-ion batteries [45, 104].

To explain the functionality of hard-carbon based anodes, a falling-card model was proposed [102], see Fig. 13.22. Accordingly, the carbon material is characterized by graphene-like segments of several nanometers in size, stacked in a somewhat random fashion like a pile of falling cards. This random stacking gives rise to small regions where multiple layers (two to three) are parallel to each other. It also results in regions of nanoscale porosity where layers are randomly oriented with respect to each other, making it suitable for adsorbing Na ions.

In addition to the hard-carbon material, PBA were used as the anode material in one of the first large-scale commercialization of Na-ion batteries. According to the report [38], Natrion utilized one PBA $Na_xMn_yFe(CN)_6$ as the cathode material, and another PBA $Na_xMn_yMn(CN)_6$ as the anode material. The nominal voltage (1.6 V) and the specific energy (67 Ah/kg) are low. However, due to the cubic structure of the material with high porosity, the power density is the highest among several popular rechargeable batteries. Because the manufacturing cost of those Ni-ion batteries is very low, for stationary energy storage, it has great commercial value.

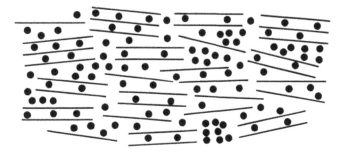

Figure 13.22 The falling-cards model of hard carbon. The hard carbon material is characterized by graphene-like segments of several nanometers in size, stacked in a somewhat random fashion like a pile of falling cards [102]. Sodium ions can easily move and then adsorb on the surfaces.

13.3.3　Rest of the System

From the point of view of research, development, and mass production, the similarity of Na-ion batteries and Li-ion batteries makes the transition much easier. Except for the cathode and anode materials, the rest of the system is very similar to the Li-ion batteries. The organic solvents listed in Table 13.4 and Fig. 13.11 are applicable. The sodium salt, $NaPF_6$, is the same as in Fig. 13.12, except Li^+ is replaced by Na^+. The separator and the packaging are identical. Therefore, a production line designed for Li-ion batteries can be adapted to produce Na-ion batteries.

For many years, because of the intrinsic lower energy density, the Na-ion batteries were considered to be stationary energy storage. Nevertheless, in recent years, the energy density of Na-ion batteries has been shown to be close to that of Li-ion counterparts. With the advantage of safety, fast charging, and especially much lower cost, Na-ion batteries are now applied to vehicles. In early 2023, two of the largest manufacturers of Li-ion batteries, CATL and BYD, announced that in late 2023, they will manufacture and install Na-ion batteries in passenger vehicles.

13.4　Traditional Rechargeable Batteries

13.4.1　Lead–Acid Batteries

To date, the most widely used rechargeable battery is the lead–acid battery. Every automobile should have one with six cells. For a fully charged lead–acid battery, the PE is made of PbO_2, and the NE is made of pure lead. The electrolyte is diluted sulfuric acid. After discharging, both the PE and the NE becomes $PbSO_4$. Sulfuric acid is thus consumed. By measuring the specific gravity of the electrolyte, and thus the concentration of sulfuric acid, the state of discharging, thus the energy remaining, can be determined.

The electrochemistry is as follows. During discharging, at the PE, PbO_2 is reduced,

$$PbO_2 + H_2SO_4 + 2e^- \longrightarrow PbSO_4 + 2OH^-. \tag{13.8}$$

At the NE, lead is oxidized,

$$Pb + H_2SO_4 \longrightarrow PbSO_4 + 2H^+ + 2e^-. \tag{13.9}$$

During charging, at the PE, $PbSO_4$ is oxidized,

$$PbSO_4 + 2OH^- \longrightarrow PbO_2 + H_2SO_4 + 2e^-. \tag{13.10}$$

At the negative electrode, $PbSO_4$ is reduced,

$$PbSO_4 + 2H^+ + 2e^- \longrightarrow Pb + H_2SO_4. \tag{13.11}$$

Lead is a heavy and toxic metal. Sulfuric acid is a dangerous liquid. The lifetime is short (300 cycles). Therefore, it is not suitable for portable devices and automobiles.

However, since lead is intrinsically an inexpensive metal and can be recycled, the overall cost is low. It is expected that such batteries will still be extensively used in the foreseeable future, for example, as energy storage devices in remote areas or in the basements of residential buildings to store solar electricity.

13.4.2 Nickel Metal Hydride Batteries

In recent decades, nickel metal hydride rechargeable batteries have been widely used in automobiles and relatively large portable electronic devices. The PE is nickel hydroxide, and the NE is an intermetallic compound. The most common metal has the general form AB_5, where A is a mixture of rare earth elements, lanthanum, cerium, neodymium, and praseodymium, and B is nickel, cobalt, manganese, and aluminum.

The electrochemistry is as follows. During discharging, at the PE, NiOOH is reduced,

$$NiOOH + H_2O + e^- \longrightarrow Ni(OH)_2 + OH^-. \tag{13.12}$$

At the negative electrode, metal hyride is oxidized,

$$MH + OH^- \longrightarrow M + H_2O + e^-. \tag{13.13}$$

During charging, at the positive electrode, $Ni(OH)_2$ is oxidized,

$$Ni(OH)_2 + OH^- \longrightarrow NiOOH^+H_2O + e^-. \tag{13.14}$$

At the negative electrode, metal is reduced,

$$M + H_2O + e^- \longrightarrow MH + OH^-. \tag{13.15}$$

When overcharged at low rates, the oxygen produced at the PE passes through the separator and recombines at the surface of the negative. Hydrogen evolution is suppressed, and the charging energy is converted to heat. This process allows NiMH cells to remain sealed in normal operation and to be maintenance free.

Figure 13.1(a) shows the discharging process. By connecting the cell to an external load, electrons flow from the NE, the anode, which is oxidized, through the external load to the PE, the cathode, where the electrons are accepted and the cathode material is reduced. The electric circuit is completed in the electrolyte by the flow of anions (negative ions) and cations (positive ions) to the anode and cathode, respectively.

Figure 13.1(b) shows the charging process. An external DC power supply is connected to the battery. The external electrical field forces electrons to flow into the NE, where reduction takes place. On the other hand, oxidation takes place at the PE, where the electrons flow out from. As the anode is, by definition, the electrode at which oxidation occurs and the cathode the one where reduction takes place, the PE is the anode and the NE is the cathode.

Problems

13.1. Each mole of electrons has a electric charge of 96500 C. Show it in terms of ampere-hours (Ah).

13.2. For a battery of nominal voltage of 3.5 V, with an active material of singly charged ion, how much energy can be stored in a mole of active material? Write the answer in watt-hours (Wh).

13.3. A Li-ion battery uses $LiCoO_3$ as cathode and LiC_6 as anode and have a nominal voltage of 4.0 V. What is the theoretical limit of specific energy in terms of Wh/kg?

13.4. In the previous problem, a Li-ion battery uses $LiCoO_2$ as cathode and LiC_6 as anode and have a nominal voltage of 4.0 V. The density of $LiCoO_2$ is 4.82 g/cm^3. The density of graphite is 2.26 g/cm^3. What is the theoretical limit of energy density in terms of Wh/L?

13.5. A Li-ion battery uses $LiFePO_4$ as cathode and LiC_6 as anode and have a nominal voltage of 3.3 V. What is the theoretical limit of specific energy in terms of Wh/kg?

13.6. In the previous problem, a Li-ion battery uses $LiFePO_4$ as cathode and LiC_6 as anode and have a nominal voltage of 3.3 V. The density of lithium iron phosphate is 3.6 g/cm^3. The density of graphite is 2.26 g/cm^3. What is the theoretical limit of energy density in terms of Wh/L?

13.7. A Na-ion battery uses Prussian blue as cathode with a general formula of $NaFe[Fe(CN)_6]$, uses hard carbon as anode with a general formula of NaC_6, and has a nominal voltage of 3.2 V. What is the theoretical limit of specific energy in terms of Wh/kg?

13.8. In the previous problem, a Na-ion battery uses Prussian blue as cathode with a general formula of $NaFe[Fe(CN)_6]$, uses hard carbon as anode with a general formula of NaC_6, and has a nominal voltage of 3.2 V. Purssian blue has a density of 1.8 g/cm^3. Hard carbon has a density of 1.5 g/cm^3. What is the theoretical limit of energy density in terms of Wh/L?

Chapter 14

Building with Sunshine

In an interview with *U.S. News and World Report* in March 2009, Steven Chu, the U.S. Energy Secretary, said this about the importance of improving building design to conserve energy [29]:

> People sometimes say energy efficiency and conservation are not sexy; they're low tech. That's actually not true. They can be very sexy and very high tech. ... Say you're building a new home. McKinsey has done a study that says for $1,000 extra in material and labor that investment could pay for itself in one and a half years and you could save a tremendous amount of energy. In terms of new homes and buildings, part of it is regulatory. But you've also got to convince people — they've got to believe in their heart and soul — that a small, minor upfront cost will actually decrease their monthly bill.

The details are in two reports by McKinsey published in 2009, entitled *"Unlocking Energy Efficiency in the U.S. Economy"* [68] and *"China's Green Revolution"* [67], which addressed the problems and suggestions for the two largest fossil fuel energy users in the world. The central conclusion is [68]:

> Energy efficiency offers a vast, low-cost energy resource for the U.S. economy — but only if the nation can craft a comprehensive and innovative approach to unlock it. Significant and persistent barriers will need to be addressed at multiple levels to stimulate demand for energy efficiency and manage its delivery across more than 100 million buildings and literally billions of devices. If executed at scale, a *holistic approach* would yield gross energy savings worth more than $1.2 trillion, well above the $520 billion needed for upfront investment in efficiency measures....

In fact, the most efficient approach to save energy is to apply passive design of the buildings into the architecture. In general, buildings consume about 40% of the world's energy. Using the holistic approach of house design, up to 50% of such energy can be saved. The principles include the following [67]:

1. Orient or position the building to absorb solar heat in cold regions and optimize solar heat in hot regions.

2. Judicially place windows to reduce the need of air conditioners, fans, and heaters.

3. Emphasize passive design to enable to use of smaller heaters or coolers.

There are two challenges for such an approach [67]. The first is to embed the approach in the mindset of professionals in the building sector. Currently, active heating and cooling systems, often using fossil fuels, are still considered the panacea. The second is to overcome a lack of effective dialogue between architects and civil engineers. Energy efficiency is still seldom taught in the curriculum of architects. Civil engineers, who may be more aware of energy efficiency considerations, tend to focus on executing a design and rarely get involved in the decisions that led to it.

The *holistic approach* to building design utilizing solar energy is not new. Many ancient cultures practiced it, but it has almost been forgotten in the developed world. To build a better future in the 21st century, we need a renaissance of the art of holistic building design.

14.1 Early Solar Architecture

14.1.1 Ancient Solar Architecture

Archeological evidence has shown that many ancient cultures built their houses according to the principles of passive solar design. According to Socrates, the ideal house should be cool in the summer and warm in the winter [17]. He also suggested that this goal can be achieved partially by orienting the opening of the house toward the south and providing a portico to create shade in the summer. Almost all important ancient cultures were located in the Northern hemisphere, and the houses were designed with large windows to the south but full walls in the north.

Holistic design principles, which were well-documented in the ancient Chinese literature since the West Zhōu dynasty (11th century *B.C.* to 771 *B.C.*), known as Fēngshŭi, meaning "wind and water," but reference to the Sun is one of the cardinal principles. In recent decades, the discipline of Fēngshŭi has become popular in the Western world. Determining true south is of fundamental importance in Fēngshŭi. In *Code of Zhōu Regulations* (Zhōu Lǐ), written c. 11th century *B.C.*, a method of determining true south was documented as follows: "by marking the point of sunrise and the point of sunset, then taking the middle point, the south point can be found." It is much more accurate than the magnetic compass.

14.1.2 Holistic Architecture in Rural China

Figure 14.1 shows a traditional peasant house in rural northern China which embodies some principles of holistic design utilizing solar energy. It was designed over 1000 years ago and is still used in less developed regions of northern China. I have personally lived in such houses for several years and have been impressed by its comfort.

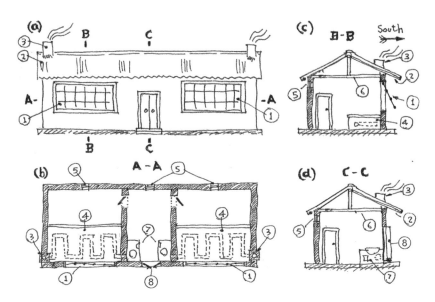

Figure 14.1 Traditional peasant house in rural northern China. (a) the front view. (b) the A-A cross sectional view. (c) the B-B cross sectional view. (d) the C-C cross secional view. The long axis of the house is east–west. Windows (1) are on the south side. Due to the long eaves (2), it has full sunlight in the winter and is in the shade in summer. The heat from the oven (7) is stored in the adobe beds (4), which can stay warm for the entire night. The wind from the front window to the small north windows (5) keeps it cool in the summer.

The long axis of the house is east–west. Large glass windows (1) are always on the south side. The windows can be swung partially open from the lower side. The eaves (2) are designed such that in the winter the windows exposes to full sunlight and in the summer the windows are in shade. Near the top of the northern wall, there are small windows (5). In the summer, by opening the small northern windows and the south window partially, wind will automatically blow from the lower opening of the south window to the small north windows, which can cool the house by ventilation. In the winter, the smoke from the oven (7) flows to the chimney (3) through the zigzagged tunnels in the adobe bed (4), called *kàng*, which can stay warm for the entire night. In the summer, cooking is often done in the sheds outside the house; thus the adobe bed, directly connected to the ground, can remain cool. The space between the ceiling (6) and the roof also helps to keep a stable temperature. To keep the rooms warm, in the winter, the front door (8) is often equipped with a heavy curtain.

14.2 Building Materials

The design of a building involves an interplay of architectural and civil engineering. In this section, some basic concepts of civil engineering for building design are presented.

14.2.1 Thermal Resistance

The thermal resistance, or the R-value, is a measure of the insulation property of a panel or sheet of building material, such as a wall panel, a door, or a window. It is the ratio of temperature difference on the two surfaces and the heat flux (heat flow per unit area). In SI unit, the R-value (RSI) is the temperature difference (in Kelvin) required to leak heat energy in watts per unit area (W/m^2). Therefore, the unit of RSI is $K \cdot m^2/W$. In British units, the R-value is the temperature difference (in Fahrenheit) required to leak heat energy in Btu per hour per unit area. Therefore, the unit of R-value is $F \cdot h \cdot f^2/Btu$). The conversion relation is 1 RSI = 5.678R, 1 R = 0.1761RSI.

For a wall consisting of several layers, the R-value is additive:

$$R = \sum_i R_i, \tag{14.1}$$

where R_i is the R-value of the i-th layer of the material. Following are some examples:

- Single-pane glass window: R-1 (RSI = 0.18)

- Double-pane glass window: R-2 (RSI = 0.35)

- Above window with low-emissivity coating: R-3 (RSI = 0.52)

14.2.2 Specific Thermal Resistance

The insulation property of a material, for example, concrete, fiberglass wool, or foam polyurethane, is measured by its *specific thermal resistance*, or the r-value. In SI units (rSI), it is defined as the temperature difference (in Kelvin) per unit thickness (in meter) required to leak heat energy in watts per unit area (W/m^2). In British unit (r), it is defined as the temperature difference (in Fahrenheit) per unit thickness (in inches) required to leak heat energy in Btu per hour per unit area ($Btu/h \cdot f^2$). The conversion relation is 1 rSI = 0.125r, 1 r = 8rSI.

Table 14.1: Specific thermal resistance

Material	r-Value (Btu)	rSI (SI unit)
Wood panels	2.5	20
Fiberglass or rock wool	3.1–3.6	25–30
Icynene spray	3.6	30
Polyurethane rigid panel	7	56
Poured concrete	0.08	0.64

For a wall consisting of several layers, the R-value is the sum of the product of the thickness and the r-value of each layer:

$$R = \sum_i t_i r_i, \tag{14.2}$$

where t_i is the thickness of the ith layer and r_i is the specific thermal resistance of the material of the ith layer.

Table 14.1 lists the r-values of some common building materials. In the United States, the most commonly used insulation material is fiberglass wool. It has a reasonably high r-value and is lightweight, inexpensive, chemically inert, and nonflammable. Foam polyurethane has a much higher r-value. It is supplied as either rigid panels or as a spray material that fills the space of the wall, for example, between the plywood panel and the sheetrock panel of a wall. The commercial name of the spray is Icynene. In terms of thermal insulation, concrete is much worse than those standard insulation materials.

The R-values of some typical wall insulation panels can be calculated using Eq. 14.2 and the data in Table 14.1. Some typical results are shown in Table 14.2.

14.2.3 Heat Transfer Coefficient: The U-Value

To calculate the total heat loss of a building or a room, the R-value is not convenient. The inverse of R-value, the heat transfer coefficient, or the U-value, is used. In the SI unit system, it is defined as the heat loss in watts per unit square meter per Kelvin temperature difference (W/K·m^2). In the British unit system, it is defined as the heat loss in Btu per hour per unit square foot per degree Fahrenheit temperature difference (Btu/h·f^2). The relation of conversion is 1U = 5.678 USI, 1 USI = 0.1761U. The relation with the R-value is

$$U = \frac{1}{R}. \tag{14.3}$$

For a room or building with walls, windows, doors, and so on, the heat loss Q can be computed as the sum of the product of the area and the U-value of each component:

$$Q = \Delta T \sum_i A_i U_i = \Delta T \sum_i \frac{A_i}{R_i}, \tag{14.4}$$

Table 14.2: Typical R-values of wall insulation materials

Material	R-Value (Btu)	RSI (SI unit)
4-in. fiberglass butts	R-13	2.3
6-in. fiberglass butts	R-19	3.3
12-in. fiberglass butts	R-39	6.7
6-in. icynene	R-22	3.8
10-in. concrete wall	R-0.8	0.14

where A_i is the area and U_i is the U-value of the ith component and ΔT is the temperature difference.

14.2.4 Thermal Mass

The stability of room temperature depends on how much heat can be retained inside the room. It is the sum of of the thermal masses of the components,

$$M = \sum_j V_j \rho_j c_{pj}, \qquad (14.5)$$

where V_j is the volume and $\rho_j c_{pj}$ is the specific heat capacity per volume of the jth component. The rate of temperature decline is

$$\frac{dT}{dt} = -\frac{Q}{M} = -\frac{\Delta T}{M} \sum_i A_i U_i = -\frac{\Delta T}{M} \sum_i \frac{A_i}{R_i}. \qquad (14.6)$$

Therefore, the greater the thermal mass, the slower the temperature decline.

Tables 14.1 and 14.2 list thermal properties of commonly used building materials. From those tables we find that water is by far the best material for sensible energy storage. Concrete and bricks are the worst, but unfortunately, are the materials often used.

14.2.5 Glazing

Glass windows let sunlight into a room, and thus provide warmth in the winter. However, as seen in Table 14.2, even high-grade windows are poor insulators. Double-pane glass windows have an RSI-value of 0.35, but the mediocre 6-in. fiberglass butts provides RSI = 3.3. Therefore, there is a one order of magnitude difference.

It has become fashionable to use a lot of glass in modern buildings. However, this requires a lot of electricity in the summer to cool the building and a lot of natural gas

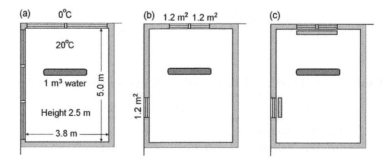

Figure 14.2 Effect of glazing on insulation. (a) All external walls are glazed with double-pane glass windows. (b) With three 1.2–m^2 double-pane glass windows. (c) After the windows are covered with polyurethane panels.

(or worse electricity) in winter to keep it warm. Too much glazing results in a waste of energy.

Figure 14.2 shows an example. A room at the corner of the house has a 1 m³ water tank as the thermal storage medium. The size of the room is 3.8 m × 5.0 m × 2.5 m. Assuming that the room temperature is 20°C and the external temperature is 0°C. The thermal mass $M = 4.19 \times 10^6$ J/K. The heat loss through the internal walls can be neglected. As shown in Fig. 14.2(a), if all external walls are glazed with double-pane glass windows, then the total area of glass is 22 m². The rate of heat loss is

$$Q = \frac{20 \times 22}{0.35} = 1257 \text{ W}. \tag{14.7}$$

The rate of temperature drop is

$$\frac{dT}{dt} = -\frac{1257}{4.19 \times 10^6} = -3 \times 10^{-4} \text{ K/s} = -1.08 \text{ K/h}. \tag{14.8}$$

Therefore, even with such a large thermal mass, without heating, the room will become very cold after several hours.

Figure 14.2(b) shows a house with three 1.2-m² double-pane windows. The rest of the wall is insulated by conventional 6-in. (150 mm) fiberglass butts. The rate of heat loss is

$$Q = 20 \times \left(\frac{22 - 3.6}{3.3} + \frac{3.6}{0.35} \right) = 317 \text{ W}. \tag{14.9}$$

The rate of temperature drop is

$$\frac{dT}{dt} = -\frac{317}{4.19 \times 10^6} = -7.6 \times 10^{-5} \text{ K/s} = -0.27 \text{ K/h}. \tag{14.10}$$

After 8 hours, the temperature will drop by 2.2°C.

At night there is no sunlight, and the windows become solely a drain of energy. By covering the windows with 25-mm-thick polyurethane rigid panels, with $R = 6.5$, the R-value of windows is increased to 8.5, or RSI = 1.5. The rate of heat loss is

$$Q = 20 \times \left(\frac{22 - 3.6}{3.3} + \frac{3.6}{1.5} \right) = 159 \text{ W}. \tag{14.11}$$

The rate of temperature drop is

$$\frac{dT}{dt} = -\frac{159}{4.19 \times 10^6} = -3.32 \times 10^{-5} \text{ K/s} = -0.136 \text{ K/h}. \tag{14.12}$$

Then, the temperature drop is about 1°C overnight. If there are people living in that room, the temperature drop will be even less because the heat generated by each person is roughly equal to an incandescent lightbulb, that is, 40–60 w. It will partially compensate for the heat loss through the walls and windows and make the temperature virtually constant.

The windows in this room are already large enough to admit sufficient sunlight to warm the room in the winter. From Fig. 4.7, in locations near latitude $40°$, from November to February, the daily solar radiation on a south-facing window is greater than 6.5 kWh/m^2. The two 1.2-m^2 windows could admit more than 10 kWh of thermal energy on a sunny day. The solar thermal power is much greater than the thermal loss, so the room would be kept warm from just the sunlight through the windows. Excessive glazing would probably make the room too hot during the day.

14.3   Example of Holistic Design

According to Steven Chu [29] and McKinsey [68, 67], when one tries to build a new home, with a little extra material and labor, a large amount of energy – and consequently a large amount of money – could be saved. I tested this statement by building my own home. I found that even the extra upfront investment money is not necessary. The key is to make a design that maximizes the benefit of sunlight. Since I appoint

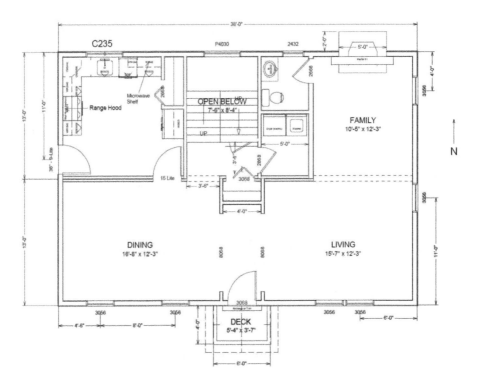

Figure 14.3 Design of a solar house: First floor. All large windows are facing south, and the entire grand hall is in the south. Less important rooms, such as the stairs and the bathroom, are in the north.

myself as the architect and chief civil enginer, the extra cost was my time.

Before design, I read almost every book and paper I could find about solar-energy house. I was somewhat disappointed that many solar energy house designs were either too exotic, requiring unconventional materials and building technique, or emphasized energy saving to demonstrate a house that solely depends on solar energy while sacrificing comfort. In addition, the planning board and architecture review board of the town required that the house be in harmony with other houses around it, and the construction company insisted on using readily available building materials.

In addition to building a comfortable home, I had some additional requirements. First, I have a Hamburg Steinway Model B, which requires a rather large room with good resonance and stable temperature and humidity. Second, I wanted to build a professional-grade home movie theater which could be operated any time of the year. Third, I want a playroom of more than 60 m^2 for my grandchildren.

Fortunately, I found an empty lot in North White Plains which fit the requirements: a 0.89 acre (3500 m^2) land sloping towards south. Because of the slope, in spite of several diligent attempts, the previous owners were unable to present a design that satisfied the requirements of the town planning board. By reviewing the previous design documents and drawings, I found that the previous owners wanted to carve out

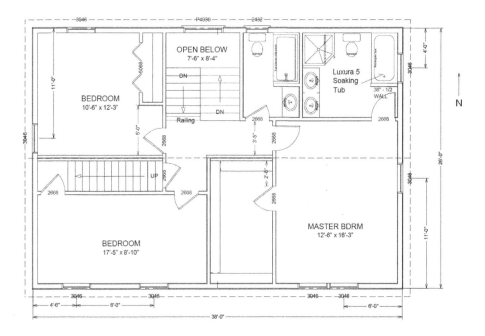

Figure 14.4 Design of a solar house: Second floor. All large windows are facing south, and the two most used rooms are in the south. Less important rooms, such as the stairs, the guest room and the bathrooms are in the north.

Figure 14.5 External view of the solar house. With the large souce-facing windows, the rooms are full of sunlight in the winter and no sunlight in the summer. The 4.2 kW solar panels on the roof drives two large central air conditing systems to keep all rooms cool. The solar panels has worked for 14 years with no problem and does not deteriorate significantly.

a large flat space through the slope, which required extensive exploding and excavation plus an 8-meter-high retaining wall, which would be very dangerous.

I decided to take advantage of the slope and bury part of the house beneath the surface of the earth. Therefore, part of the structure can utilize the shallow geothermal energy – the solar energy stored in the ground. The front plane of the house is designed to face true south, with nothing blocking the sunlight. All large windows face south to admit as much sunlight as possible. The north face of the house is close to the hill. Very few windows are needed. The idea was quickly approved by the planning board and architectural review board of the Town of North Castle.

The design drawings are shown in Figs. 14.3 and 14.4. From outside to inside, it looks like a normal central-hall colonial. But solar energy is extensively utilized. The first floor, with a high ceiling, hosts a Steinway B as its focal point. All large windows face south. The less important rooms, such as the scissor stairs and bathrooms, are in the north. On the second floor, the two most used rooms, the master bedroom and the office (bedroom 2) are in the south, and all large windows are facing south. Less important rooms, such as the guest room (bedroom 3) are in the north.

Westchester Module Home Construction Company worked out the details and built the house using some of the best conventional materials and building methods. All windows are double-pane airtight Anderson windows. The insulation for the ceiling is R39, and others are R19. It was finished in August 2009. See Fig. 14.5. On cold and sunny winter days, sunlight from the large south windows often bring the temperature

Figure 14.6 Internal view of the solar house. With the large souce-facing windows, the rooms are full of sunlight for a sunny day in the winter. The room temperature often exceeds 22 °C, and the thermostat for the heating system triggers off. There is almost no sunlight in the summer.

of the two most used rooms to above 22°C, with the thermostat almost always turned off, see Fig. 14.6. This is reasonable by counting solar radiation. As shown in Fig. 4.7, near latitude 40°, on a sunny day in winter, through the two south-facing windows each has an area of one square meter, sunlight can bring in more than 10 kWh of thermal energy. In the summer, because of the eaves, no sunlight comes in. Furthermore, there is a solar-powered attic fan on the roof to assist cooling the entire house.

The movie theater was placed in the far corner of the foundation, which is deepest in the ground. Utilizing the solar energy stored in the ground, the temperature remains comfortable throughout the year.

The 4.2-kW photovoltaics system also worked as planned. Because the orientation of the roof faces exactly south, the tilt angle equals the latitude, and sunlight is unobstructed, the efficiency is high. The total electricity purchased from Con Edison for a whole year was about 1600 kWh, roughly one-tenth of a regular house of similar size. Fourteen years after it was built, the solar photovoltaic system has worked smoothly without any problem, and the efficiency shows no visible deterioration.

Problems

14.1. Consider a house of dimension 12 m long, 8 m wide, and 5 m high, with typical insulation of 6-in. fiberglass with R-19 on peripherals and 12-in. fiberglass with R-39 on the ceiling and floor. In the summer, the average external temperature is 80°F (26.67°C), but the comfortable internal temperature is 68°F (20°C). By using the latent heat of the tank of ice to cool the house (equipped with a thermostat and circulating mechanism), for how long can the tank of ice keep the room temperature constant?

14.2. In the New York area (latitude 40°47′), on the south side of a house the is a window of height $H = 1.67$ m, see Fig. 14.7. In order to fully utilize the sunlight at the winter solstice and fully block the sunlight at the summer solstice, what should be the length of overhang L and the distance D from the top of the window to the base of the overhang?

14.3. A typical room in a well-insulated house has a dimension of 20 ft in length, 16 ft in width, and 8 ft in height. All walls, including ceiling and floor, are insulated by R39 material. In the winter, the outside temperature is 0°C and the room temperature is 20°C. What is the thermal loss in watts? If the room contains one ton of water as thermal mass, what is the rate of temperature loss in °C per hour?

Figure 14.7 Calculation of eaves.

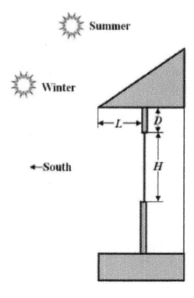

Appendix A

Energy Unit Conversion

Because energy is one of the most important quantities, there are many energy units which often creates confusion. Throughout this book, we used the SI units for all physical quantities. The SI unit of energy is joule, defined as the energy capable of pushing an object with one newton of force by one meter:

$$J = N \cdot m. \tag{A.1}$$

The basic SI unit of power, the watt, equals one joule per second,

$$W = J/s. \tag{A.2}$$

The most frequently used units of energy and power are listed in Table A.1.

Electrical power is the product of the current in amperes and voltage in volts:

$$W = A \cdot V. \tag{A.3}$$

Utility companies often use an energy unit derived from electrical power, the kilowatt-hour, or kWh:

$$1 \text{ kWh} = 3600 \text{ kJ} = 3.6 \text{ MJ}. \tag{A.4}$$

Table A.1: Energy and power units

Name	Symbol	Equals
Kilojoule	kJ	10^3 J
Megajoule	MJ	10^6 J
Gigajoule	GJ	10^9 J
Exajoule	EJ	10^{18} J
Kilowatt	kW	10^3 W
Megawatt	MW	10^6 W
Gigawatt	GW	10^9 W
Terawatt	TW	10^{12} W

In microscopic physics, the electron-volt (eV) is often used as a unit of energy. Because the charge of an electron is $e = 1.60 \times 10^{-19}$ C, 1 eV $= 1.60 \times 10^{-19}$ J. Another unit frequently found in the literature is the calorie, abbreviated as cal. It is defined as the energy required to raise the temperature of one gram of water by one degree Celsius. Because at different temperatures the amount of energy required is different, the definition of the unit is ambiguous. A well-defined and widely accepted definition of the calorie is

$$1 \text{ cal} = 4.184 \text{ J.} \tag{A.5}$$

One kilocalorie (kcal) is defined exactly as 4184 J. Therefore, unless necessary, SI units are much preferred.

In the United States, the British thermal unit (Btu) is often used as the unit of energy. It is defined as the energy to raise the temperature of one pound water by one degree Fahrenheit. Similar to calorie, the exact value of the unit is ambiguous. The International Table defines the Btu as 1055.06 J, which is very close to 1 kJ. In dealing with renewable energy problems, we can take the approximation

$$1 \text{ Btu} \approx 1 \text{ kJ.} \tag{A.6}$$

For large amounts of energy, a frequently used unit is the exajoule (EJ), which is defined as 10^{18} J. In the United States, the corresponding unit for large amounts of energy is quadrillion Btu, abbreviated as quad, which is defined as 10^{15} Btu. Because 1 Btu is very close to 1 kJ, for practical purposes, we can consider the two units to be almost equivalent.

$$1 \text{ quad} \approx 1 \text{ EJ.} \tag{A.7}$$

Another unit for large amounts of energy is terawatt-hour (TWh). It equals one billion kilowatt-hours, or 0.0036 EJ. One terawatt-hour equals 3.6×10^{15} J. One exajoule equals 277.8 TWh.

In the utility industry, the gigawatt (GW) and terawatt (TW) are often used. One gigawatt-year equals 3.156×10^{16} J, roughly equals one 1/32 EJ. One terawatt-year equals 3.156×10^{19} J, roughly equals 32 EJ. In 2007, the world's energy consumption of energy is approximately 500 EJ, or an average power of 15 TW; the energy consumption of the United States is approximately 100 EJ, or an average power of 3 TW. In 2008, the energy consumption of the United States is slightly less than 100 EJ.

Several important approximate relations are worth noting:

- One barrier of crude oil is defined as 5.8×10^6 Btu.

- One cubic foot of natural gas is approximately 1000 Btu, or 1 MJ.

- One therm of natural gas is defined as 100,000 Btu, approximately 100 MJ.

- One kilojoule per mole is 0.01036 eV, approximately 10 meV.

- One kilocalorie per mole is 0.0434 eV.

Appendix B

Spherical Trigonometry

When we look into the sky, it seems that the Sun and all the stars are located on a *sphere* of a large but unknown radius. In other words, the location of the Sun is defined by a point on the *celestial sphere*. On the other hand, the surface of Earth is, to a good approximation, a sphere. Any location on Earth can be defined by a point on the *terrestrial sphere*; namely by the *latitude* and the *longitude*. In both cases, we are dealing with the geometry of spheres.

To study the location of the Sun with respect to a specific location on Earth, we will correlate the coordinates of the location on the terrestrial sphere of Earth with the location of the Sun on the celestial sphere. The mathematical tool of this study is *spherical trigonometry*. In this Appendix, we will give a brief introduction to spherical trigonometry, sufficient to deal with the problem of tracking the sunlight.

B.1   Spherical Triangle

A plane passing through the center of a sphere O cuts the surface in a circle, which is called a *great circle*. For any two points A and B on the sphere, if the line AB does not pass the center O, there is one and only one great circle which passes both points. The angle $\widehat{AOB}$, chosen as the one smaller than $180°$ or π in radians, is defined as the length of the arc AB. Given three points A, B, and C on the sphere, three great circles can be defined. The three arcs AB, BC, and CA, each less than $180°$ or π in radians, form a *spherical triangle*; see Fig. B.1.

Following standard notation, we denote the sides BC, CA, and AB by a, b, and c, respectively. The length of side a is defined as the angle $\widehat{BOC}$, the length of side b is defined as the angle $\widehat{COA}$, and the length of side c is defined as the angle $\widehat{AOB}$. The vertex angles of the triangle are defined in a similar manner: The vertex angle A is defined as the angle between a straight line AD tangential to AB and another straight line AE tangential to AC, and so on.

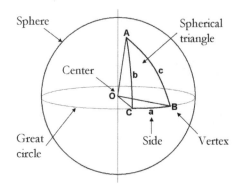

Figure B.1   The spherical triangle. A *great circle* is defined by the intersection of a plane that cuts the sphere in two equal halves through the center O. Given three points A, B, and C on the sphere, three great circles can be defined. The three arcs AB, BC, and CA, each less than $180°$ or π in rad, form a *spherical triangle*.

B.2   Cosine Formula

In planar trigonometry, there is a cosine formula

$$a^2 = b^2 + c^2 - 2bc \cos A. \tag{B.1}$$

In spherical trigonometry, a similar formula exists:

$$\cos a = \cos b \cos c + \sin b \sin c \cos A. \tag{B.2}$$

When the arcs are short and the spherical triangle approaches a planar triangle, Eq. B.2 reduces to Eq. B.1. In fact, for small arcs,

$$\cos b \approx 1 - \frac{1}{2}b^2, \tag{B.3}$$

and

$$\sin b \approx b, \tag{B.4}$$

and so on. Substituting Eqs B.3 and B.4 into Eq. B.2 reduces it to Eq. B.1.

Here we give a simple proof of the cosine formula in spherical trigonometry by an analogy to that in planar trigonometry, see Fig. B.2. To simplify notation, we set the radius of the sphere $OA = OB = OC = 1$. By extending line OB to intersect a line tangential to AB at a point D, we have

$$AD = \tan c; \quad OD = \sec c. \tag{B.5}$$

Similarly, by extending line OC to intersect a line tengential to AC at at a point E, we have

$$AE = \tan b; \quad OE = \sec b. \tag{B.6}$$

From the planar triangle DAE, using the planar cosine formula,

$$DE^2 = AD^2 + AE^2 - 2\,AD \cdot AE \cos \widehat{DAE}$$
$$= \tan^2 c + \tan^2 b - 2\tan b \tan c \cos A. \tag{B.7}$$

Figure B.2   Derivation of cosine formula. The derivation is based on the projection of a spherical triangle onto a plane and the cosine formula in planar trigonometry. A straight line tangential to arc AB intersects the extension of line OB at D. Another straight line tangential to arc AC intersects the extension of line OC at E. By applying the cosine formula in planar trigonometry on triangles ODE and ADE, after some brief algebra, the corresponding cosine formula is obtained.

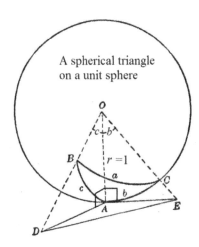

A spherical triangle on a unit sphere

Similarly, for the planar triangle DOE,

$$DE^2 = OD^2 + OE^2 - 2\,OD \cdot OE \, \cos \widehat{DOE}$$
$$= \sec^2 c + \sec^2 b - 2\sec b \, \sec c \, \cos a. \tag{B.8}$$

Subtrating Eq. B.7 from B.8, and notice that

$$\sec^2 b = 1 + \tan^2 b, \quad \sec^2 c = 1 + \tan^2 c, \tag{B.9}$$

we obtain

$$2 - 2\sec b \, \sec c \, \cos a = -2\tan b \, \tan c \, \cos A. \tag{B.10}$$

Multiplying both sides with $\cos b \, \cos c$ yields

$$\cos a = \cos b \, \cos c + \sin b \, \sin c \, \cos A. \tag{B.11}$$

Similarly, for vertices B and C,

$$\cos b = \cos c \, \cos a + \sin c \, \sin a \, \cos B; \tag{B.12}$$

$$\cos c = \cos a \, \cos b + \sin a \, \sin b \, \cos C. \tag{B.13}$$

B.3   Sine Formula

In planar trigonometry, there is the sine formula

$$\frac{a}{\sin A} = \frac{b}{\sin B} = \frac{c}{\sin C}. \tag{B.14}$$

In spherical trigonometry, a similar formula exists,

$$\frac{\sin a}{\sin A} = \frac{\sin b}{\sin B} = \frac{\sin c}{\sin C}. \tag{B.15}$$

Obviously, for small arcs, Eq. B.15 reduces to Eq. B.14.

To prove Eq. B.15, we rewrite Eq. B.2 as

$$\sin b \sin c \cos A = \cos a - \cos b \cos c. \tag{B.16}$$

Squaring, we obtain

$$\sin^2 b \sin^2 c \cos^2 A = \cos^2 a - 2\cos a \cos b \cos c + \cos^2 b \cos^2 c. \tag{B.17}$$

The left-hand side can be written as

$$\sin^2 b \sin^2 c \ - \sin^2 b \sin^2 c \sin^2 A, \tag{B.18}$$

or

$$1 - \cos^2 b - \cos^2 c + \cos^2 b \cos^2 c \ - \sin^2 b \sin^2 c \sin^2 A. \tag{B.19}$$

Hence,

$$\sin^2 b \sin^2 c \sin^2 A = 1 - \cos^2 a - \cos^2 b - \cos^2 c + 2\cos a \cos b \cos c. \tag{B.20}$$

Define a quantity symmetric to a, b, and c,

$$Z = \frac{\sin^2 a \sin^2 b \sin^2 c}{1 - \cos^2 a - \cos^2 b - \cos^2 c + 2\cos a \cos b \cos c}, \tag{B.21}$$

equation B.20 can be written as

$$\frac{\sin a}{\sin A} = \pm\sqrt{Z}. \tag{B.22}$$

By definition, in a spherical triangle, the sides and the vertical angles are always smaller than $180°$. Therefore, in Eq. B.22, only a positive sign is admissible. Because Z is symmetric to A, B and C, we obtain

$$\frac{\sin a}{\sin A} = \frac{\sin b}{\sin B} = \frac{\sin c}{\sin C}. \tag{B.23}$$

B.4   Formula C

We rewrite the cosine formula B.11 in the following form and use Eq. B.13:

$$\sin b \, \sin c \, \cos A = \cos a - \cos b \, \cos c$$
$$= \cos a - \cos b \, (\cos a \, \cos b + \sin a \, \sin b \, \cos C) \qquad \text{(B.24)}$$
$$= \cos a \, \sin^2 b - \cos b \, \sin a \, \sin b \, \cos C.$$

Dividing both sides by $\sin b$, one obtains *formula C*

$$\sin c \, \cos A = \cos a \, \sin b - \sin a \, \cos b \, \cos C. \qquad \text{(B.25)}$$

Similarly,

$$\sin a \, \cos B = \cos b \, \sin c - \sin b \, \cos c \, \cos A, \qquad \text{(B.26)}$$
$$\sin a \, \cos C = \cos c \, \sin b - \sin c \, \cos b \, \cos A, \qquad \text{(B.27)}$$

and so on.

Problems

B.1. Show that if one of the arcs is $180°$, then no spherical triangle can be constructed.

B.2. If one of the vertex angles of a spherical triangle, for example, C, is a right angle, show that for small arcs the cosine formula leads to the Pythagorean theorem.

B.3. Using the cosine and sine formulas, show that

$$\cot a \, \sin b = \cot A \, \sin C + \cos b \, \cos C, \tag{B.28}$$

$$\cot c \, \sin a = \cot C \, \sin B + \cos a \, \cos B, \tag{B.29}$$

and so on.

B.4. For a rectangular spherical triangle, where $C{=}90°$, show that

$$\sin a = \sin c \, \sin A, \tag{B.30}$$

$$\sin b = \sin c \, \sin B, \tag{B.31}$$

$$\tan a = \tan c \, \cos B, \tag{B.32}$$

$$\tan b = \tan c \, \cos A, \tag{B.33}$$

$$\tan a = \sin b \, \tan A, \tag{B.34}$$

$$\tan b = \sin a \, \tan B. \tag{B.35}$$

Appendix C

Vector Analysis and Determinants

C.1   Vector Analysis

For handling elements in a three-dimensioanl field, especially electromagnetics, *vector analysis* is a convenient matiematical tool. The central component is a differential operator usually called a *del operator*, defined as

$$\nabla = \mathbf{i}\frac{\partial}{\partial x} + \mathbf{j}\frac{\partial}{\partial y} + \mathbf{k}\frac{\partial}{\partial z},$$ (C.1)

where $\mathbf{i}$, $\mathbf{j}$, and $\mathbf{k}$ are unit vectors in the x, y, and z directions, respectively.

There are three commonly used operations. The gradient of a scalar function $\varphi(\mathbf{r})$, a vector function, is defined as

$$\operatorname{grad}\varphi \equiv \nabla\varphi \equiv \mathbf{i}\frac{\partial\varphi}{\partial x} + \mathbf{j}\frac{\partial\varphi}{\partial y} + \mathbf{k}\frac{\partial\varphi}{\partial z}.$$ (C.2)

The divergence of a vector function $\mathbf{A}(\mathbf{r})$, a scalar function, is defined as

$$\operatorname{div}\mathbf{A} \equiv \nabla \bullet \mathbf{A} \equiv \frac{\partial A_x}{\partial x} + \frac{\partial A_y}{\partial y} + \frac{\partial A_z}{\partial z}.$$ (C.3)

And the curl of a vector function $\mathbf{A}(\mathbf{r})$, a vector function, is defined as

$$\operatorname{curl}\mathbf{A} \equiv \nabla \times \mathbf{A} \equiv \begin{vmatrix} \mathbf{i} & \mathbf{j} & \mathbf{k} \\ \dfrac{\partial}{\partial x} & \dfrac{\partial}{\partial y} & \dfrac{\partial}{\partial z} \\ A_x & A_y & A_z \end{vmatrix}.$$ (C.4)

the mathematical operation in Eq. C.4 is the *determinant*, and we will present it later. The dot product of two del operators is a Laplacian,

$$\nabla^2 \equiv \nabla \bullet \nabla \equiv \frac{\partial^2}{\partial x^2} + \frac{\partial^2}{\partial y^2} + \frac{\partial^2}{\partial z^2}.$$ (C.5)

There a series of identitites of interest. Here are some examples:

$$\nabla(\varphi + \psi) \equiv \nabla\varphi + \nabla\psi. \tag{C.6}$$

$$\nabla \bullet (\varphi \mathbf{A}) \equiv \varphi(\nabla \bullet \mathbf{A}) + \mathbf{A} \bullet (\nabla\varphi). \tag{C.7}$$

$$\nabla \times (\varphi \mathbf{A}) \equiv \varphi(\nabla \times \mathbf{A}) + (\nabla\varphi) \times \mathbf{A}. \tag{C.8}$$

$$\nabla \times (\mathbf{A} \times \mathbf{B}) \equiv \mathbf{A}(\nabla \bullet \mathbf{B}) - \mathbf{B}(\nabla \bullet \mathbf{A}) + (\mathbf{B} \bullet \nabla)\mathbf{A} - (\mathbf{A} \bullet \nabla)\mathbf{B}. \tag{C.9}$$

$$\nabla \bullet (\mathbf{A} \times \mathbf{B}) \equiv \mathbf{B} \bullet (\nabla \times \mathbf{A}) - \mathbf{A} \bullet (\nabla \times \mathbf{B}). \tag{C.10}$$

$$\nabla \bullet (\nabla \times \mathbf{A}) \equiv 0. \tag{C.11}$$

$$\nabla \times (\nabla\varphi) \equiv 0. \tag{C.12}$$

$$\nabla \times (\nabla \times \mathbf{A}) \equiv \nabla(\nabla \bullet \mathbf{A}) - \nabla^2\mathbf{A}. \tag{C.13}$$

C.2 Determinants

For a square matrix A of n×n elements, a scalar as a function of the matrix can be defined, which is an important characterstics of the matrix. For $n = 2$, the determinant of a matrix

$$A = \begin{bmatrix} a & b \\ c & d \end{bmatrix} \tag{C.14}$$

is defined as

$$\det(A) \equiv \begin{vmatrix} a & b \\ c & d \end{vmatrix} = ad - bc. \tag{C.15}$$

For $n = 3$, a good example is the volume of a parallelepiped constructed from three vectors $\mathbf{r}_1 = (x_1, y_1, z_1)$, $\mathbf{r}_2 = (x_2, y_2, z_2)$, and $\mathbf{r}_3 = (x_3, y_3, z_3)$. By writing the three vectors as a 3×3 matrix,

$$A = \begin{bmatrix} x_1 & y_1 & z_1 \\ x_2 & y_2 & z_2 \\ x_3 & y_3 & z_3 \end{bmatrix}, \tag{C.16}$$

the volume is

$$V = \mathbf{r}_1 \cdot (\mathbf{r}_2 \times \mathbf{r}_3) = \det(A) \equiv \begin{vmatrix} x_1 & y_1 & z_1 \\ x_2 & y_2 & z_2 \\ x_3 & y_3 & z_3 \end{vmatrix}. \tag{C.17}$$

The determinant is deined as

$$\det(A) = x_1 y_2 z_3 + x_2 y_3 z_1 + x_3 y_1 z_2 - x_1 y_3 z_2 - x_2 y_1 z_3 - x_3 y_2 z_1. \tag{C.18}$$

The definition of curl in Eq. C.4 is then

$$\nabla \times \mathbf{A} \equiv \mathbf{i} \left(\frac{\partial A_z}{\partial y} - \frac{\partial A_y}{\partial z} \right) + \mathbf{j} \left(\frac{\partial A_x}{\partial z} - \frac{\partial A_z}{\partial x} \right) + \mathbf{k} \left(\frac{\partial A_y}{\partial x} - \frac{\partial A_x}{\partial y} \right). \tag{C.19}$$

In general, for a $n \times n$ matrix

$$A = \begin{bmatrix} a_{1,1} & a_{1,2} & \cdots & a_{1,n} \\ a_{2,1} & a_{2,2} & \cdots & a_{2,n} \\ \cdots & \cdots & \cdots & \cdots \\ a_{n,1} & a_{n,2} & \cdots & a_{n,n} \end{bmatrix}, \tag{C.20}$$

the definition of a determinant of A is

$$\det(A) = \sum_{i_1, i_2, \ldots i_n} \epsilon_{i_1, i_2, \ldots i_n} \, a_{1,i_1} a_{2,i_2} \cdots a_{n,i_n}. \tag{C.21}$$

where the sum is taken over all permutations of integers in $(1, 2, \ldots, n)$. The Levi-Civita symbol $\epsilon_{i_1, i_2, \ldots i_n}$ is defined on a permutation of integers in $(1, \ldots n)$. It is 0 if two of the integers are equal. If $i_k = k$, it is $+1$. If a pair of the integers i_{k1} and i_{k2} is swapped, the sign of the Livi-Civita symbol is reversed.

Problems

C.1. Prove Eq. C.9.

C.2. Prove Eq. C.10.

C.3. Prove Eq. C.13.

Appendix D

Real Spherical Harmonics

D.1   The Spherical Coordinate System

To make an analysis of the hydrogen atom problem, we first write the Schrödinger equation in spherical coordinates (r, θ, ϕ), defined as

$$
\begin{aligned}
x &= r \sin \theta \cos \phi, \\
y &= r \sin \theta \sin \phi, \\
z &= r \cos \theta.
\end{aligned}
\tag{D.1}
$$

Instead of the unit vectors in rectangular coordinates, those in spherical coordinates $\mathbf{i}_r, \mathbf{i}_\theta$, and $\mathbf{i}_\phi$ are used, see Fig. D.1. The gradient of wavefunction ψ is

$$
\mathbf{A} = \nabla p = \frac{\partial \psi}{\partial r} \mathbf{i}_r + \frac{1}{r} \frac{\partial \psi}{\partial \theta} \mathbf{i}_\theta + \frac{1}{r \sin \theta} \frac{\partial \psi}{\partial \phi} \mathbf{i}_\phi.
\tag{D.2}
$$

The divergence of $\mathbf{A}$ is the total outgoing flow of $\mathbf{A}$ in three directions divided by the element volume. As shown in Fig D.1, the three sides are dr, $rd\theta$, and $r \sin \theta d\phi$. Therefore, the elementary volume is

$$
dv = dr \cdot rd\theta \cdot r \sin \theta d\phi = r^2 \sin \theta dr d\theta d\phi.
\tag{D.3}
$$

Figure D.1 Laplace operator in spherical coordinates. In order to derive the expression of the Laplace operator in spherical coordinates, the geometry of the spherical coordinate system and the elemental vectors are shown as $\mathbf{i}_r$, $\mathbf{i}_\theta$, and $\mathbf{i}_\phi$. The scale factors in the directions of (r, θ, ϕ) are 1, r, and $r \sin \theta$, respectively.

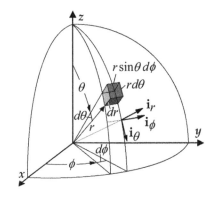

The flow in the r direction is

$$df_r = \frac{\partial(r^2 \sin\theta A_r)}{\partial r} dr\, d\theta\, d\phi = \sin\theta \frac{\partial}{\partial r}\left(r^2 \frac{\partial\psi}{\partial r}\right) dr\, d\theta\, d\phi, \tag{D.4}$$

the flow in the θ direction is

$$df_\theta = \frac{\partial(r \sin\theta A_\theta)}{r\partial\theta} dr\, d\theta\, d\phi = \frac{\partial}{\partial\theta}\left(\sin\theta \frac{\partial\psi}{\partial\theta}\right) dr\, d\theta\, d\phi, \tag{D.5}$$

and the flow in the ϕ direction is

$$df_\phi = \frac{\partial(A_\phi)}{r \sin\theta\partial\phi} dr\, d\theta\, d\phi = \frac{1}{\sin\theta}\frac{\partial}{\partial\phi}\left(\frac{\partial\psi}{\partial\phi}\right) dr\, d\theta\, d\phi. \tag{D.6}$$

Adding together, the Laplacian of a wavefunction ψ is

$$\begin{aligned}
\nabla^2\psi &= \frac{1}{dv}\left(df_r + df_\theta + df_\phi\right) \\
&= \frac{1}{r^2}\frac{\partial}{\partial r}\left(r^2 \frac{\partial\psi}{\partial r}\right) + \frac{1}{r^2 \sin\theta}\frac{\partial}{\partial\theta}\left(\sin\theta \frac{\partial\psi}{\partial\theta}\right) + \frac{1}{r^2 \sin^2\theta}\frac{\partial^2\psi}{\partial\phi^2}.
\end{aligned} \tag{D.7}$$

The Schrödinger equation, Eq. 7.75 is then

$$-\frac{\hbar^2}{2m_e\, r^2}\left[\frac{\partial}{\partial r}\left(r^2 \frac{\partial\psi}{\partial r}\right) + \frac{1}{\sin\theta}\frac{\partial}{\partial\theta}\left(\sin\theta \frac{\partial\psi}{\partial\theta}\right) + \frac{1}{\sin^2\theta}\frac{\partial^2\psi}{\partial\phi^2}\right] - \frac{K}{r}\psi = E\psi. \tag{D.8}$$

Using an *angular momentum operator* $\mathbf{L}^2$ defined as

$$\mathbf{L}^2\psi \equiv -\frac{1}{\sin^2\theta}\left[\sin\theta\frac{\partial}{\partial\theta}\left(\sin\theta\frac{\partial\psi}{\partial\theta}\right) + \frac{\partial^2\psi}{\partial\phi^2}\right]. \tag{D.9}$$

The Schrödinger equation becomes

$$-\frac{\hbar^2}{2m_e}\left[\frac{1}{r^2}\frac{\partial}{\partial r}\left(r^2 \frac{\partial\psi}{\partial r}\right) - \frac{1}{r^2}\mathbf{L}^2\psi\right] - \frac{K}{r}\psi = E\psi. \tag{D.10}$$

D.2 Spherical Harmonics

To understand the hydrogen atom and other atoms, spherical harmonics is a necessary mathematical tool. In many quantum mechanics textbooks, the general formulas of the spherical harmonics are introduced in terms of associate Legendre functions and complex exponentials. Nevertheless, for almost all applications in condensed-matter physics and chemistry, only nine real spherical harmonics are necessary. Furthermore, the representations of those spherical harmonics in cartesian coordinates are quite useful in the understanding of physicsl phenomena. Here we show that those real spherical harmonics can be introduced using elementary calculus without the general theory of associate Legendre functions and complex numbers.

For the excited states of the hydrogen atom, the wavefunctions $\psi(\mathbf{r})$ are written as a product of a function of radius $R(r)$ and a function of angle variables, $Y(\theta, \phi)$, called *spherical harmonics*:

$$\psi(\mathbf{r}) = R(r)Y(\theta, \phi). \tag{D.11}$$

Insert Eq. D.11 into Eq. D.8, we find

$$\frac{1}{R(r)}\frac{d}{dr}\left(r^2\frac{dR(r)}{dr}\right) + k^2\,r^2 = \frac{1}{Y(\theta, \phi)}\mathbf{L^2}Y(\theta, \phi). \tag{D.12}$$

The left-hand side of the equation depends only on r, and the the right-hand side of the equation of the spherical harmonics depends only on angular variables θ and ϕ. Therefore, both sides must be a constant.

In the following, using the differential equation for the function $Y(\theta, \phi)$ from Eq. D.9, we look for the explicit value of constant λ:

$$-\frac{1}{\sin^2\theta}\frac{\partial^2 Y(\theta, \phi)}{\partial\phi^2} - \frac{1}{\sin\theta}\frac{\partial}{\partial\theta}\left(\sin\theta\frac{\partial Y(\theta, \phi)}{\partial\theta}\right) = \lambda Y(\theta, \phi). \tag{D.13}$$

The angular variables can be separated by writing $Y(\theta, \phi)$ as a product,

$$Y(\theta, \phi) = \Theta(\theta)\Phi(\phi). \tag{D.14}$$

Insert Eq. D.14 into Eq. D.13, we obtain

$$\frac{\sin^2\theta}{\Theta}\left[\frac{1}{\sin\theta}\frac{d}{d\theta}\left(\sin\theta\frac{d\Theta}{d\theta}\right) + \lambda\Theta\right] = -\frac{1}{\Phi}\frac{d^2\Phi}{d\phi^2}. \tag{D.15}$$

The left-hand side only depends on θ, and the right-hand side only depends on ϕ. Both sides must be a constant. To ensure continuity, the function $\Phi(\phi)$ should be a sinusoidal function of an integer multiple of ϕ, either

$$\Phi(\phi) = \cos m\phi \tag{D.16}$$

or

$$\Phi(\phi) = \sin m\phi, \tag{D.17}$$

to satisfy the ordinary differential equation

$$\frac{d^2\Phi}{d\phi^2} + m^2\,\Phi = 0. \tag{D.18}$$

The index m is related only to the azimuthal variable ϕ. Therefore, it is approrpiate to called it the *azimuthal quantum number*. The right-hand side of Eq. D.15 is then

$$\frac{1}{\sin\theta}\frac{d}{d\theta}\left(\sin\theta\frac{d\Theta}{d\theta}\right) - \frac{m^2}{\sin^2\theta}\Theta + \lambda\Theta = 0. \tag{D.19}$$

To solve Eq. D.19, we use a trail function

$$\Theta(\theta) = \sin^m \theta. \tag{D.20}$$

Insert Eq. D.20 into Eq. D.19, after a short algebra, we obtain

$$-m(m+1)\Theta + \lambda\Theta = 0. \tag{D.21}$$

Therefore, it is a good solution with

$$\lambda = m(m+1). \tag{D.22}$$

Here λ is a parameter for Θ, a function of polar angle θ. In the standard literature, an *orbital quantum number* l is defined,

$$\lambda = l(l+1). \tag{D.23}$$

Table D.1: Spherical harmonics

Mathematical notation	Chemist's Name	Formula in angular variables	In cartesian coordinates
$Y_{00}(\theta, \phi)$	s	$\dfrac{1}{\sqrt{4\pi}}$	$\dfrac{1}{\sqrt{4\pi}}$
$Y_{10}(\theta, \phi)$	p_z	$\sqrt{\dfrac{3}{4\pi}}\cos\theta$	$\sqrt{\dfrac{3}{4\pi}}\dfrac{z}{r}$
$Y_{11}^g(\theta, \phi)$	p_x	$\sqrt{\dfrac{3}{4\pi}}\sin\theta\cos\phi$	$\sqrt{\dfrac{3}{4\pi}}\dfrac{x}{r}$
$Y_{11}^u(\theta, \phi)$	p_y	$\sqrt{\dfrac{3}{4\pi}}\sin\theta\sin\phi$	$\sqrt{\dfrac{3}{4\pi}}\dfrac{y}{r}$
$Y_{20}(\theta, \phi)$	d_{z^2}	$\sqrt{\dfrac{5}{4\pi}}\left(\dfrac{3}{2}\cos^2\theta - \dfrac{1}{2}\right)$	$\sqrt{\dfrac{5}{4\pi}}\left(\dfrac{3}{2}\dfrac{z^2}{r^2} - \dfrac{1}{2}\right)$
$Y_{21}^g(\theta, \phi)$	d_{xz}	$\sqrt{\dfrac{15}{4\pi}}\cos\theta\sin\theta\cos\phi$	$\sqrt{\dfrac{15}{4\pi}}\dfrac{zx}{r^2}$
$Y_{21}^u(\theta, \phi)$	d_{yz}	$\sqrt{\dfrac{15}{4\pi}}\cos\theta\sin\theta\sin\phi$	$\sqrt{\dfrac{15}{4\pi}}\dfrac{zy}{r^2}$
$Y_{22}^g(\theta, \phi)$	$d_{x^2-y^2}$	$\sqrt{\dfrac{15}{16\pi}}\sin^2\theta\cos 2\phi$	$\sqrt{\dfrac{15}{16\pi}}\dfrac{x^2-y^2}{r^2}$
$Y_{22}^u(\theta, \phi)$	d_{xy}	$\sqrt{\dfrac{15}{16\pi}}\sin^2\theta\sin 2\phi$	$\sqrt{\dfrac{15}{16\pi}}\dfrac{2xy}{r^2}$

For the case of $l = m$, the expression of $\Theta(\theta)$ in Eq. D.20 is correct. In general, m could be smaller than l. For the general case, the expressions are complicated. However, we need only three more cases. Here is a complete list:

$$\Theta_{10}(\theta) = \cos\theta, \qquad l = 1, \quad m = 0; \tag{D.24}$$

$$\Theta_{20}(\theta) = 3\cos^2\theta - 1, \qquad l = 2, \quad m = 0; \tag{D.25}$$

and

$$\Theta_{21}(\theta) = \cos\theta\sin\theta, \qquad l = 2, \quad m = 1. \tag{D.26}$$

The correctness of those solutions can be verified by Eq. D.19,

$$\frac{1}{\sin\theta}\frac{d}{d\theta}\left(\sin\theta\frac{d\Theta_{lm}(\theta)}{d\theta}\right) - \frac{m^2}{\sin^2\theta}\Theta_{lm}(\theta) + l(l+1)\Theta_{lm}(\theta) = 0, \tag{D.27}$$

which is left as an exercise.

For applications in condensed-matter physics and chemistry, expressions of spherical harmonics in Cartesian coordinates are useful. Table D.1 lists the first nine spherical harmonics in real variables. In the Table, superscript g means symmetric versus x, and superscript u means antisymmetric versus x. The names of those spherical harmonics by chemists are shown in the second column. The numerical constant such as $1/\sqrt{4\pi}$ is to normalize the spherical harmonics over a spherical surface of radius 1.

Appendix E

Complex Numbers

E.1 Definition of Complex Numbers

Complex number is a powerful and useful mathematical tool in physics and engineering. The basic definition and properties are taught in standard high-school algebra courses. The quadratic equation

$$x^2 + 1 = 0 \tag{E.1}$$

has no real solutions. The solutions are the imaginary unit

$$x = \pm\sqrt{-1}. \tag{E.2}$$

There are two solutions. Either one is valid. In physics and in electrical engineering, for historical reasons, the notations and definitions of the imaginary unit are different. In physics, the imaginary unit is denoted as

$$i = \sqrt{-1}, \tag{E.3}$$

and the time evolution of a quantity is defined as

$$f(t) = e^{-i\omega t}. \tag{E.4}$$

In electrical engineering, the imaginary unit is denoted as

$$j = \sqrt{-1}, \tag{E.5}$$

and the time evolution of a quantity is defined as

$$f(t) = e^{j\omega t}. \tag{E.6}$$

Although this textbook is designed for both physicsts and electrical engineers, we will follow the physicsits convention, Eqs. E.3 and E.4.

E.2 The Euler Formula

The most important formula for the application of complex number in quantum mechanics is the Euler formula. A simple proof can be done by taking a derivative to the expression $f(x) = e^{-ix}(\cos x + i \sin x)$. Using the chain rule of differentiation, one finds

$$\frac{df(x)}{dx} = e^{-ix}(i \cos x - \sin x) - ie^{-ix}(\cos x + i \sin x) = 0. \tag{E.7}$$

Therefore, the function $f(x)$ is a constant. Because $f(0) = 1$, for all real x, $f(x) = 1$. The following Euler's formula is valid for all real x:

$$e^{ix} = \cos x + i \sin x. \tag{E.8}$$

Using Euler's formula, many equalities in plannar geometry and plannar trigonometry can be easily proved. An example is a proof of the double-angle formulas for the sine and cosine functions.

Euler's formula for argument $2x$ is

$$e^{2ix} = \cos 2x + i \sin 2x. \tag{E.9}$$

On the other hand,

$$\begin{aligned} e^{2ix} &= e^{ix}e^{ix} \\ &= (\cos x + i \sin x)(\cos x + i \sin x) \\ &= (\cos^2 x - \sin^2 x) + 2i \cos x \sin x. \end{aligned} \tag{E.10}$$

Separating real part and imaginary part, one obtains

$$\cos 2x = \cos^2 x - \sin^2 x, \tag{E.11}$$

and

$$\sin 2x = 2 \cos x \sin x. \tag{E.12}$$

Appendix F

Statistics of Particles

There are three types of statstics: *Maxwell–Boltzmann statistics, Fermi–Dirac statistics* and *Bose–Einstein statistics*. Maxwell–Boltzmann statistics is valid for a system of *distinguishable particles* in a system of energy levels allowing unlimited occupancy; whereas Fermi–Dirac statistics is valid for a system of *indistinguishable particles* in a system of energy levels allowing limited occupancy, that is, systems satisfying the Pauli exclusion principle. Bose–Einstein statistics is valid for a system of *indistinguishable particles* in a system of energy levels allowing unlimited occupancy, such as photons. It is yet another way of deriving the blackbody radiation formula.

The starting point of the derivation is the Boltzmann expression of entropy,

$$S = k_B \ln W, \tag{F.1}$$

where k_B is Boltzmann's constant and W is the total number of all possible configurations of the system.

Consider a situation of N particles in a system consisting of n energy levels. The occupancy number of the ith level is N_i, where $i = 1, 2, 3, \ldots\ldots n$. The energy of the ith level is E_i, and the total energy of the system is E. We have the conditions

$$N = \sum_{i=1}^{n} N_i,$$

$$E = \sum_{i=1}^{n} N_i E_i. \tag{F.2}$$

The condition of equilibrium is that the entropy, Eq. F.1, reaches maximum under the two constraints in Eq. F.2.

Intuitively, the more uniform the distribution, the greater the randomness, or the greater the entropy. However, the condition of constant total energy adds another condition: It is preferable to have more particles in the levels of lower energy and less particles in the levels of higher energy. The problem can be resolved using *Fermat's theorem* together with the *Lagrange multiplier* method. Introducing two Lagrange multipliers α and β, the condition of equilibrium is

$$\frac{\partial}{\partial N_i} \left[S + \alpha \left(N - \sum_{i=1}^{n} N_i \right) + \beta \left(E - \sum_{i=1}^{n} N_i E_i \right) \right] = 0. \qquad (F.3)$$

The distribution can be found by combining the solution of Eq. F.3 with the two constraints in Eqs. F.1 and F.2.

During the calculation, one needs an approximate value of the derivative of $\ln N!$ for large N. This can be simply taken as

$$\frac{d}{dN} \ln N! \approx \ln N! - \ln(N-1)! = \ln N. \qquad (F.4)$$

F.1 Maxwell–Boltzmann Statistics

In the case of Maxwell–Boltzmann statistics, applicable to classical atomic systems, the particles are distinguishable, and the number of particles per energy level is unlimited. The number of possible configurations W can be determined as follows.

First, the total number of possibilities of placing N particles into N bins is $N!$. However, there are n energy levels. In each bin, the occupation number is N_1, N_2, ... N_i, ... and so on. Inside each bin, the number of different ways of placing is $n_i!$. The total number of possible placements of N particles into n bins is then,

$$W = \frac{N!}{\prod_i N_i!}. \qquad (F.5)$$

The entropy is

$$S = k_B \ln W = k_B \left(\ln N! - \sum_{i=1}^{n} \ln N_i! \right). \qquad (F.6)$$

The condition of thermal equilibrium gives

$$\frac{\partial}{\partial N_i} \left[S + \alpha \left(N - \sum_{i=1}^{n} N_i \right) + \beta \left(E - \sum_{i=1}^{n} N_i E_i \right) \right] = 0. \qquad (F.7)$$

The result is

$$k_B \ln N_i = -\alpha - \beta E_i, \qquad (F.8)$$

The meaning of the parameter β can be interpreted based on thermodynamics. Because the system is under the condition of constant temperature and constant volume, according to Eq. 6.28, we have

$$dE = T \, dS. \qquad (F.9)$$

By treating S and E as variables, comparing Eq. F.3 and Eq. F.9, we find, heuristically,

$$\beta = \frac{1}{T}. \tag{F.10}$$

The constant α can be determined by the condition that the total number of particles is N. Equation F.8 can be rewritten as

$$N_i = \frac{N}{Z} \exp\left(\frac{-E_i}{k_B T}\right), \tag{F.11}$$

where Z is determined by the condition that the total number of particles in is N,

$$\sum_{i=1}^{n} N_i = N, \tag{F.12}$$

in other words,

$$Z = \sum_{i=1}^{n} \exp\left(\frac{-E_i}{k_B T}\right). \tag{F.13}$$

By introducing a *probability* of the ith energy level, $p_i = N_i/N$, Eq. F.13 becomes

$$p_i = \frac{1}{Z} \exp\left(\frac{-E_i}{k_B T}\right). \tag{F.14}$$

Obviously, the sum of all probabilities is 1,

$$\sum_{i=1}^{n} p_i = 1. \tag{F.15}$$

F.2 Fermi–Dirac Statistics

Electrons are fermions obeying the Pauli exclusion principle. Each state can only be occupied by one electron. The electrons satisfies Fermi–Dirac statistics.

For each energy value, there can be multiple states. For example, each electron can have two spin states that have the same energy level. Let the *degeneracy*, that is, the number of states at energy E_i, be g_i. The number of electrons staying at that energy level, N_i, should not exceed g_i. The number of different ways of occupation in the g_i states is

$$W_i = \frac{g_i!}{(g_i - N_i)!\, N_i!}, \tag{F.16}$$

Following Eq. F.3, we obtain

$$k_B \left[\ln(g_i - N_i) - \ln N_i\right] = \alpha + \beta E_i, \tag{F.17}$$

or

$$N_i = \frac{g_i}{\exp\left(\dfrac{\alpha + \beta E_i}{k_B}\right) + 1}. \tag{F.18}$$

Using Eq. F.10 and introducing the probability $p_i = N_i/g_i$,

$$p_i = \frac{1}{\exp\left(\dfrac{E_i - E_F}{k_B T}\right) + 1}. \tag{F.19}$$

Equation F.19 is called the *Fermi function*.

F.3   Bose–Einstein Statistics

Photons are bozons. The particles are indistinguishable, but each state can have unlimited number of particles. Assuming that energy level i has a degeneracy of g_i. Assuming that the occupation number of ith state is N_i. The number of possible configuration can be estimated as placing $g_i - 1$ lines among $N_i + g_i - 1$ places, which is

$$W_i = \frac{(N_i + g_i - 1)!}{(g_i - 1)! \, N_i!}. \tag{F.20}$$

Because g_1 is a large number, the difference between $g_i - 1$ and g_1 is insignificant. Eq. F.20 can be simplified to

$$W_i = \frac{(N_i + g_i)!}{g_i! \, N_i!}. \tag{F.21}$$

For bosons, the total number of particles is not a constant. The condition of thermal equilibrium is

$$k_B \left[\ln(g_i + N_i) - \ln N_i\right] + \beta E_i = 0, \tag{F.22}$$

or

$$N_i = \frac{g_i}{\exp\left(\dfrac{E_i}{k_B T}\right) - 1}. \tag{F.23}$$

The probability, defined as $p_i = N_i/g_i$, is

$$p_i = \frac{1}{\exp\left(\dfrac{E_i}{k_B T}\right) - 1}. \tag{F.24}$$

Appendix G

Measurement in Quantum Mechanics

Quantum mechanics is the centerpiece of modern physics. It underlies much of modern science and technology, including transistors and integrated circuits, solar cells, light-emitting diodes, lasers, all of chemistry, materials science, and molecular biology. Nevertheless, for many decades, quantum mechanics was enshrouded in a thick cloud of mystery. Teaching and learning are significantly impeded. Richard Feynman famously said, "I can easily say that nobody understands quantum mechanics."

The 2023 Nobel Prize in Physics recognizes the experimental methods for producing attosecond light pulses and using them to provide images of processes inside atoms and molecules [3, 74]. The images of the electrons inside atoms and molecules are real wavefunctions in three-dimensional space [43, 112]. Furthermore, using scanning tunneling microscopy (STM), which was awarded the 1986 Nobel Prize in Physics, the wavefunctions including their nodal structures are observed and mapped at a subpicometer resolution, see Section 7.2.4. Both experiments showed that at a subatomic scale, wavefunction is an observable physical reality. Starting with the experimental facts that wavefunction is an observable physical reality, the thick cloud used to enshroud the crown jewelry of modern science can be removed, to open up quantum mechanics to everyone who is interested in and deserves to know.

G.1 The Measurement Postulate

Quantum mechanics was established in the first half of the twentieth century. At that time, a great majority of physicists thought of physical reality as material points, or point particles. According to a popular quantum mechanics textbook, "After all, a particle, by its nature, is localized at a point [37]." The basic dynamic variables or observables of a point particle are position and momentum. Measurement is to obtain the value of position or momentum of a point particle through experiments.

After Schrödinger published his papers in 1926, it was clear that the central concept of quantum mechanics, the wavefunction, is a continuous function spread out in space. A basic question facing physicists in 1926 was the relation between the state of a point particle, that is the wavefunction, and the observables of the point particle, that are position and momentum. John von Neumann, then a 22-year-old mathematician working under David Hilbert at University of Göttingen, ventured to create a mathematically rigorous theoretical system of quantum mechanics. After publishing three papers in 1927, he completed a monograph in 1932, *Mathematische Grundlagen der Quantunmechanik* [111]. In that monograph, the axiomatic formulation of quantum mechanics was defined. The mathematical tool, the Hilbert space, is also defined in a 161-page chapter. His axiomatic presentation of quantum mechanics in terms of Hilbert space was followed by many textbook authors [37, 44, 98]. The first three axioms, following a succinct language from Shankar [98], are as follows:

1. The state of the particle is represented by a vector $|\psi(t)\rangle$ in a Hilbert space.

2. The independent variables, for example coordinate x, is represented by a Hermitian operator $\hat{x}$ with the eigenvectors satisfying

   $\langle x|\hat{x}|x_0\rangle = x_0\delta(x - x_0)$.

3. If the particle is in a state $|\psi(t)\rangle$, a measurement of $\hat{x}$ yields one of the eigenvalues x_0 with the probability $p(x_0) \propto |\psi(x_0)|^2$. The state of the system $|\psi(t)\rangle$ then collapses to an eigenvector of $\hat{x}$, which is $|x_0\rangle = \delta(x - x_0)$.

A state $|\psi(t)\rangle$ of a particle can be represented by a wavefunction in three-dimensional space. For example, the most well-known wavefunction is the ground state of hydrogen atom, see Eq. 7.87 and Fig. 7.6. Because the size of that wavefunction is about one Bohr radius, $a_0 = 53$ picometers, to make a meaningful measurement of the position of an electron in the $1s$ wavefunction of the hydrogen atom, the resolution should be of the order of a few picometers, see Fig G.1(a). After the measurement, the $1s$ wavefunction collapses to an eigenfunction of the coordinate, a δ-function centered at the outcome of the measurement x_0, see Fig G.1(b).

There are two extensions to the measurement postulate which are accepted by some physicists. First, whereas a stand-alone quantum system evolves causally following the Schrödinger equation, the measuring process is not. Measuring represents an unpredictable jump of the quantum state, not governed by the Schrödinger equation. Second,

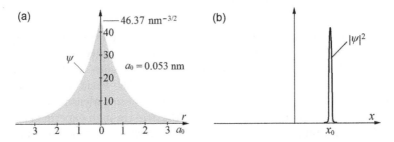

Figure G.1 Electron in the wavefunction of hydrogen atom (a) the wavefunction of the ground state of hydrogen atom. (b) After a position measurement, the wavefunction collapses to a δ-function centered at the outcome of the measurement, x_0, adapted from Fig. 1.3 in [37].

measurement always requires the participation of an observer. Consequently, in some sense, the result of a measurement is *created* by the observer.

Nevertheless, in the entire history of science, there has never been any measurement of the position of an electron with a resolution finer than the size of a hydrogen atom. The resolution is always much greater than the size of any atom, typically on the scale of micrometers. There is no experimental evidence to support the measurement postulate in the von Neumann axiomatic system at an atomic scale.

G.2 Experiments in Position Detection

In this section, we show that all experiments ostensibly considered as detecting the position of a particle are actually the interaction of a continuous field with a number of independent detectors, each one is a quantum-mechanical system with many atoms. A detector can be excited or ionized by the continuous field, then emits radiation. The radiation thus emitted can be detected by macroscopic instruments. Such a measurement process can be precisely described by the Schrödinger equation. The resolution of such a position measurement is limited by the size of the detector.

The central concept, the Golden Rule, is presented in Section 7.5. A detector as a quantum mechanical system with many atoms interacts with the incoming radiation. The radiation causes the detector to undergo a transition from a state of lower energy $|i\rangle$ to a state of higher energy $|f\rangle$. The transition rate is proportional to the square of the matrix element $\langle f|v|i\rangle$. Because the interaction potential v is proportional to the field amplitude *at the location of the detector*, the transition rate is proportional to *the square of the field amplitude* at the location of the detector.

Take an example of the double-slit experiment, see Fig. G.2. An array of detectors

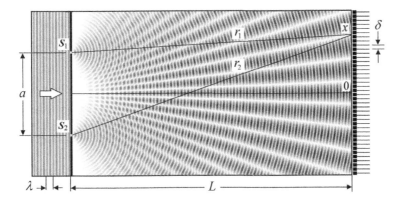

Figure G.2 Double-slit experiment with single-photon detectors A plane wave of wavelength λ falls on a screen with two slits S_1 and S_2 of distance a. An interference pattern is formed on an array of single-photon detectors each has a width of δ. The probability of photon detection is proportional to the square of the field amplitude at that detector.

Figure G.3 Building-up of the interference pattern. The screen is a two-dimensional array of single-electron detectors, each is made of many atoms. Each detector is interacting individually with the interferring electron wavefunction. The probability of excitation of a detector is proportional to the square of the wavefunction at the detector. When the number of detections increases from (a) to (d), the pattern of electron detection exhibits the interference pattern. *Source*: Tonomura *et al.* [109].

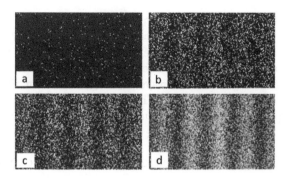

of width δ are on the detector side, D. Each detector is made of many atoms and has many energy levels. An electromagnetic wave causes a transition from a state of lower energy $|i\rangle$ to a state of higher energy $|f\rangle$. According to the Golden Rule, the transition rate is proportional to the square of the field amplitude at the location of that detector. As a result, an interference pattern is observed.

The building-up process of interference patterns of single-electron detection devices, shown in Fig. G.3, can also be understood using the Golden Rule. The screen is a two-dimensional array of macroscopic single-electron detectors. Each detector is individually interacting with the incoming wavefunction. The rate of excitation of a detector is proportional to the square of the wavefunction at the location of that detector. Therefore, as the number of detections increases from (a) to (d), the pattern of detection approaches the intensity distribution of interference.

Taking another example. In Section 30 of Leonard Shiff's *Quantum Mechanics* [94], the process of creating a track in a Wilson chamber by a plane wave of electron was treated using Schrödinger's equation. The initial state is a planal wavefunction and an atom in its ground state. The final state is an outgoing wavefunction and the atom at an excited state. Using time-dependent perturbation theory, Schiff showed that the outgoing wave is concentrated in a cone with the apex at the detector and the axis along the wavevector of the incoming plane wave $\mathbf{k}_0$. The cone angle is

$$\theta \lesssim \frac{1}{k_0\,a_0}, \tag{G.1}$$

where k_0 is the wavevector of the incoming plane wave, and a_0 is the dimension of the wavefunction of the atom. Detailed mathematics was worked out for a hydrogen atom, where a_0 is the Bohr radius, 52.9 pm. Outside the cone, the outgoing wavefunction decays rapidly as θ^{-12}; see Fig. G.4.

The process of generating a tract in a Wilson chamber is elucidated by Schiff as follows. The chamber is filled with a large number of detectors, each one is a quantum-mechanical system of many atoms, and interacting with the incoming wavefunction individually. According to the Golden Rule, the transition rate of a detector is proportional to the square of the incoming wavefunction at the location of the detector. When a detector is excited, the outgoing wavefunction is concentrated in a narrow cone

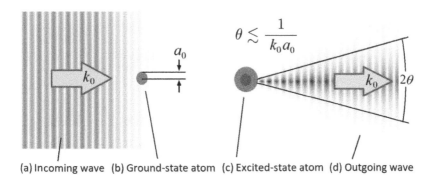

(a) Incoming wave (b) Ground-state atom (c) Excited-state atom (d) Outgoing wave

Figure G.4 Creating a Wilson-chamber track by a plane wave. (a) and (b), The initial state $|i\rangle$ consists of the incoming plane wave and a detector that is a ground-state atom. (c) and (d), After interaction, the final state $|f\rangle$ emerges that consists of an outgoing wavefunction and an excited-state atom. The entire process can be precisely described by Schrödinger's equation.

with the apex at the detector and the axis along the wavevector $\mathbf{k}_0$ of the incoming plane wave. Also according to the Golden Rule, on the right-hand side of the excited atom, the atoms along the line of the wavevector have the highest probability of being excited. An atom thus excited then becomes the apex of a new outgoing wave.

According to Eq. G.1, for a hydrogen atom and an electron of 3 keV, the angle of the cone is about one radian. Therefore, for fast electrons with energy greater than 100 keV, the cone of the outgoing wavefunction is extremely narrow.

In Section 30 of Schiff's book, the Wilson-chamber track is alternatively studied using the Ehrenfest theorem. The center of wavefunction follows Newton's second law of mechanics. The size of the wavefunction is often greater than the wavelength of the visible light. Nevertheless, from naked human eyes, it could be misinterpreted as a track of subatomic dimension. The scattered wave in a narrow cone could be misinterpreted as the collapse of the wavefunction by a mysterious process. According to Schiff's rigorous treatment, the process of creating Wilson-chamber tracks is at a macroscopic scale which can be precisely described by Schrödinger's equation.

G.3 Tomographic Imaging of Wavefunctions

As we have presented, at a subatomic scale, single electrons and single photons as point particles have never been experimentally observed. On the other hand, using attosecond light pulses, atomic and molecular wavefunctions were imaged in real space using a method similar to the tomographic imaging of medical subjects [3, 74, 43, 112]. The truthfulness of the medical imaging method is verified by in-touch examination and autopsy. Similarly, the truthfulness of tomographic imaging of wavefunctions is verified by scanning tunneling microscopy, see Section 7.2.4.

The principle of the topographical imaging of molecular wavefunctions is, the wave-

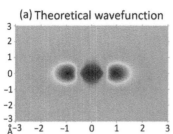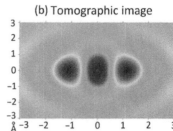

Figure G.5 Tomographic image of a nitrogen molecule (a) Theoretical wavefunction of the HOMO of a N_2 molecule from quantum mechanical computation. (b) Tomographic image of the HOMO wavefunction of a N_2 molecule. *Source*: After Itatani *et al.* [43].

lengths of higher harmonics of the attosecond light pulses are about one angstrom. The scattered waves from the higher harmonics of attosecond light pulses resemble a Fourier transform of the wavefunctions at an angstrom resolution [3, 74].

Figure G.5 shows a tomographic image of the highest occupied molecular orbital (HOMO) of an N_2 molecule using the attosecond light pulses [43]. Especially, the phase contrast, that is, the lobes with opposite signs, are detected experimentally, showing that the phase contrast is an objective reality.

Figure G.6 shows a tomographic image of the HOMO of the CO_2 molecule, using the attosecond light pulses [112]. Similarly, the phase contrast is detected experimentally, showing that the phase contrast is an objective reality.

Because the tomographic images of the molecular wavefunctions are obtained with attosecond light pulses, the process of wavefunction evolution can be observed with a high temporal resolution. Therefore, the chemical process, that is, the evolution of frontal wavefunctions in real time, can be examined continuously.

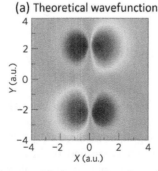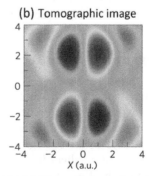

Figure G.6 Tomographic image of carbon dioxide (a) Theoretical wavefunction of the HOMO of a CO_2 molecule from quantum mechanical computation. (b) Tomographic image of the HOMO wavefunction of CO_2. The atomic unit a.u. is 53 pm. *Source*: After Vozzi *et al.* [112].

G.4   Einstein's Opinion on Quantum Mechanics

There is a popular misunderstanding that Einstein was an advocator of the hidden-variable theory, that underneath the wavefunction, there are trajectories of electrons as point particles, and the hidden trajectories follow a causal pattern. And the negative results of the experiments with the Bell inequality rejected his opinion. However, Einstein's serious publications, often overlooked, revealed otherwise.

In an article for the centenary of Maxwell's birth, *Maxwell's Influence on the Development of the Conception of Physical Reality*, Einstein wrote [66]:

> Before Maxwell, people thought of physical reality—in so far represented events in nature—as material points, whose changes consist only in motions which are subject to total differential equations. After Maxwell, they thought of physical reality as represented by continuous fields, not mechanically explicable and subject to partial differential equations. This change in the conception of reality is the most profound and the most fruitful that physics has experienced since Newton.

In that article about Maxwell's fields, Einstein criticized the probabilistic interpretation of quantum mechanics, and advocated a viewpoint in terms of fields [66]:

> Nevertheless, I am inclined to think that physicists will not be satisfied in the long run with this kind of indirect description of reality, even if an adaptation of the theory to the demand of general relativity can be achieved in a satisfactory way. Then they must surely be brought back to the attempt to realize the program which may suitably be designated as Maxwellian: a description of physical reality in terms of fields which satisfy partial differential equations in a way that is free from singularities.

Clearly, according to Einstein's opinion, wavefunctions should be interpreted as continuous physical fields, the same as electromagnetic fields. Quantum mechanics should conform to causality the same as electromagnetics does. The recent experiments of topographical imaging and the real-space STM imaging of wavefunctions as physical fields fulfilled Einstein's prophecy. The very fast and non-destructive imaging of wavefunctions means that quantum mechanics follows the causality rule.

G.5   A Modern View of Schrödinger's Cat

The cat paradox was devised by Schrödinger in 1935 [96], more than three quarters of a century ago. At that time, measurement technology was not well developed. Schrödinger's satirical story was even mistreated as a serious theory of physics. In the stone age of technology, by enclosing a cat in a box, no information can be obtained unless someone opens the box. And the only information may be obtained after opening the box is a binary determination whether the cat is alive or dead.

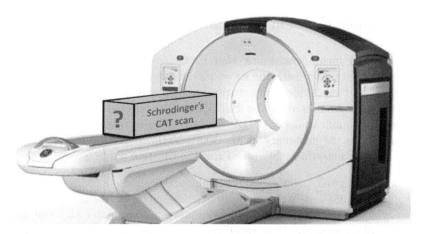

Figure G.7 Schrödinger's CAT scan. Using computer assisted tomography (CAT scan), detailed images of a cat in a box can be obtained every few minutes without opening the box.

Nevertheless, during the three quarters of a century, measurement technology has advanced tremendously. One example is, in October 1979, the Royal Swedish Academy of Sciences announced that

> The Nobel Prize in Physiology or Medicine 1979 was awarded jointly to Allan M. Cormack and Godfrey N. Hounsfield "for the development of computer assisted tomography."

The technology is alternatively called computerized axial tomography [31], also abbreviated as CAT scan. Using CAT scan, three-dimensional images of a cat in a box or the details of the brain inside a skull can be obtained without opening it.

Therefore, using modern experimental methods, three-dimensional images of a cat in a box can be obtained every few minutes. And the result of measurements is not a binary determination of life or death, but three-dimensional images of the cat as a continuous field with great details. In quantum mechanics, using the tomographic imaging method presented in Section G.3 [3, 74, 43, 112], Schrödinger's wavefunction can be observed and mapped in detail with a high temporal resolution.

G.6 A Natural Presentation of Quantum Mechanics

Starting from the experimental fact that wavefunction is a physical field similar to vector potential and scalar potential in electromagnetics, and treating Schrödinger's equation as a field equation similar to Maxwell's equations, a quantum mechanics primer is presented in Chapter 7. It includes the following topics:

1. A heuristic derivation of the static Schrödinger's equation from the classical energy integral and de Broglie's postulate.

2. A complete solution of the harmonic oscillator problem using the real creation operator and the real annihilation operator.

3. A detailed solution of the hydrogen-atom problem for K, L, and M shells.

4. The chemical bond, analytic theory, and examples.

5. A heuristic derivation of many-electron Schrödinger's equation. The Stern-Gerlach experiment. Pauli exclusion principle and Slater determinants.

6. Highest occupied molecular orbital (HOMO) and lowest unoccupied molecular orbital (LUMO).

7. A conceptual introduction to density functional theory.

8. Bloch waves and an introduction to the solid state.

9. A heuristic derivation of the dynamic Schrödinger's equation from the classical energy integral, de Broglie's postulate, and the Planck–Einstein relation.

10. Time-dependent perturbation theory and a derivation of the Golden Rule.

The thick cloud enshrouding quantum mechanics is removed. Especially, the following topics are excluded:

1. Concept of material points or point particles.

2. Complex number as an intrinsic feature of quantum mechanics.

3. Hilbert space and Hermitian operators.

4. Born's statistical interpretation on a microscopic scale.

5. The Copenhagen interpretation and wave-particle duality.

6. Heisenberg's uncertainty principle.

7. Von Neumann's axioms especially the measurement postulate.

8. Hidden variables and underlying particle trajectories.

9. Bell's inequality (an experimental test of the existence of hidden variables).

10. Schrödinger's cat (it is continuously observable using CAT scan).

This appendix is an abridged text of a plenary speech at the Third International Forum on Physics and Astronomy, December 11, 2023, with the same title. The video is in http://www.columbia.edu/~jcc2161/documents/WF2023.mp4.

Bibliography

[1] W. Adams. Cooking by Solar Heat. *Scientific American*, 1878:376, 1878.

[2] International Energy Agency. Trends in batteries. *Global Electrical Vehicle Outlook*, 2023, 2023.

[3] C. Altucci and R. Velotta. Ultra-fast dynamic imaging: An overview of current techniques, their capabilities and future prospects. *Journal of Modern Optics*, 57:916–952, 2010.

[4] P. Arora and Z. J. Zhang. Nonaqueous liquid electrolytes for lithium-based rechargeable batteries. *Chemical Reviews*, 104:4419–4462, 2004.

[5] J. Asenbauer, T. Eisenmann, M. Kuenzei, A. Kazzazi, Z. Chen, and D. Bresser. The success story of graphite as a lithium-ion battery anode materails – fundamentals, remaining challanges, and recent developments. *Sustainable Energy Fuels*, 4:5387, 2020.

[6] D. Banks. *An Introduction to Thermogeology: Ground Source Heating and Cooling*. Blackwell Publishing, Oxford, UK, 2008.

[7] A. J. Bard and M. A. Fox. Artificial photosynthesis solar splitting of water to hydrogen and oxygen. *Acc. Chem. Res.*, 28:141–145, 1995.

[8] J. K. Beatty, C. C. Peterson, and A. Chaokin. *The Solar System*. Cambridge University Press, Cambridge, UK, 1999.

[9] H. A. Bethe. Energy Production in Stars. *Physical Review*, 55:434–456, 1939.

[10] R. E. Blankenship. *Molecular Mechanisms of Photosynthesis*. Blackwell Science, Oxford, UK, 2002.

[11] J. O'M. Bockris and A. K. N. Reddy. *Modern Electrochemistry*. Plenum Rosetta, 1977.

[12] J. B. Bolton and D. O. Hall. Photochemical conversion and storage of solar energy. *Ann. Rev. Energy*, 4:353–401, 1979.

[13] M. Born and E. Wolf. *Principles of Optics*. Seventh Edition, Cambridge University Press, Cambridge, 1999.

[14] G. Boyle. *Renewable Energy*. Second Edition, Oxford University Press, Oxford, UK, 2004.

[15] C. J. Brabec, N. S. Sariciftci, and J. C. Hummelen. Plastic solar cells. *Advanced Functional Materials*, 11:15–26, 2001.

[16] J. Britt and O. Ferekides. Thin-film CdS/CdTe solar cell with 15.8% efficiency. *Applied Physics Letters*, 62:2851–2852, 1993.

[17] K. Butti and J. Perlin. *A Golden Thread*. Marion Boyers, London Boston, 1980.

[18] C. J. Chen. *Introduction to Scanning Tunneling Microscopy, Third Edition*. Oxford University Press, Oxford, UK, 2021.

[19] K. L. Chopra, P. D. Paulson, and V. Dutta. Thin-film solar cells: an overview. *Progress in Photovoltaics*, 12:69–92, 2004.

[20] R. J. Damburg and R. K. Propin. On asymptotic expansions of electronic terms of the molecular ion H_2^+. *J. Phys. B, Ser. 2*, 1:681–691, 1968.

[21] C. Darwin. *On the Origin of Spicies*. Oxford University Press, Oxford, 1859.

[22] P. A. M. Dirac. Quantum mechanics of many-electron systems. *Proceedings of the Royal Society A*, 123:714–456, 1929.

[23] A. Duffie and W. A. Beckman. *Solar Energy Thermal Processes*. John Wiley and Sons, New York, 1974.

[24] A. Duffie and W. A. Beckman. *Solar Engineering of Thermal Processes*. Third edition, John Wiley and Sons, Hoboken, NJ, 2006.

[25] US DoE EERE. *National Algal Biofuels Technology Roadmap*. US Department of Energy, 2010.

[26] A. Einstein. Ist die Trägheit eines Körpers von seinem Energieinhalt abhängig? *Annalen der Physik*, 18:639–641, 1905.

[27] A. Einstein. Über einen die Erzeugung und Verwandung des Lichts betreffenden heuristischen Gesichtspunkt. *Annalen der Physik*, 17:132–148, 1905.

[28] M. M. Farid, A. M. Khudhair, and S. A. K. Razack S. Al-Hallaj.

[29] K. Garber. Steven Chu, Obama's Point Man on Energy, Says Conservation Is 'Sexy'. *U.S. News and World Report*, March 2009, 2009.

[30] H. P. Garg, S. C. Mullik, and A. K. Bhargava. *Solar Thermal Energy Storage*. D. Reider Publishing Company, Dordricht, 1985.

[31] L. P. Gerson. Computerized axial tomography: a brief survey. *Cardiovascular Diseases*, 4:237–239, 1977.

[32] M. Grätzel. Photoelectrochemical cells. *Nature*, 414:338–344, 2001.

[33] M. Grätzel. Dye-sensitized solar cells. *Journal of Photochemistry and Photobiology C: Photochemistry Reviews*, 4:145–153, 2003.

[34] M. A. Green. *Solar cells*. Prentice-Hal, Englewood Cliffs, NJ, 1982.

[35] M. A. Green. Limits on the open-circuit voltage and efficiency of silicon solar cells imposed by intrinsic auger processes. *IEEE Transactions on Electron Devices*, 31:671–678, 1996.

[36] M. A. Green, J. Zhao, A. Wang, and S. R. Wenham. Progress and outlook for high-efficiency crystalline silicon solar cells. *Solar Energy Materials and Solar Cells*, 65:9–16, 2001.

[37] D. J. Griffiths and D. F. Schroeter. *Introduction to Quantum Mechanics*. Cambridge University Press, 2018.

[38] M. He, R. Davis, D. Chartouni, M. Johnson, M. Abplanalp, P. Troendle, and R.-P. Suetterlin. Assessment of the fist commertial prussian blue based sodium-ion batteries. *Journal of Power Sources*, 548:232036, 2022.

[39] P. Heremans, D. Cheyns, and B. P. Rand. Strategies for increasing the efficieny of heterojunction organic photovoltaic cells: Material selection and device architecture. *Accounts of Chemical research*, 42:1740–1747, 2009.

[40] A. Home. Column: Lithium slump puts China's spot price under spotlight. *Reuters*, page May 21, 2023.

[41] G. Hoover. Tech wars: RCA and the television industry. *American Business History Center*, Feb. 5, 2021.

[42] H. C. Hottel and A. Whillier. Evaluation of flat-plate solar collector performance. *Transcaction of Conference on the Use of Solar Energy*, II:74–104, 1958.

[43] J. Itatani, L. Levesque, D. Zeidler, H. Niikura, H. Pepin, J. C. Kieffer, P. B. Corkum, and D. M. Villeneuve. Tomographic imaging of molecular orbitals. *Nature*, 432:867–871, 2004.

[44] M. Jammer. *The Philosophy of Quantum Mechanics*. John Wiley and Sons, New York, 1974.

[45] C. Karuppiah, D. O. Jaiihindh, and C.-C. Yang. Oxides free materails as anodes for sodium-ion batteries. *Journal of the Electrochemical Society*, 147:1271–1273, 2022.

[46] C. E. Kennedy. Reiew of mid- to high-temperature absorber materials. *National Renewable Energy Laboratory report*, 520:31267, 2002.

[47] C. E. Kennedy and H. Price. Progress in development of high-temperature solar-selective coating. *Proceedings of ISEC 2005*, 520:36997, 2005.

[48] R. R. King, D. C. Law, K. M. Edmondson, C. M. Fetzer, G. S. Kinsey, H. Yoon, R. A. Sherif, and N. H. Karam. 40% efficient metamorphic GaInP/GaInAs/Ge multijunction solar cells. *Appl. Phys. Lett.*, 90:183516, 2007.

[49] S. A. Klein. Calculation of flat-plate collector loss coefficients. *Solar Energy*, 17:79–80, 1975.

[50] S. A. Klein. Why wind power works for Denmark. *Proc. ICE Civil Engineering.*, 2005.

[51] S. A. Klein, W. A. Beckman, and A. Duffie. A design procedure for solar heating systems. *Solar Energy*, 18:113–127, 1975.

[52] M. M. Koltun. *Selective optical surfaces for solar energy converters.* Allerton Press, Inc. New York, 1981.

[53] P. Kusuma, B. Fatzinger, B. Bugbee, W. Soer, and R. Wheeler. LEDs for extraterrestrial agriculture: tradeoffs between color perception and photon efficacy. *NASA*, TM-20210016720, 2021.

[54] Sandia National Laboratory. Sandia, Stirling Energy Systems set new world record for solar-to-grid conversion efficiency. *News Release*, February 12, 2008.

[55] A. D. Leite. *Energy in Brazil.* Earthscan, London, 2009.

[56] P. Lenard. Über die lichtelektrische Wirkung. *Annalen der Physik*, 8:149–170, 1902.

[57] D. A. Levine. *Quantum Chemistry.* Viva Books, 2003.

[58] J. Lund, B. Sanner, L. Rybach, R. Curtis, and G. Hellstrom. Geothermal (ground-source) heat pumps: a world overview. *GHC Bullitin*, September:31267, 2004.

[59] P. J. Lunde. *Solar Thermal Engineering.* John Wiley and Sons, New York, 1980.

[60] C. Lyell. *Elements of Geology.* C. H. Key & Co., Pittzburgh, 1839.

[61] H. A. Macleod. *Thin Film Optical Filters.* American Elsevier Publishing Company, Inc., New York, 2005.

[62] H. P. Maruska and W. C. Rhines. A modern perspective on the history of semiconductor nitride blue light sources. *Solid State Electrronics*, 111:32–41, 2015.

[63] H. P. Maruska, D. A. Stevenson, and J. I. Pankove. Violet luminescence of Mg-doped GaN. *Applied Physics Letters*, 22:303–305, 1973.

[64] H. P. Maruska and J. J. Tietjens. The preparation and properties of vapor-deposited single-crystalline GaN. *Applied Physics Letters*, 15:327–329, 1969.

[65] J. C. Maxwell. *A Dynamic Theory of the Electromagnetic Field.* Reprinted by Wipf and Stock Publishers, 1996, 1864.

[66] J. C. Maxwell. *The Dynamic Theory of the Electromagnetic Field.* Einstein's article is on pp. 29-32. Wipf and Stock Publishers, Eugene, Oregon, 1996.

[67] McKinsey&Company. China's Green Revolution. *McKinsey Global Energy and Materials,* July 2009:1–136.

[68] McKinsey&Company. Unlocking Energy Efficiency in the U.S. Economy. *McKinsey Global Energy and Materials,* July 2009:1–144.

[69] Y. S. Meng, V. Srinivasan, and K. Xu. Design better electrolytes. *Science,* 378:1065, 2022.

[70] R. A. Millikan. A Direct Photoelectric Determination of Planck's h. *Physical Review,* 7:355–388, 1916.

[71] K. Mizushima, P. C. Jones, P. J. Wiseman, and J. B. Goodenough. LixCoO2: A new cathode material for batteries of high energy density. *Mat. Res. Bull.,* 15:783–789, 1980.

[72] D. Moché. *Astronomy.* 7th Ed, John Wiley and Sons, Hoboken, 2009.

[73] S. Nakamura and M. R. Kramers. History of gallium-nitride based light-emitting diodes for illumination. *Proceedings of the IEEE,* 101:2211–2220, 2013.

[74] H. Niikurai and P. B. Corkum. Attosecond and angstrom science. *Advances in Atomic, Molecular and Optical Physics,* 54:511–548, 2007.

[75] P. U. Nzereogu, A. D. Omah, F. I. Ezema, E. I. Iwuoha, and A. C. Nwanya. Anode materails for lithium-ion batteries: A review. *Applied Surface Science Advances,* 9:100233, 2022.

[76] Royal Swedish Academy of Sciences. The Nobel Prize in Chemistry. 2019.

[77] B. O'Regan and M. Grätzel. A low-cost, high-efficiency solar cell based on dye-sensitized colloidal TiO_2 films. *Nature,* 353:737–740, 1991.

[78] A. K. Pahdi, K. S. Nanjundaswamy, and J. B. Goodenough. Phospho-olivines as positive-electrode for rechargeable lithium batteries. *Journal of the Electrochemical Society,* 144:1188–1194, 1997.

[79] J. I. Pankove. *Optical processes in semiconductors.* Cover Publications, Inc. New York, 1971.

[80] L. Pauling. *General Chemistry.* Dover Publications, Inc., 1970.

[81] R. Pauncz. *Spin Eigenfunctions.* Plenum Press, New York and London, 1979.

[82] J. P. Peixoto and A. H. Oort. *Physics of Climate*. Third Edition, John Wiley and Sons, New York, 1992.

[83] J. Peng, W. Zhang, Q. Liu, and S. Chou. Prussian blue analogues for sodium-ion batteries: past, present, and future. *Advanced Materials*, 34:2108384, 2022.

[84] J. Perlin. *From Space to Earth*. Aatec Publications, Ann Arbor, Michigan, 1999.

[85] J. H. Phillips. *Guide to the Sun*. Cambridge University Press, Cambridge, UK, 1992.

[86] T. Placke, R. Kloepsch, S. Durhnen, and M. Winter. Lithium ion, lithium metal, and alternative rechargeable battery technologies: The odyeesy for high energy density. *J. Solid State Electrochem*, 21:1939–1964, 2017.

[87] M. Powalla and D. Bonnet. Thin-film solar cells based on polycrystalline compound semiconductors CIS and CdTe. *Advances in Optoelectronics*, 2007:97545, 2007.

[88] Jr. R. W. Bliss. The derivations of several plate efficiency factors in the design of flat-plate solar collectors. *Solar Energy*, 3:55–64, 1959.

[89] S. Rehman, L. M. Al-Hadhrimi, and Md. Mahbub Alam. Pumped hydro energy storage system: A technological review. *Renewable and Sustainable Energy*, 44:486–598, 2015.

[90] H. H. Rogner. *World Energy Assessment: Energy and the Challenge of Sustainability*. United Nations Development Programme, 2000.

[91] M. Romero, R. Buck, and J. E. Pacheco. Water-in-glass evacualte tube solar water heaters. *Journal of Solar Energy Engineering*, 124:98–108, 2002.

[92] H. J. Sauer and R. H. Howell. *Heat Pump Systems*. John Wiley and Sons, New York, 1983.

[93] B. Sayahpour, H. Hirsh, and Y. S. Meng. Perspective: design of cathode materials for sustainable sodium-ion batteries. *MRS Energy & Sustainability*, 9:183 – 197, 2022.

[94] L. Schiff. *Quantum Mechanics, Second Edition*. McGraw-Hill, 1955.

[95] E. Schrodinger. *Collected Papers on Wave Mechanics*. Blackie & Son Limited, London and Glascow, 1929.

[96] E. Schrodinger. Die gegenwartige situation in der quantenmechanik. *Naturwissensachaften*, 23:807–812, 1935.

[97] E. F. Schubert. *Light Emitting Diodes, Second Edition*. Cambridge University Press, 2006.

[98] R. Shankar. *Principles of Quantum Mechanics, Second Edition.* Springer, 1994.

[99] J. L. Shay and S. Wagner. Efficient $CuInSe_2$/CdS solar cells. *App. Phys. Lett.*, 27:89–90, 2007.

[100] W. Shockley and H. J. Queisser. Detailed balance limit of efficiency of p-n junction solar cells. *Journal of Applied Physics*, 32:510–519, 1961.

[101] B. J. Stanbery. Copper indium selenides and related materials for photovoltaic devices. *Critical Reviews in Solid State and Materials Sciences*, 27:73–117, 2002.

[102] D. A. Stevens and J. R. Dahn. High capacity anode materails for rechargeable sodium-ion batteries. *Journal of the Electrochemical Society*, 147:1271–1273, 2000.

[103] M. Stix. *The Sun, an Introduction.* Second Edition, Springer, New York, 2002.

[104] S. Tan, H. Yang, Z. Zhang, X. Xu, J. Zhou, X. Zhou, Z. Pan, X. Rao, Y. Gu, Z. Wang, Y. Wu, X. Liu, and Y. Zhang. The progress of hard carbon as an anode material in sodium-ion batteries. *Molecules*, 28:3134, 2023.

[105] C. W. Tang. Two-layer organic photovoltaic cell. *Appl. Phys. Lett.*, 48:183–185, 2004.

[106] W. Thompson. On the Age of Sun's Heat. *Macmillan's Magazine*, 5:388–393, 1862.

[107] W. Thompson. Nineteenth Century Clouds over the Dynamical Theory of Heat and Light. *Royal Institution Proceedings*, 16:363–397, 1900.

[108] T. Tiedje, E. Yablonovitch, G. D. Cody, and B. G. Brooks. Limiting efficiency of silicon solar cells. *IEEE Transactions on Electron Devices*, 31:711–716, 1996.

[109] A. Tonomura, J. Endo, T. Matsuda, and T. Kawasaki. Demonstration of single-electron buildup of an interference pattern. *American Journal of Physics*, 57:117–120, 1989.

[110] D. Voet and J. D. Voet. *Biochemistry.* John Wiley and Sons, Hoboken, NJ, 2004.

[111] J. von Neumann. *Mathematical Foundations of Quantum Mechanics.* JPrinceton University Press, 1955.

[112] C. Vozzi, M. Negro, F. Calegari, G. Sansone, M. Nisoli, S. De Silvestri, and S. Stagira. Generalized molecular orbital tomography. *Nature Physics*, 7:822–826, 2004.

[113] S. Wagner, L. J. Shay, P. Migliorato, and H. M. Kasper. $CuImSe_2$/CdS heterojunction photovoltaic detectors. *Applied Physics Letters*, 25:434–435, 1974.

[114] M. S. Whittingham. Electrical energy storage and intercalation chemistry. *Science*, 192 (4244):1126–1127, 1976.

[115] M. S. Whittingham. Chemistry of intercalation compounds: metal guests in chalcogenide hosts. *Progress of Solid State Chemistry*, 12:41–99, 1978.

[116] M. S. Whittingham. Lithium batteries and cathode materails. *Chemical Reviews*, 104:4271–4301, 2004.

[117] M. S. Whittingham. Ultimate limits to intercalation reactions for lithium batteries. *Chemical Reviews*, 114:11414–11443, 2014.

[118] K. Xu. Nonaqueous liquid electrolytes for Li-based recgargeable batteries. *Chemical Reviews*, 104:4303–4417, 2004.

[119] K. Xu. Electrolytes and interphases in Li-ion batteries and beyond. *Chemical Reviews*, 114:11504–11618, 2014.

[120] W. Xu, J. Wang, F. Ding, X. Chen, E. Nasybulin, and J.-G. Zhang. Lithium metal anodes for rechargeable batteries. *Energy & Environmental Science*, 2014:513, 2014.

[121] N. Yabuuchi, K. Kubota, M. Dabbi, and S. Komaba. Rearch development on sodium-ion batteries. *Chemical Reviews*, 114:11636–11682, 2014.

[122] A. Yoshino. The lithium-ion batteries: Two breakthroughs in development and two reasons for the nobel prize. *Bulletin of the Chemical Society of Japan*, 95:195–197, 2022.

[123] H. Zhang, Y. Yang, D. Ren, L. Wang, and X. He. Graphite as anode materails: fundamental mechanism, recent progress and advances. *Energy Storage Materials*, 36:147–170, 2021.

[124] J. Zhao, A. Wang, P. P. Altermatt, S. R. Wenham, and M. A. Green. 24% efficient perl silicon solar cell: Recent improvements in high-efficiency silicon solar cell rezsearch. *Solar Energy Materials and Solar Cells*, 41–42:87–99, 1996.

Index

Printed and bound by CPI Group (UK) Ltd, Croydon, CR0 4YY

16/04/2025

14658352-0003